MOLECULAR COMPLEXES IN EARTH'S, PLANETARY, COMETARY, AND INTERSTELLAR ATMOSPHERES

MOLECULAR COMPLEXES IN EARTH'S, PLANETARY, COMETARY, AND INTERSTELLAR ATMOSPHERES

Editors

Andrei A. Vigasin

Obukhov Institute of Atmospheric Physics, Russia

Zdenek Slanina

Toyohashi University of Technology, Japan

World Scientific

Singapore • New Jersey • London • Hong Kong

Published by

World Scientific Publishing Co. Pte. Ltd.
5 Toh Tuck Link, Singapore 596224
USA office: 27 Warren Street, Suite 401-402, Hackensack, NJ 07601
UK office: 57 Shelton Street, Covent Garden, London WC2H 9HE

Library of Congress Cataloging-in-Publication Data
Vigasin, Andrei A.
Molecular complexes in earth's planetary, cometary and interstellar atmosphere / Andrei A. Vigasin, Zdenek Slanina.
p. cm.
Includes bibliographical references.
ISBN-13 978-981-02-3211-5
ISBN-10 981-02-3211-X
1. Molecular astrophysics. 2. Atmospheric chemistry. 3. Intermolecular forces. I. Slanina, Zdenek, 1948– II. Title.
QB462.6.V54 1998
551.51'1--dc21 98-6454
CIP

British Library Cataloguing-in-Publication Data
A catalogue record for this book is available from the British Library.

Preface

Everybody needs the atmosphere for breathing. The global contamination of the Earth's atmosphere is becoming so significant that even breathing may cause danger in certain regions of the world. It has become commonplace to say that Earth's fragile atmosphere needs to be preserved from possible unfavorable global changes. Infernal perspectives are presented by the mass-media under the titles of the "ozone hole", "greenhouse effect", and "nuclear winter", among others. Ironically it is primarily politicians and journalists, not scientists, who benefit from these public discussions. Science is becoming incorporated in the free market and certainly needs to be advertised, although true science can rarely be put into a nicely painted package for the common use. Everyday scientific achievements in laboratories, which by themselves may be minor, may allow the scientific community to forecast transformations of our atmosphere and suggest methods of remediation.

The goal of this volume is to draw attention to weak atmospheric effects which may be of significant value for scientific forecasts of the state of the atmosphere. We believe that this book will serve different audiences. Professional scientists will find here new threads for creating better models for atmospheric physics and chemistry, climate theory, and natural data retrieval. Those who are studying molecular interactions will be interested to find an exciting area of application for data taken in their laboratory. Some articles in this volume are well suited to educational programs, while others will interest lay persons.

Atmospheres are generally represented in terms of a mixture of ideal gases in which the molecules do not interact with each other. To make this model more precise, one has to account for intermolecular forces which give rise to a complicated interplay and to transformations of atmospheric constituents. Weak intermolecular forces may cause the formation of bound or quasibound quantum states between various groups of chemically distinguishable species. Temporary formation of bound and quasibound states results in thermodynamic non-ideality of real gases. This can be represented in terms of a virial expansion of thermodynamic properties in a power series in the gas density. Similarly the spectroscopic non-ideality may be introduced and many-particle effects can be traced in the gaseous spectra via a density expansion of the correlation function.

This book is an attempt to systematically involve the concept of weakly bound molecular complexes into atmospheric research. The gas-phase compositions of the atmospheres of the planets in our Solar system are well known. The data from ground- and space-based observations demonstrate a wide range of thermodynamic conditions and gaseous composition of planetary and cometary atmospheres. Molecular association effects may occur in various atmospheres as the association is facilitated either by higher pressure or lower temperature.

Typically, all atmospheres contain a variety of minor gaseous constituents which play key roles in radiative processes. In the case of the Earth's atmosphere, these gases are water vapor, carbon dioxide, ozone and others, which are frequently called radiatively active minor or trace atmospheric constituents. Their concentration and properties are thoroughly documented in the atmospheric texts and databases, such as HITRAN or GEISA. It is noteworthy, however, to comment that no systematic data on the atmospheric abundancies of the weakly bound species are available at present, although these species are readily formed by conventional 'parent' monomeric constituents.

This book is a collection of papers focused on a variety of relevant atmospheric applications, written by an international team of theoreticians and experimentalists. A number of important questions are overlooked or discussed only superficially. For instance we do not discuss here the problem of chemical reactivity of associated species, kinetics of dimer formation, dimer enhancement of the energy relaxation in the atmosphere, gas-phase aerosol growth, etc. This is due in part to the fragmentary nature of the current state of the knowledge and in part to the professional bias of the editors. Therefore, this volume should be considered an introduction to the problem rather than its solution.

A. Vigasin owes much to the late Prof. V. I. Dianov-Klokov who introduced him to the world of atmospheric science. Many colleagues kindly supported this project at different stages. Not having the possibility to identify and to thank all of them by name, however, it is a pleasure to acknowledge the inestimable assistance of Profs. F. Huisken, L. Trafton, L. Frommhold and R. Gamache whose inspiring councils and advice enabled the author to overcome a number of difficult situations.

Z. Slanina equally warmly thanks Profs. P. J. Crutzen and K. Heinzinger for their frequent friendly words of encouragement during the years the project was pursued, and the Max-Planck-Institut für Chemie (Otto-Hahn-Institut), Mecca of atmospheric science, for generous technical support of the extended computational studies.

The editors of this volume thank Dr. J.-F. Crifo for his valuable help on the initial stage of this project. Our thanks are also due to Prof. A. Borysow who supported this project. Their friendly advice and assistance helped to shape its profile.

December 1997
A. A. Vigasin, Z. Slanina

Contents

Introduction

General Concept

The concept of 'weakly bound' complexes has no simple definition at the present time. In what follows, species formed by neutral atomic or molecular particles held together by very weak interatomic (intermolecular) forces are addressed. Very roughly, these are characterized by weak bonds with dissociation energies that do not exceed 5 to 10 kcal per mole. Traditionally these weakly bound complexes are called van der Waals molecules or, in some specific cases, hydrogen-bonded species. It is spectacular that all known atomic and molecular particles are capable of forming weakly bound pairs. The corresponding dissociation energies range over several orders of magnitude from tenths of kcal per mole for the rare gas dimer to tens of kcal per mole for strongly bound complexes. It is for this reason that the individual characteristics and macroscopic manifestations of the weakly bound species are so diverse.

In most cases intermolecular potential energy surface is so shallow that moderate heating may result in the thermal excitation of the high-lying vibro-rotational intermolecular quantum states. This easy excitation of intermolecular states has at least three consequences. First, the concept of a rigid complex, possessing definite structure, is no longer valid and should be considered only in terms of the low-temperature ground-state limit of a rich spectrum of allowed coupled states. Generally, the so-called resonance states, which are quasistable complexes temporarily held together, have to be accounted for. These states are capable of spontaneous decay caused by intramolecular energy redistribution. Second, the local minima, which are situated on the potential energy surface of the weakly interacting monomers, are easily accessible and can be significantly populated. These local minima account for the structural isomers which play an important role in thermodynamics and kinetics of dimer formation. Third, molecular complexes are apt to undergo internal motions with large amplitudes, such as anharmonic intermolecular vibrations, internal rotations over low barriers, tunneling permutations of light nuclei and so forth. Theoretical treatment based on conventional models adapted to rigid molecules often do not yield valid results for these loosely bound species.

Ab initio quantum chemical methods have progressed significantly over the past few decades. The characterization of global minima of weakly bound molecular systems using quantum chemical calculations is becoming routine. However, to reveal the details of the entire potential energy surface and to predict the excited states with sufficient accuracy is much more complicated. At this point the focus of laboratory studies and atmospheric applications diverge in a certain sense. In their investigations of the weakly bound molecules, laboratory studies frequently deal with cold media, such as adiabatically cooled supersonic flows. Applications, however, demand that the media to be model be at significantly higher temperatures, where the role of highly excited intermolecular states drastically increases. This fact is why the study of the ground states represents only a primary step towards elaboration of truly applied models.

The low temperatures reached in adiabatic expansions, down to tenths of a degree Kelvin, favor molecular association in rarefied gas flows. Also, weakly bound complexes are studied by virtue of deposition of a gas on a low-temperature inert host material. The other extreme in laboratory techniques that promotes the formation of aggregated species is the use of an equilibrium gas, which in most cases is at high density. By contrast to free jet expansion or matrix isolation techniques, a gas at equilibrium can hardly be cooled to the point where the contribution from states other than the ground state is insignificant. Instead, the role of quasistable complexes increases at elevated temperatures and the issue of molecular interaction increasingly resembles pure elastic scattering. The relative contribution due to a manifold of free states increases as the temperature rises, but the number of collisions at higher gas density favors the formation of a substantial quantity of aggregated species, although not necessarily in their ground states. Ultimately the distribution of coupled and free states is formed as a result of a complicated interplay between collisional excitation, relaxation, and spontaneous decay of resonant states.

In most cases, as far as spectroscopic, thermodynamic, and other macroscopic properties of associating gases are concerned, one has to admit that the general theoretical approaches fall short of the mark. Vast amounts of experimental data, which characterize structural and spectroscopic features of hundreds of the weakly bound molecular systems, have been obtained. Unfortunately, the state of the knowledge has not been presented in a single review. There are a number of articles that are recommended. These include a number of monographs and papers, [1–6] feature articles, [7–10] and bibliography database. [11]

Pair states are known to manifest themselves spectroscopically by deviating from molecular absorption that depends linearly on density. Also, forbidden transition may become allowed due to intermolecular interactions. An advanced area of research, collision-induced absorption (CIA) in gases, is closely related to formation of molecular complexes in the gas-phase. The bibliography on collision-induced absorption amounts to several hundreds of publications since its first observation in gases by H. Welsh *et al.* in 1949. [12] For a comprehensive review of the CIA theory the excellent book by L. Frommhold[13] is recommended. The bibliography of collision-induced phenomena in gases is thoroughly documented in. [14–17]

The first two chapters of this book aim to introduce the general theoretical features of the weakly bound species. In Chapter 1, the statistical physics of dimer formation in equilibrium gases is reviewed with particular attention to the partitioning of truly bound, metastable or quasibound, and free pair states. Relations, which make it possible to estimate the fractional abundance of dimers at equilibrium from simplified model intermolecular potentials, are given and discussed. The law of mass-action for weakly bound systems has to be modified to account for a substantial decrease in the effective dissociation energy due to thermal excitation. For most applications it is necessary to apply a statistical averaging procedure over the domains of states in the phase-space of the interacting monomers. This procedure is obtained from classical statistical physics and is discussed in this Chapter. Examples of statistical averaging procedure over the domains in the phase space of interacting particles are given. The average thermal excitation energy of a dimer is calculated as a function of temperature and the partitioning of the integrated collision-induced absorption for symmetrical linear molecules is considered.

Chapter 2 focuses primarily on the problem of non-rigidity in the interpretation of the spectroscopic data of weakly bound species. Advanced group theory as well as *ab initio* quantum chemical calculations are used to reveal the nature of non-rigid behavior of complexes. It turns out that accurate spectroscopic identification of a number of the van der Waals molecules of atmospheric interest is only possible due to the floppy nature of these species, namely tunnel splitting. Typically large-amplitude motions in van der Waals molecules give rise to vibrational averaging effects, which affect the rotational constants and may significantly complicate the assignment of spectra taken *in situ*. The state of the art and relevant theoretical approaches to effects of non-rigidity are discussed for SO_2 and C_2H_2 dimers, polymers and complexes with Ar as examples.

Atmospheric Applications

The next four chapters describe various aspects of spectroscopic features attributable to weakly bound absorptions in the spectra of terrestrial, planetary, cometary, and interstellar atmospheres. Chapters 3 and 4 are somewhat overlapping in their topics, although the presentation is quite different. Much attention is paid to the calculation of dimers profiles in the planetary and cometary atmospheres. The relation between dimeric and collision-induced absorption in the atmosphere is also discussed. The properties of dimers and mixed intermolecular complexes of water, carbon dioxide, nitrogen, oxygen, ozone, hydrogen and some other molecules are briefly considered and their possible role in various atmospheres is discussed. As far as the terrestrial atmosphere is concerned the presence of molecular complexes or collisional pairs may, in some cases, affect the remote sensing and spectroscopic retrieval of the column content of atmospheric gases. The radiative transfer and heating of the Earth's troposphere and middle mesosphere is believed to be influenced by absorption of the oxygen couples. The water dimer is both of importance for the propagation of radiation through the Earth's atmosphere and for the kinetics of condensation in expanding cometary atmospheres, namely for the comet Halley. Besides $(H_2)_2$ dimers detected spectroscopically in the atmospheres of giant planets, a number of weakly bound dimers and complexes are discussed in the literature as candidate systems responsible for the absorption and emission of radiation in the planetary atmospheres. Some individual molecular characteristics and thermodynamics of these complexes are also briefly discussed in Chapters 3 and 4.

Chapter 5 addresses the C_{60} fullerene and other carbon aggregates in space. It is emphasized that the carbon aggregates might be involved in the formation of diffuse interstellar bands (DIBs), and in fact two DIBs indeed coincide with spectral features of C_{60}^+. Strong evidence has been found that relates some characteristics of the observed DIB emissions to carbon-rich molecules, although their nature is still not completely understood. In recent years, fullerenes and smaller carbon aggregates, both pure carbon and C_nH_m chains, have been considered as a possible carriers of these bands. The DIB problem is closely linked to the fullerene discovery and synthesis and it has helped to fuel real research avalanche in this area, which was highlighted by the 1996 Nobel Prize in chemistry. In fact the 1995 and 1996 Nobel Prizes in Chemistry were awarded for investigations derived from or related to atmospheric studies in the general sense.

In Chapter 6, the problem of collision-induced absorption (CIA) in planetary atmospheres is considered. In some cases the spectroscopic manifestations of dimers can be masked by unresolved diffuse contours of collision-induced absorption in the atmosphere. This paper updates the contemporary knowledge in this area, thus continuing a series of previous publications (see, e.g. the papers by L. Trafton,[18] R. Tipping,[19] A. Borysow,[20]) devoted to different aspects of this problem. The problem of collision-induced absorption is of great importance for the radiative processes in the atmospheres of giant planets, as first shown in the pioneering works by L. Trafton. Moreover, it has recently been demonstrated (see e.g. A. Borysow [20]) that CIA phenomenon might be of importance for the atmospheres of cold stars, which possess rather dense atmospheres at temperatures of about 2000–3000 K.

Relevant Laboratory Studies

Chapters 7 and 8 of this book aim to illustrate the application of contemporary laboratory experimental techniques to the study of molecular association in atmospheric gases. Certainly the domain of laboratory investigations of dimers, binary complexes and small clusters, which might be relevant to the atmospheric problems, is vast. We have chosen for this book only two types of experiments that might shed new light on the mechanisms of photochemical transformations in the stratosphere.

In Chapter 7, the use of matrix isolation techniques is reviewed with respect to studies concerning spectroscopy and photochemistry of atmospheric gases. Infrared and Raman spectroscopy of species trapped in the noble gas or other inert matrices is a powerful method to reveal the conformations of reacting species, which occur frequently via the formation of van der Waals complexes. The problem of isomeric equilibrium and photoisomerisation is discussed and several examples are given of its variations caused by the influence of the host material. This review paper can serve as a reference database for subsequent photochemical studies of molecular transformations in the atmosphere. Infrared predissociation of the hydrogen-bonded species as well as UV and visible photochemistry are also discussed.

Impressive experimental breakthroughs have been made recently by the use of molecular beam techniques in conjunction with powerful laser spectroscopy methods. Chapter 8 describes the application of molecular beam infrared predissociation spectroscopy to the study of water cluster formation. Water clusters have had a significant role in studies of the Earth's and cometary atmospheres. [21] For the latter, the kinetics of recondensation of water in cometary coma is of importance following the evaporation from the solid ice nuclei. For the former, the absorption of infrared radiation by aggregating atmospheric water vapor has been discussed in the scientific literature for a long time. Heterogeneous photochemical processes are now recognized to be responsible for ozone depletion cycles in the stratosphere. Tiny water crystals forming polar stratospheric clouds are a prominent candidate system enhancing the freeing of the halogen atoms, which attack the ozone molecules. To model this system in laboratory controlled conditions, the spectroscopic investigations of small clusters deposited on the surface of host clusters are very promising and fruitful. This technique employs the method of vibrational infrared predissociation spectroscopy in crossed atomic and molecular beams. These experiments open a direct way to reveal leading mechanisms of heterogeneous gas-phase reactions which are recognized to be of crucial importance for the photochemistry of the upper atmosphere.

References

1. J. O. Hirschfelder, C. F. Curtiss, and R. B. Bird, *Molecular Theory of Gases and Liquids* (Wiley, N. Y., 1954).
2. R. S. Mulliken and W. B. Person *Molecular Complexes* (Wiley-Interscience, N. Y., 1969).
3. *Dynamics of Polyatomic van der Waals Complexes*, ed. N. Halberstadt and K. C. Janda (Plenum, N. Y., 1990).
4. *Structure and Dynamics of Weakly Bound Molecular Complexes*, ed. A. Weber (Reidel, Dordrecht, 1987).
5. E. R. Bernstein, *Atomic and Molecular Clusters. Studies in Physical Theoretical Chemistry* (Amsterdam, Elsevier, 1990).
6. *Clusters of Atoms and Molecules, Part I*, ed. H. Haberland (Berlin, Springer, 1995).
7. F. G. Celii and K. C. Janda, *Chem. Rev.* **285**, 120 (1987).
8. M. C. Heaven, *Ann. Rev. Phys. Chem.* **43**, 283 (1992).
9. A. Vigasin, *J. Struct. Chem.* **28**, 735 (1987).
10. R. J. Le Roy and J. S. Carley in *Potential Energy Surfaces*, ed. K. P. Lawley (Wiley, N. Y., 1980), p. 353.
11. S. E. Novick, *Bibliography of Rotational Spectra of Weakly Bound Complexes*, (1997). Electronic updates are available from the author upon request at snovick@wesleyan.edu and on the web at http://www.wesleyan.edu/chem/bios/vdw.html.
12. M. F. Crawford, H. L. Welsh, and J. L. Locke, *Phys. Rev.* **75**, 1607 (1949).
13. L. Frommhold, *Collision-induced Spectroscopy in Gases* (Cambridge University Press, Cambridge, 1993).
14. N. Rich and A. R. W. McKellar, *Can. J. Phys.* **54**, 486 (1976).
15. J. L. Hunt and J. D. Poll, *Mol. Phys.* **59**, 163 (1986).
16. A. Borysow and L. Frommhold in *Adv. Chem. Phys.*, V.**75**, ed. I. Prigogine and S. A. Rice (1989), p. 439.
17. L. Ulivi and L. Frommhold, update 1993 of.[16]
18. L. Trafton in *Collision and Interaction-Induced Spectroscopy*, ed. G. C. Tabisz and M. N. Neuman (Kluwer Academic Press, Dordrecht, 1995), p. 517.
19. R. Tipping in *Phenomena Induced by Intermolecular Interactions*, ed. G. Birnbaum (Plenum Press, New York, 1985), p. 727.

20. A. Borysow in *Phenomena Induced by Intermolecular Interactions*, ed. G. Birnbaum (Plenum Press, New York, 1985), p. 529.
21. Z. Slanina, F. Uhlík, and J.-F. Crifo, *J. Mol. Struct.* **270**, 1 (1992).

STATISTICAL PHYSICS AND THERMODYNAMICS OF BIMOLECULAR COMPLEXES

A.A. VIGASIN
Obukhov Institute of Atmospheric Physics,
Russian Academy of Sciences,
Pyzhevsky per. 3, Moscow 109017 Russia

Basic ideas are summarized which refer to molecular association thermodynamics and statistical physics of molecular pairs. The principles and methods often used in the calculation of equilibrium composition of associating gases are commented on. Attention is mainly focussed on the rigorous definition of tightly bound, metastable and free states in the phase space of interacting pairs.

1 Introduction

The formation of dimers in equilibrium gases obeys in its general outline the known principles of chemical thermodynamics which are familiar to every chemist. There are, however, some peculiarities which should be accounted for while calculating thermodynamic functions and equilibrium composition of associating gases. One of the aims of the present Chapter is to emphasize the most important differences in the treatment of thermodynamic behaviour of the weakly bound species as compared to conventional chemically bound compounds. Most complications result from the fact that intermolecular dissociation energy of a complex with weak intermolecular van der Waals or hydrogen bonding may be several tens or even hundreds times lower than the value typical for 'normal' molecules. This results in highly excited rovibrational states of these dimers being readily populated even at moderate temperatures. It then follows that thermodynamic behaviour of the weakly bound species is strongly influenced by the population of high-lying states situated in the vicinity of, or even above, the dissociation limit. Moreover, the concept of a newly created (e.g., dimer) molecule as an entity which may be rigorously distinguished from a pair of its monomeric progenitors, is no longer valid, and an explicit and well grounded definition of dimeric states is needed. At least for bimolecular pairs, such a definition can be formulated in terms of a thorough statistical physics examination of the interaction of the monomers. This implies the consideration in phase space of interacting species to be made, i.e. making allowance for spatial positions, relative impulses and end-over-end rotational moments of molecular pair constituents. In doing so the tightly bound dimeric states are bridged with the states of isolated free monomers via a manifold of intermediate quasibound pair states. These latter states are frequently called metastable states since they are capable of spontaneously decaying to independent parent monomers. They can be considered also in terms of the scattering resonances (rotational, Feshbach, shape resonances etc.) and as such are treated by advanced quantum scattering theory. We shall see below that metastable dimeric states can be unexpectedly important for a number of applications, namely in their application to atmospheric spectroscopy. This situation makes weakly bound dimers noticeably at variance with any conventional atmospheric gaseous constituents, since highly excited metastable states of 'normal' molecular gases at equilibrium can hardly be reached at reasonable temperatures. As a rare example, however, note that the problem of accounting for the metastable states of conventional molecules like N_2 or O_2 arises in the chemical kinetics calculations of the heating experienced by a space vehicle re-entering the upper reaches of the atmosphere. [1,2] Dissociation occurs under nonequilibrium conditions with respect to ro-vibrational and translational degrees of freedom and the temperature may reach 30–50 thousands of degrees K. This illustrative example shows clearly the difference between conventional and weakly bound molecules since, for the latter, an impact of the metastable states formation on the macroscopic gas-phase properties is well pronounced, even near room temperature. In what follows we first comment on the general principles and methods employed in the calculations of the equilibrium composition of associating gases. Then the problem of subdivision of phase space into tightly bound, metastable and free contributions will be discussed as applied to a couple of atomic (structureless) or polyatomic interacting particles. Finally, two examples are given which concern the statistical physics average of the dissociation energy and the partitioning of collision-induced intensity of absorption in a gas consisting of symmetric linear molecules.

2 Equilibrium Composition of Associating Gas

2.1 Mass-Action Law

The simplest and most frequently used assumption on the composition of associating gases consists of interpreting the gas as a mixture of monomers and a set of clusters (dimers, trimers, tetramers etc.) which are in thermodynamic equilibrium. By accepting this model for the present section, we postpone a discussion of the explicit definition of bimolecular states to Sec. 3.

In a mixture of reacting gases equilibrium is governed by conventional mass-action laws. Briefly, this may be summarized as follows. For the sake of simplicity let us consider only one association reaction

$$A + B + M \Longleftrightarrow C + M, \tag{1}$$

in which monomers A and B form a complex C = A...B in the course of a three-body collision with particle M. Thermodynamic equilibrium of this system is known[3] to be characterized by the Gibbs free energy G being a minimum. This Gibbs energy $G(N,P,T)$ (P and T being, respectively, pressure and temperature) is an additive function of the number of particles N and, as such, can be represented as

$$G = \mu \frac{N}{N_0} = -\frac{N}{N_0} RT \ln \frac{Q}{N}. \tag{2}$$

Here μ stands for the chemical potential of the gas, N_0 is Avogadro's number, R is the gas constant, and Q is the partition function. A minimum of G for reaction (1) at fixed P and T can be found from the following relationship

$$\frac{\partial G}{\partial N} + \frac{\partial G}{\partial N_A}\frac{\partial N_A}{\partial N_C} + \frac{\partial G}{\partial N_B}\frac{\partial N_B}{\partial N_C} = \frac{\partial G}{\partial N_C} - \frac{\partial G}{\partial N_A} - \frac{\partial G}{\partial N_B}. \tag{3}$$

Substituting (2) in (3) results in a relation which is conventionally called the mass-action law:

$$\frac{N_C}{N_A N_B} = \frac{Q_C}{Q_A Q_B} = K_N(T,P). \tag{4}$$

The function $K_N(T,P)$ is the equilibrium constant for the reaction (1).

It is convenient to express the ideal gas equilibrium constant in terms of partial pressures instead of partial number densities. This converts Eq. 4 to the form:

$$K_P = \frac{P_C}{P_A P_B} = \frac{N}{P} K_N(T,P). \tag{5}$$

Here $P = P_A + P_B + P_C$ and $N = N_A + N_B + N_C$. A remarkable feature of K_P constant is that it is a function only of the gas temperature. Formulae (2)–(5) enable one to obtain the standard Gibbs free energy change $\Delta G_r^0 = G_C^0 - G_A^0 - G_B^0$ for reaction (1) in the form:

$$\Delta G_r^0 = -RT \ln K_P. \tag{6}$$

Zero upper index indicates the standard pressure state of 1 atm (P^0) for each component. In case of a dimer D forming from monomers A

$$A + A + M \Longleftrightarrow D + M \tag{7}$$

the Dalton law $P = P_A + P_D$ and mass action law

$$K_P = \frac{P_D}{P_A^2} \tag{8}$$

make it possible to determine all unknown partial pressures in a simple way. Let this system of equations be expressed in terms of molar fractions of the reagents $x_1 = N_A/N = P_A/P$ and $x_2 = N_D/N = P_D/P$:

$$\begin{aligned} x_1 + x_2 &= 1, \\ K_P &= \frac{x_2}{P\, x_1^2}. \end{aligned} \tag{9}$$

The solution of these equations yields

$$x_2 = \frac{1 + 2PK_P - \sqrt{1 + 4PK_P}}{2PK_P}. \tag{10}$$

In the low pressure limit $x_2 \approx PK_P$. This enables one to obtain the following relationship for the dimer number density of

$$n_D = \frac{N}{V} x_2 = \frac{1}{kT} P^2 K_P. \tag{11}$$

Thus in the low pressure limit the molar fraction of dimers in a gas at equilibrium varies linearly with pressure and the number density of dimers is proportional to the pressure squared. Formula (10) provides an exact description of the dimer mole fraction behaviour versus gas pressure, however, it must be kept in mind that gas compression progressively causes more and more pronounced aggregation of molecular species. As a consequence, a condition for the formation of only dimers is justified only in the limit of initial stage of association, that is one has to assume that the relationship $x_2 \ll x_1$ should be fulfilled to approximate a gas in terms of a mixture of hypothetical ideal monomeric and dimeric gases. In some cases it is convenient to introduce the definitions of the component's fractional abundances which are at variance with molar fractions defined above. For instance we can introduce mass fractions given by

$$a_1 = x_1 \frac{M_A}{\overline{M}}; \qquad a_2 = x_2 \frac{M_D}{\overline{M}}.$$

Here M_A and M_D denote, respectively, molecular masses of a monomer and a dimer, $\overline{M}$ is mean molecular mass of a gas particle. In the case of dimerization, the relationships of the molar and mass fractions have the following form

$$x_1 = \frac{2a_1}{1 + a_1}; \qquad x_2 = \frac{1 - a_1}{1 + a_1}. \tag{12}$$

When $x_2 \ll x_1$, it can be assumed that $a_2 \approx 2x_2$. The formula (10) expressed in terms of mass fractions reduces to:

$$a_1 = 1 - a_2 = \frac{1}{\sqrt{1 + 4PK_P}}. \tag{13}$$

To establish the extent of association dependence upon temperature and pressure, the calculations of the standard Gibbs free energy G_C^0 for the complex should be carried out as a function of temperature. The corresponding standard Gibbs energies for the majority of interesting individual practical substances are collected in the reference editions.[4,5] This makes it possible to establish the Gibbs energy change in association reactions and then to use the mass-action law through relationships (6), (10).

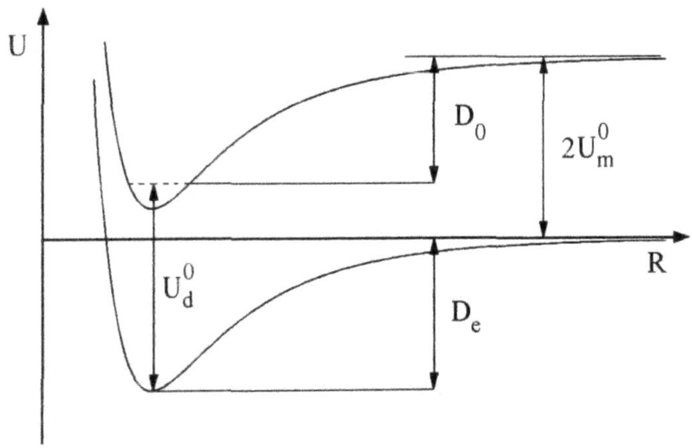

Figure 1: Potential energy curves for the dimer with 'allowed' (upper curve) and 'frozen' (lower curve) vibrations.

According to classical thermodynamics, the relationship $G = H - TS$ is valid between the Gibbs free energy G, the enthalpy H and the entropy S of a gas. It then follows that any partial standard Gibbs energy G^0 is ambigious due to the arbitrary energy scale. To overcome this difficulty in the calculations of the Gibbs energy change, it is conventionally assumed that the energy of dissociation of a dimer $D_0 = D_e + 2U_m^0 - U_d^0$ should be subtracted from the calculated G_d^0 (see Fig. 1).[a] Here D_e stands for the total potential well depth,

[a]We shall see below that for van der Waals molecules the decrease of the average dissociation energy upon thermal excitation may become significant and has to be accounted for. This can be done roughly by the use of the effective temperature dependent function $\langle D(T)\rangle$ instead of true dissociation energy D_0 as it is described in detail in Sec. 4.1.

U_m^0 and U_d^0 are zero-point energies for the monomer and dimer, respectively. This will result in the correspondence of the energy scales for the dimer and two monomers taken at infinite separation from each other.

2.2 Partition Functions

The basis for the calculations of thermodynamic functions is provided by the concept of the partition function

$$Q = \sum_{j=1} g_j exp(-\frac{e_j}{kT}). \tag{14}$$

Here the summation is carried out over all quantum states j of the species, having energy e_j and degeneracy g_j, k is the Boltzmann constant and T is temperature. It is generally assumed that individual motions within a molecule are independent. It then follows that the total partition function Q can be factorized in terms of individual partition functions corresponding to different types of motion:

$$Q = Q_{tr}\, Q_{rot}\, Q_{vibr}\, Q_{el}\, Q_{ns}. \tag{15}$$

Here Q_{tr}, Q_{rot}, Q_{vibr}, Q_{el} are the partition functions, respectively, over translational, rotational, vibrational, and electronic degrees of freedom, and Q_{ns} accounts for the nuclear spin. In case of a monatomic gas of identical atoms, $Q_{ns} = 2s + 1$ where s is the spin of the nucleus. In case of polyatomics Q_{ns} can be determined, provided nuclear spins of atoms and the symmetry point group of a molecule are given. Electronic excitations can be neglected in most cases, thus giving $Q_{el} = const$. It is worthy to note however that some molecules such as O_2, NO_2, ClO_2 are characterized by degenerate electronic ground states. The calculation of partition functions for such species has to account for the electronic ground state statistical weight and the excitation of the low-lying electronic levels.

The rigid rotator–harmonic oscillator (RRHO) approximation is the most frequently used assumption in the calculations of the partition and thermodynamic functions for polyatomics. As applied to weakly bound species this assumption is very poor, since these species are extremely floppy and capable of having internal movements with large amplitudes. Note that the RRHO approximation, however, makes it possible to perform thermodynamic calculations on the basis of minimal input information relevant to the details of the potential energy surface of a complex. Attempts to account for the effects of nonrigidity are currently still limited due to the lack of reliable data on the intermolecular potential energy surfaces.

For practical use, the formulae given in the references, such as *JANAF*,[4] or[5] can be recommended. For example, the following relationship can be used to calculate the Gibbs free energy for a nonlinear polyatomic complex in the RRHO approximation:

$$\begin{aligned}\frac{G^0 - H^0}{RT} &= -1.5\ln M - 4\ln T + \ln\sigma - 0.5\ln(I_A I_B I_C) \\ &\quad + \sum_{i=1}^{s} \ln\left(1 - \exp\left(-\tfrac{h\nu_i}{kT}\right)\right) - 129.519.\end{aligned} \tag{16}$$

Here M is the molecular mass of a complex in g/mole, σ is the symmetry number, $(I_A I_B I_C)$ is the product of principal moments of inertia in $(\mathrm{g}\cdot\mathrm{cm}^2)^3$, s is the number of vibrational degrees of freedom, and ν_i are the vibrational frequencies in cm^{-1}, $h/k = 1.4388$ cm K.

Equations 16, 6 and 10 form a basis for the calculation of the equilibrium composition of an associating gas provided that the complex can be characterized by a set of molecular parameters appropriate to rigid or quasi-rigid molecular structure. This implies that the complex should be near its ground state so that the effect of large amplitude internal movements can be treated as minor corrections. To deviate from this assumption, one has to account for the complete intermolecular potential energy surface. This will also permit the role played by metastable dimeric states in the overall equilibrium gas properties to be elucidated.

2.3 Predictions Based on Intermolecular Potentials

Model intermolecular potentials provide an efficient tool for the calculation of statistical averages needed to ensure better comprehension of the molecular association processes. From a thermodynamic viewpoint, the effect of molecular association in a gas can be considered to result from its nonperfect behaviour. One of the simplest ways to account for the effect of nonideality of a gas consists in representation of the equation of state of a dilute gas in terms of the virial expansion

$$\frac{PV}{RT} = 1 + \frac{B(T)}{V} + \frac{C(T)}{V^2} + \dots \tag{17}$$

Here $B(T)$, the second virial coefficient, can be easily expressed via the pair intermolecular interaction energy U, which is a function of the intermolecular separation and the relative orientation of pairs of molecules. The case of a spherically-symmetric interaction potential $U = U(R)$ is realized very rarely.

This approximation can be applied, strictly speaking, to atoms in their S-states, but not to typical atmospheric molecular constituents, which possess an anisotropic intermolecular potential. Nevertheless, the assumption that the potential U is independent of molecular orientation is widely used since its simplicity enables one to apply an approximate but unified approach to a variety of species possessing very complicated potentials. Although, as we shall see in the Section below, this approach is not well-grounded nor justified in all cases, it seems worthwhile to initially assume a spherically symmetric potential. Then the following relationship is valid for the second virial coefficient:

$$B(T) = 2\pi N \int_0^\infty [1 - \exp(-\frac{U(r)}{kT})] R^2 \, dR. \tag{18}$$

The second virial coefficient is often represented in terms of contributions from free and coupled pair states

$$B(T) = B_f(T) + B_c(T). \tag{19}$$

Here subscript f refers to the pair states which experience purely elastic collision, whilst the states denoted by subscript c are those which relate to the formation of either firmly bound or metastable dimers. The latter may be thought to be identical with the shape resonance states arising in the course of collision due to the end-over-end rotation of a pair against its center of mass. The distinction between 'tightly bound' and 'metastable' dimers is explained by Stogryn and Hirshfelder[6] in terms of the so-called effective potential energy. Let an effective potential Φ_{eff} be a sum of potential energy U and centrifugal energy L

$$\Phi_{eff} = U(R) + L. \tag{20}$$

Here L can be represented via the parameters of collision as $L = Kb^2/R^2$ where b is the impact parameter and K is the initial relative kinetic energy. The term L is known as the centrifugal energy and can assume the following form: $L = \hbar^2 J(J+1)/2I$ where $I = \mu R^2$ is moment of inertia of a pair, μ being a reduced mass, J is the overall angular momentum quantum number, $\hbar$ is Planck's constant. Fig. 2 illustrates the radial dependence of the effective potential parametrised by growing rotational energy.

It is seen that three separate regions can be unequivocally characterized. For values of L exceeding certain L_{max}, there are no extrema on the curves of $\Phi_{eff}(R)$. For values of L less than L_{max}, each effective potential curve has maximum and minimum points. It is apparent that the states with Φ_{eff}

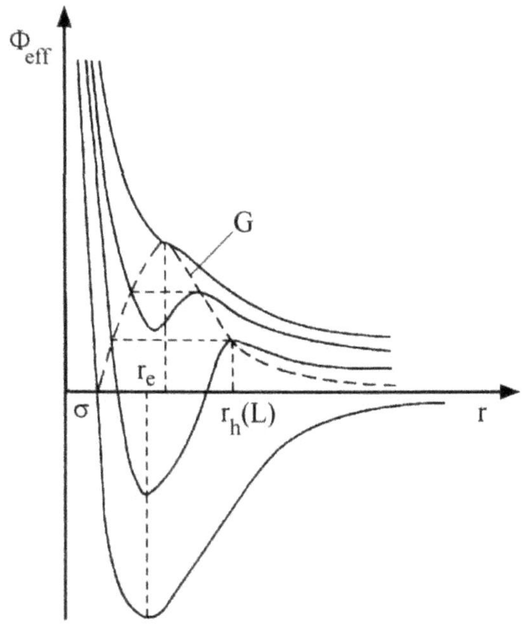

Figure 2: Radial dependence of the effective intermolecular potential energy. G is an auxiliary function.

below zero belong to the tightly bound dimeric manifold of states. The region of states situated at $\Phi_{eff} > 0$ and below the hump of the potential curve with the selected L may be called metastable, since from the standpoint of quantum mechanics these states are capable of dissociating through the potential barrier. Note that from the classical viewpoint, these states are indistinguishable from tightly bound states, since in both cases the parent molecules can only be freed by a collision with a third body. Numerical estimates made by Stogryn and Hirshfelder[6] show that in the case of argon dimers, the life-time of metastable states varies from 10^{10} to 10^{-11} sec depending on the excitation energy and the total angular momentum. The mean time between subsequent collisions amounts to 10^{-9} or 10^{-10} sec at normal conditions and the average time of flight of molecules through the region of effective interaction is of the order of 10^{-12} sec. It is apparent that most metastable argon dimers do not dissociate spontaneously before undergoing a collision. Therefore, the metastable dimers can be thought of as molecular pairs similar to highly excited dimers and having a life-time to spontaneous decay in excess of the duration of an elastic collision experienced by the monomers.

Stogryn and Hirshfelder[6] have show that the equilibrium constant entering the mass-action law (5) can be broken into contributions from the tightly bound (b) and metastable (m) pairs $K = K_b + K_m$. This representation is in accord with the subdivision of the second virial coefficient $B_c(T) = B_b(T) + B_m(T)$, if both terms in this formula are associated with the corresponding partial equilibrium constants through the use of the general relationship $B(T) = -RTK(T)$ for the so-called 'bound' and 'metastable' parts of the second virial. It then follows from (9) that the dimer mole fraction can be represented in terms of the 'coupled' second virial:

$$x_2 = -\tilde{n}B_c x_1^2 = -\tilde{n}(B_b + B_m)x_1^2 \; . \tag{21}$$

Here $\tilde{n} = n/N_0$ is the monomer number of moles in a volume V, $n = N/V$, N_0 is Avogadro's number, and x_1 stands for the mole fraction of monomers. In the case of heteromolecular A ... B complex formation, x_1^2 should be replaced by $2x_A x_B$. Equation 21 is employed frequently to estimate the equilibrium composition in associating gases, namely in the atmosphere.

Numerical evaluations of B_b and B_m parts of second virial coefficient as functions of the reduced temperature $T^* = kT/\epsilon$ were performed by Stogryn and Hirshfelder [6] for several intermolecular potentials. Anderson *et al.* [7] made a least square fitting of $B_b(T^*)$ calculated in [6] for the Lennard-Jones (6-12) interparticle potential. This resulted in the following expression often employed to calculate the dimer mole fraction:

$$x_2 = b_0 \frac{P}{RT} \exp\left[0.105 + \frac{0.364}{T^*} - 1.46 \ln T^*\right]. \tag{22}$$

Here $b_0 = \frac{2}{3}\pi N_0 \sigma^3$, σ is the hard sphere diameter of the molecule, R is the gas constant.

It seems instructive to reveal the physical meaning of the x_2 temperature dependence yielded by Eq. 22. Rigorous definitions introduced in Sec. 2.1 enable us to express x_2 in terms of the partition functions Q_1 and Q_2 for isolated monomer and dimer, respectively. In the low-pressure limit the use of Eqns. 4, 5, and 10 results in:

$$x_2 = \frac{P}{kT}\frac{VQ_2}{Q_1^2} \exp\left(\frac{D_0}{kT}\right). \tag{23}$$

Here D_0 is the dimer dissociation energy; the dimer partition function Q_2 should be taken over tightly bound dimeric states $Q_2 = Q_{2b}$. Formula (22) corresponds to the structureless monomers forming a dimer. The monomer partition function Q_1 reduces in this case to the translational partition function only $Q_1 = Q_1^{tr}$ which, in the classical limit, is known to be proportional to $VT^{3/2}$.

Generally the temperature dependence of Q_{2b} is determined by several contributions:

$$Q_{2b} = Q_{2b}^{tr}\, Q_{2b}^{rot}\, Q_{2b}^{vibr}. \tag{24}$$

Here $Q_{2b}^{tr} \sim VT^{3/2}$, Q_{2b}^{rot}, and Q_{2b}^{vibr} are translational and truncated rotational and vibrational partition functions, respectively. To allow for tightly bound states formation, the sum of the overall rotational and vibrational energies of a dimer has to be limited by the dissociation threshold. Obviously in the high-temperature classical limit $kT \gg D_0$ all ro-vibrational levels in the truncated potential well are equally populated. This results in both Q_{2b}^{rot} and Q_{2b}^{vibr} being temperature independent.

At relatively high temperature one has to account for the J-dependence effect of the state-selected dissociation energy in the effective potential well (see Fig. 2). Detailed consideration of this effect is given in the Sec. 4.1 below. It is shown (see Eq. 55) that at high temperature the thermally averaged state-selected dissociation energy is roughly $\langle D_0 \rangle \approx 0.333 D_0$. This value has to be inserted in (23) together with the classical partition functions for the bound dimer and monomer, which reduce in our case to translational partition functions. As a result, the following temperature dependence of the dimer mole fraction x_2 stems directly from (23):

$$x_2 \sim \frac{P}{kT} \exp\left[\frac{0.333 D_0}{T} - 1.5 \ln T\right]. \tag{25}$$

It is seen that this dependence is almost identical to that yielded by Eq. 22 and derived from purely numerical calculations by Stogryn and Hirshfelder.[6]

Comparing this expression with (22) enables us to deduce that the restrictions to validity of the latter are not due to the choice of the model potential function (the Lennard-Jones potential). The deficiency of the formula (22) is due to the suppositions that, first, the monomers are structureless and, second, the thermal energy is well in excess of the dissociation limit. It is worthy to note that expressions (22) and (25) seem to be valid even at $kT \approx D_0$. The neglect of anisotropy of interaction, as we shall see in the Sections below, may result in more serious consequencies. These shortcomings are caused by the loss of significant domain of the quasibound states in the phase space of the interacting particles as true anisotropic potential function is imitated by the isotropic one. Although this approach is physically inconsistent, this way of accounting for the anisotropy is widely used and deserves to be briefly commented below.

In the works by Calo [8,9] it was suggested to extend the relationship (22) to anisotropically interacting particles by reducing the problem to the above discussed spherically-symmetric one. The spherically symmetric potential may be constructed through the angular average of an anisotropic, e.g. Stockmayer potential:

$$\begin{aligned} U(R,\theta_1,\theta_2,\phi) &= 4\epsilon\left[(\tfrac{\sigma}{R})^{12}-(\tfrac{\sigma}{R})^{6}\right]-\tfrac{\mu^2}{R^3}f(\theta_1,\theta_2,\phi), \\ f(\theta_1,\theta_2,\phi) &= 2cos\theta_1 cos\theta_2 - sin\theta_1 sin\theta_2 cos\phi. \end{aligned} \tag{26}$$

Here r is the intermolecular separation, ϵ and σ are conventional Lennard-Jones potential parameters, and the angles θ_1, θ_2, and ϕ determine the relative orientation of the dipole moment μ appropriate to monomeric units. According to Calo, [9] the equilibrium dimer mole fraction can be found in this case using the following relationship:

$$x_2 = b_0\frac{P}{RT}\exp\left[0.105+\frac{0.364}{T^*}-1.46\ln T^*+A(1.655+\frac{1.295}{T^*}+0.418\ln T^*)\right]. \tag{27}$$

Here $A = 1-\exp(-\delta^2/2.806T^*)$, $\delta = \mu^2/2\epsilon\sigma^3$, and $T^* = kT/\epsilon$ is the reduced temperature.

Equation 21 has been obtained under more general assumptions than Eqns. 22, 27. These latter imply that the anisotropy of intermolecular interaction, if any, is to be effectively averaged over the relative orientations of the monomeric units. This means that the monomers are allowed to rotate freely within a dimer molecule. Note that the heat capacity of these internal rotations is not accounted for neither in (22), nor in (27). Rigorous calculation of the Ar ... N_2 energy levels position made by Brocks [10] shows that a transition to free internal rotation occurs in the vicinity of the dissociation level, or even higher. It looks like, in most cases, free internal rotation within a complex is a privilege of metastable rather than tightly bound stable dimers. That is why the estimates given by Eqns. 22, 27 are not applicable to those dimers and temperatures which favor an excess of metastable states over tightly bound states. Model calculations of simple systems such as He...LiH made by Certain and Moiseyev, [11] show that nearly isotropic interaction results in the formation of rotational (shape) resonances only. Levine [12] showed that for two spherical molecules forming a dimer the metastable contribution to far-infrared and Raman spectra is always considerably less than the bound state contribution. By contrast in a strongly anisotropic potential, the role of rotational resonances is negligible. In the Sections below it will be shown that the anisotropy of interaction gives rise to a manifold of metastable states beyond the rotational resonance states, which affects the thermodynamics and spectroscopy of molecular pairs significantly.

3 Bound and Metastable States of the Interacting Molecules

3.1 A Set of Quantum Oscillators

The goal of this Section is to give an illustrative example of how metastable states can be easily formed even in a nonrotating polyatomic complex. Let the bimolecular complex be represented as an array of s harmonic oscillators having vibrational frequencies ν_i . The energy levels of the ith oscillator are known to be described by

$$E_i = h\nu_i\left(v^{(i)} + \frac{1}{2}\right).$$

Here h is the Planck's constant, $v^{(i)}$ is vibrational quantum number. The total amount of vibrational energy of a complex equals

$$E = E_0 + \sum_{i=1}^{s} h\nu_i v^{(i)}.$$

Here E_0 stands for zero-point energy. The statistical weight of each state with fixed v_i equals unity, that is why the total vibrational partition function Q_v is a geometrical progression. Neglecting the term containing zero-point energy, the total partition function Q_v assumes the following form:

$$Q_v = \prod_{i=1}^{s}\sum_{v^{(i)}}^{\infty} \exp\left(-\frac{h\nu_i}{kT}v^{(i)}\right) = \left(1 - e^{-h\nu_i/kT}\right)^{s}. \tag{28}$$

In reality, only a fraction of states which contribute to the partition function (28) may be called true dimeric states. Molecules having a total amount of vibrational energy in excess of the dissociation limit D_0 (so that $D_0 < E < \infty$) must be called metastable since they are capable of dissociating spontaneously before undergoing a collision. Consequently, only the truncated partition function

$$Q_b = \sum_{E=0}^{D_0} f(E) \tag{29}$$

is appropriate to tightly bound dimeric states. In formula (29) the distribution function $f(E)$ is introduced for the total vibrational energy E of a complex. This function can be represented as a product of the density of states $g(E)$ and the Boltzmann factor:

$$f(E) = g(E) \exp(-E/kT). \tag{30}$$

For simplicity, let all the vibrational frequencies be equal to each other. In this case the density of states is known to assume the form:

$$g(E) = \frac{(j+s+1)!}{j!\,(s-1)!}. \tag{31}$$

Here $j = E/\theta$ and $\theta = h\nu/k$.

The qualitative behaviour of the distribution function (30) with varying temperature is illustrated by Fig. 3.

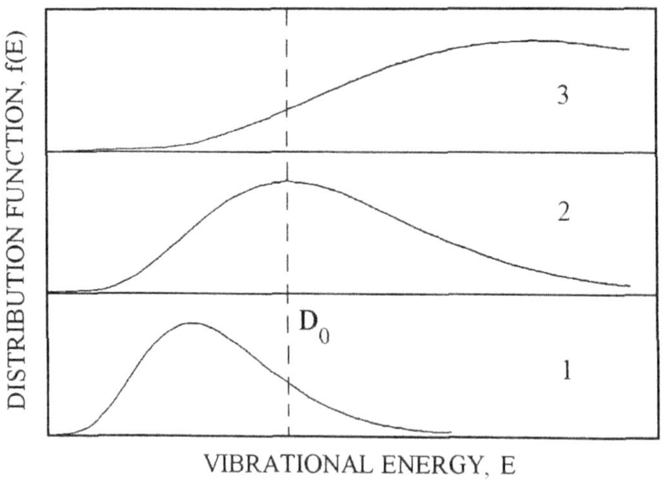

Figure 3: Qualitative variation of the distribution function over total vibrational energy at low (1), medium (2), and high (3) temperatures. 1–$T < T'$; 2–$T = T'$; 3–$T > T'$.

The characteristic temperature at which the maximum of the distribution function $f(E)$ over total vibrational energy E reaches the dissociation energy D_0 can be roughly estimated as

$$T' \approx \frac{D_0}{k(s-1)}.$$

It is seen from the Fig. 3 that at $T > T'$, the maximum of (30) shifts above the dissociation threshold and the majority of the complexes may be found in their metastable states. This statement is at variance with the widely held intuitive belief that the metastable states are capable of playing a significant role only at temperatures which are in excess of the effective dissociation threshold $T > D_0/k$ (see e.g. the review of Blaney and Ewing [13]). The manifold internal degrees of freedom in polyatomic complexes leads to an extension of the domain in phase space which may be occupied by metastable states. An explicit calculation [14] of the truncated partition function (29) results in the expression:

$$Q_b = \frac{\Gamma(s+m+1)}{\Gamma(m+1)\Gamma(s+1)}\, {}_1F_1(-m, s, s+1, 1-x). \tag{32}$$

Here, $m = D_0/\theta$, $x = \exp(-\theta/T)$, $\Gamma(\cdots)$ and ${}_1F_1(\cdots)$ are gamma- and hypergeometric functions, respectively. The truncated partition function apparently cannot be represented in terms of a product of truncated partition functions appropriate to the individual oscillators, as is erroneously asserted, e.g. in. [15]

The model we have considered in this Section is obviously oversimplified since it ignores at least three important details: First, this model does not make an allowance for the rotation of real molecular complexes in the gas phase. Second, the model assumes the harmonic oscillator approximation for the internal degrees of freedom within a complex, although the type of internal motion may easily transform to hindered or free internal rotation as the temperature rises. Third, the model does not consider free pair states which may exist in the purely repulsive segments of the intermolecular potential energy. All these defects can be corrected in the bimolecular phase space framework of classical statistical physics.

3.2 Subdivision of Phase Space for Diatomics

The problem of phase space subdivision into bound, metastable and free parts has been solved by Hill [16] and Stogryn and Hirshfelder [6] in the case of two structureless interacting particles. Hill [16] has shown that there exists a unique way to determine the classical partition function Q_b over bound states of two particles interacting with a spherically-symmetric potential $U(r)$. The total partition function can be represented as

$$Q = \frac{1}{2h^6}\int e^{-H/kT}\, dx\,dy\,dz\,d\theta\,d\phi\,dr\,dp_x\,dp_y\,dp_z\,dp_\theta\,dp_\phi\,dp_r. \tag{33}$$

Here x, y and z are the center of mass coordinates, θ, ϕ and r are conventional polar coordinates, which determine the relative position of the two centers of masses, the conjugate momenta are denoted by p_i, and H is the Hamiltonian of the system:

$$H = \frac{1}{4m}(p_x^2 + p_y^2 + p_z^2) + \frac{1}{mr^2}\left(p_\theta^2 + \frac{P_\phi^2}{\sin^2\theta}\right) + \frac{p_r^2}{m} + U(r). \qquad (34)$$

Here $m = \dfrac{m_1 m_2}{(m_1 + m_2)}$ is the reduced mass of a pair. Making the substitution of variables

$$P_\theta = \frac{p_\theta}{(mr^2 kT)^{1/2}}, \qquad P_\phi = \frac{p_\phi}{(mr^2 kT \sin^2\theta)^{1/2}}, \qquad P_r = \frac{p_r}{(mkT)^{1/2}} \qquad (35)$$

and performing an integration in (33) over the coordinates and momenta of the center of mass, one obtains

$$Q = \frac{2V}{\Lambda^6\sqrt{\pi}} \int e^{-U(r)/kT} r^2 dr \int e^{-(P_\theta^2 + P_\phi^2 + P_r^2)} dP_\theta dP_\phi dP_r. \qquad (36)$$

Here, V is gas volume and $\Lambda = \dfrac{h}{\sqrt{2\pi mkT}}$. After making an integration over the relative momenta from zero to infinity, the total partition function Q assumes the form

$$Q = \frac{2\pi V}{\Lambda^6} \int e^{-U(r)/kT} r^2 dr. \qquad (37)$$

Obviously, the condition of a pair being bound may be expressed by the following constraint:

$$P_\theta^2 + P_\phi^2 + P_r^2 \leq -\frac{U(r)}{kT}. \qquad (38)$$

This means that the sum of internal kinetic energy terms should be less than or equal to the potential energy of interaction taken with a negative sign, so that the particles are trapped in the potential well until a freeing collision occurs. Choosing the proper integration limits in (36) according to condition (38) one can obtain the following exact relationship for the truncated partition function over the bound pair states:

$$Q_b = \frac{2\pi V}{\Lambda^6} \int \frac{\gamma(\frac{3}{2}, -\frac{U(r)}{kT})}{\Gamma(\frac{3}{2})} e^{-U(r)/kT} r^2 dr. \qquad (39)$$

To reveal the true role played by the metastable states, a thorough analysis of the interaction potential is needed. As was mentioned in the previous Section, such an analysis was first made by Stogryn and Hirschfelder.[6] In what follows we shall deviate from the model of Stogryn and Hirshfelder in favor of the formalism of the energy distribution functions.[17] This will enable us to generalize the theory of Stogryn and Hirschfelder[6] to bimolecular complexes of arbitrary complexity and to obtain the results of[6] as a particular case.

3.3 General Model of the Phase Space Subdivision for Polyatomic Dimers

A detailed outline of this model is given elsewhere.[17] Briefly, it may be reduced to the following: An interaction potential is supposed to be a function of the intermolecular separation r and a set of angles $\Omega = \{\theta_1, \theta_2, \ldots, \theta_t\}$ which determine the relative orientation of the monomers. The number of angular intermolecular coordinates t equals three in the case of two linear interacting molecules; for two nonlinear molecules, $t = 5$ and so forth. The translational motion of the center of mass for molecular pairs is disregarded below. It is also assumed that different types of internal rotational and vibrational movements do not couple. Typically, the intermolecular separation within a weakly bound bimolecular complex exceeds the dimensions of the separate monomeric units. This enables one to consider a complex as a prolate symmetric top having two equal rotational constants B $\simeq$ C $\neq$ A.

The Hamiltonian for an interacting pair may be represented as:

$$H = U + L + E. \tag{40}$$

Here $U = U(r, \Omega)$ stands for the intermolecular potential surface, L is the energy of the two-dimensional rotation of a complex about the axes perpendicular to intermolecular A-axis, $L = hcB(M_2^2 + M_3^2)$, E incorporates all contributions from the of kinetic energy of internal movements and the energy of rotation against A-axis. Consequently, the number of degrees of freedom which contributes to the energy term E obviously equals to $(t + 2)$. The distribution function $f(L)$ is known[3] to assume the form appropriate to diatomics:

$$f(L) = \exp(-L/kT). \tag{41}$$

The energy term E may be thought of as a quadratic function of momenta, since the vibro-rotational interaction is neglected. In this case the distribution function over the energy E may be readily written as

$$f(E) = E^{\frac{t+2}{2}-1} \exp(-E/kT) = E^{\frac{t}{2}} \exp(-E/kT). \tag{42}$$

In the classical limit the partition function of a system with Hamiltonian (40) is equal to

$$Q = \int exp(-H/kT)dqdp. \tag{43}$$

It is worthwhile to transform the set of coordinates q and momenta p in (43) into the new variables r, Ω, E and L. Substituting (40), (41) and (42) into (43) and performing integration over E and L, one obtains the following expression for the total partition function:

$$Q = C' \int \int_0^\infty [Ja] \exp\left(-\frac{U(r,\Omega)}{kT}\right) r^2 dr d\Omega, \tag{44}$$

where

$$\int d\Omega \equiv \int d\theta_1\, d\theta_2 \cdots d\theta_t. \tag{45}$$

Here C' is a constant, $U(r,\Omega)$ is the intermolecular potential, and the Jacobian $[Ja]$ appears due to the transformation of coordinates. In general $[Ja]$ is a function of the coordinates; it is constant however when all the internal rotors are symmetrical tops.[18] For simplicity, let $[Ja] = const$ in the present Section. An exact definition is to be used in the Section below for two interacting linear molecules.

At every point of the potential energy surface (r, Ω), the bound states correspond to the negative values of Hamiltonian (40). In fact, the total kinetic energy of the motions which can cause dissociation of a complex proves to be less than the effective potential energy $\Phi_{eff} = U + L$ (cf. e.g. Hill's relationship (38)):

$$E < -\Phi_{eff} \qquad \text{or} \qquad E < -U - L.$$

Here, the effective potential Φ_{eff} must be negative; hence, $L < -U$. Let us turn to Fig. 4, from which the meaning of these limits becomes clear. Region 1 in this figure, limited by the right-angled triangle, refers to tightly bound states. Setting the proper integration limits in (43), one finds the partition function for bound states has the form:

$$Q_b \sim \int \int_\sigma^\infty e^{-U/kT} \int_0^{-U} e^{-L/kT} \int_0^{-U-L} E^{t/2} e^{-E/kT}\, dE\, dL\, r^2\, dr\, d\Omega. \tag{46}$$

Integration over E yields an incomplete gamma-function; subsequent integration over L leads to

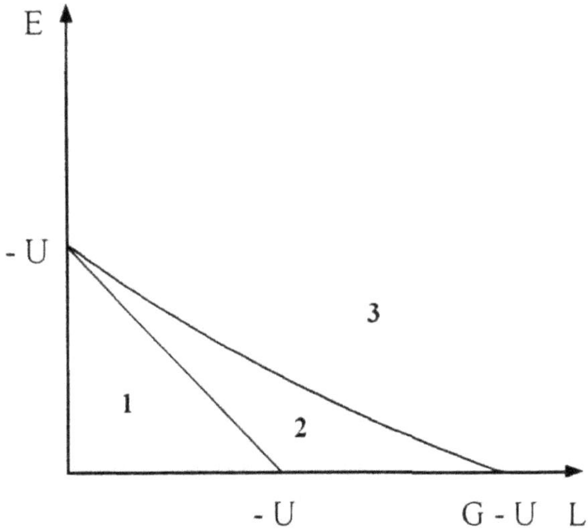

Figure 4: Domains of stable (1), metastable (2), and free (3) energy states for the structureless interacting monomers.

$$Q_b = (kT)^{(t/2)+2}\, C \int \int_{\sigma}^{\infty} \frac{\gamma(\frac{t}{2}+2, -\frac{U}{kT})}{\Gamma(\frac{t}{2}+2)} e^{-U/kT} r^2 dr\, d\Omega. \tag{47}$$

Here $C = C'[Ja] = const.$ In formulae (46) and (47) the integration is performed over a distance r starting from $r = \sigma$ at which the potential $U(r)$ is equal to zero (see Fig. 2).

Formula (47) transforms into corresponding formula (39) of Hill upon removing the integration over the angular coordinates and putting $t = -1$. The latter is necessary to rule out both internal rotations and the rotation of the complex relative to the intermolecular A-axis.

To determine explicitly the domains of metastable and free states, it is convenient to introduce an auxiliary surface $G(r, \Omega)$ which contains all the maxima points at $0 < L < L_{max}$ as well as the corresponding points on the purely repulsive branch of the effective potential energy curves (see Fig. 2). At $L > L_{max}$, the effective potential curves exhibit no extrema. This means that an interaction is repulsive and the formation of bound or quasibound states is not allowed. Figs. 2 and 4 give an idea of how the metastable states can be defined in the case of two interacting structureless particles. On the (E, L) plane metastable states are limited by the curve $E = G_h - U - L$. Here $G_h(L)$ is a function describing the dependence of the barrier height on the value of L. At any given interparticle separation $r > \sigma$, the terminal magnitude of rotational energy L, which leaves monomers within the region of stable or quasistable states equals $G - U$, where G is the magnitude of the auxiliary function $G(r)$ (see Fig. 2). The dependence of G_h on rotational energy L is determined by the shape of the selected potential curve. For instance, if one assumes that only dispersion-like forces contribute to the attractive part of the potential, i.e. $U(r) \sim 1/r^6$, then $G_h(L)$ is given by $G_h = 2L^{3/2}/3\sqrt{6D}$, where D is the dissociation energy.

Region 2 in Fig. 4 represents the domain of metastable states on the (E, L) plane. In this case the calculation of the truncated partition function is completely analogous to that of (47) and results in:

$$\begin{aligned} Q_m \sim & \int_\sigma^\infty e^{-U/kT} \int_0^{G-U} e^{-L/kT} \int_0^{G_h-U-L} E^{-1/2} e^{-E/kT}\, dE\, dL\, r^2\, dr \\ & - \int_\sigma^\infty e^{-U/kT} \int_0^{-U} e^{-L/kT} \int_0^{-U-L} E^{-1/2} e^{-E/kT}\, dE\, dL\, r^2\, dr \end{aligned} \tag{48}$$

or

$$\begin{aligned} Q_m = & (kT)^{1/2} C \int_\sigma^\infty e^{-U(r)/kT} \int_0^{G-U} \frac{\gamma(\frac{1}{2}, -\frac{G_h-U-L}{kT})}{\Gamma(\frac{1}{2})} e^{-L/kT}\, dL\, r^2\, dr \\ & -(kT)^{3/2} C \int_\sigma^\infty \frac{\gamma(\frac{3}{2}, -\frac{U(r)}{kT})}{\Gamma(\frac{3}{2})} e^{-U(r)/kT} r^2\, dr. \end{aligned} \tag{49}$$

The free states partition function Q_f can be obtained from the following obvious relationship

$$Q_f + Q_b + Q_m = Q.$$

Everythings changes principally if we turn to the case of a polyatomic complex with $(t+1)$ internal coordinates along which the complex may perform vibrational, hindered rotational or freely rotational movements. Here, metastable states are possible even at $H > G_h$ or $E > G_h - U - L$ at $\Phi_{eff} < G$ or $L < G - U$. The energy E can be arbitrarily large because the temporary redistribution of this energy is apt to take place among the internal motions which do not cause a dissociation of the complex. For instance, the excitation of a free internal rotation against the intermolecular axis will not cause dissociation at any excitation energy, provided the monomers are assumed to be rigid.

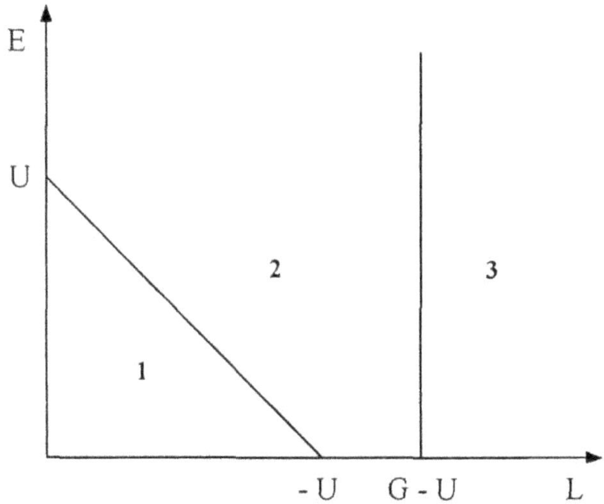

Figure 5: Domains of stable (1), metastable (2), and free (3) energy states for anisotropically interacting monomers.

Thus, integration over region 2 of the (E, L) plane (see Fig. 5) for a polyatomic complex yields an expression which may not be analogous to that of Stogryn and Hirshfelder:

$$Q_m \sim \int \int_{\sigma}^{\infty} e^{-U/kT} \int_{0}^{G-U} e^{-L/kT} \int_{0}^{\infty} E^{t/2} e^{-E/kT} \, dE \, dL \, r^2 \, dr \, d\Omega - Q_b$$

$$Q_m = (kT)^{(t/2)+2} \, C \int \left[\int_{\sigma}^{\infty} e^{-U/kT} \, r^2 \, dr - \int_{\sigma}^{\infty} e^{-G/kT} \, r^2 \, dr \right] d\Omega - Q_b. \tag{50}$$

Here Q_b is given by (47). The formulae obtained for truncated partition functions (47) and (49) or (47) and (50), respectively, for spherically symmetric and anisotropic intermolecular potentials enable one to calculate the statistical averages of any physical observable. To do this it is sufficient simply to introduce the coordinate dependent function into the corresponding integrand and to perform all the integrations over the domains of kinetic energy (E, L) as was explained above. In the following Section we shall consider two illustrative examples of this procedure.

4 Statistical Average of the Individual Characteristics of Dimers

4.1 Dissociation Energy

Weakly bound or van der Waals dimers exemplify a type of molecular system whose rotational or internal degrees of freedom can be substantially populated even at room temperature. Simple diatomics such as N_2 or O_2 are mentioned in the literature (see e.g. [1,2]) in connection with the problem of highly excited rovibrational states. This problem arises due to the crucial role played by the detailed chemistry in determining the heating load experienced by a re-entering space vehicle. The rates of dissociation of major gases govern the rate of spacecraft frictional heating in the upper reaches of the atmosphere. Dissociation rates at extremely high temperatures (up to ca. 50000 K) should account for the effective lowering of the dissociation energy due to the rotational and vibrational excitation of the reacting species. This effect holds for weakly bond dimers near room temperature since the dissociation energies of dimers amount typically to several hundreds of degrees K. Some disagreements may arise in the experimental estimates of the dissociation energy of weakly bound dimers due to this effect, until exact spectroscopic, e.g. line strength measurements are employed.

As a primary step in the modelling of this effect, the concept of the temperature dependent effective bond energy can be introduced which is an average of the state-selected energy of dissociation over the equilibrium or nonequilibrium thermal distribution. Alternatively, this value may be considered as a characteristic of the thermally averaged reservoir of energy of the statistical ensemble of diatomics or weakly bound dimers.

In most cases, the nonequilibrium regime can be characterized by a multi-temperature scheme. Rotational degrees of freedom can be considered at equilibrium with translational movement and can be described by $T_r = T_t$, whilst vibrational excitation is characterized by its own temperature $T_v \neq T_r$. Since the integration in formulae (46),(48) are performed over separate subsystems, it is sufficient to ascribe different temperatures to roto-translational and vibrational subsystems. Consequently, the following relationship can be obtained for the partition function over bound and metastable states in the two-temperature regime:

$$Q_b + Q_m = \int_0^\infty e^{-U/kT_v} \int_0^{G-U} e^{-L/kT_r} \int_0^{G_h-U-L} E^{-1/2} e^{-E/kT_v}\, dE\, dL\, r^2\, dr. \quad (51)$$

Suppose that the intermolecular potential can be approximated well by a harmonic oscillator potential. This means that an additional quadratic term involving one of the canonical variables (distance) appears in the Hamiltonian. To account for this term, it is not necessary to perform an integration over the interparticle separation; it is sufficient to augment the number of degrees of freedom by one in the expression for the distribution function (42) of the internal energy E. Hence, the formula (51) assumes the following form

$$Q_b + Q_m = \int_0^{G+D_e} e^{-L/kT_r} \int_0^{G_h+D_e-L} e^{-E/kT_v}\, dE\, dL. \tag{52}$$

Here D_e is the full depth of the potential well U of a nonrotating dimer. The use of the harmonic approximation makes it possible to obtain simple estimates valid irrespective of the model adapted for the shape of intermolecular potential well.

To calculate the temperature dependence of the effective dissociation energy, one should introduce the function $D(L) = G_h + D - E - L$ into the integrand of (52) and normalize the integral obtained over the corresponding partition function given by (52). Here $D = D_e - E_0$, where E_0 is the zero-point energy. It has been shown in [2] that the metastable states contribute insignificantly to the average dissociation energy. Provided the metastable states are neglected, that is $G_h = G = 0$, the integrations in (52) may be carried out and one can get an expression for the average decrease of dissociation energy $\Delta D = D - \langle D \rangle$ in the form:

$$\Delta D = \frac{k(T_v^2 - T_r^2) + T_r(D + kT_r)exp(-D/kT_r) - T_v(D + kT_v)exp(-D/kT_v)}{(T_v - T_r) + T_r\, exp(-D/kT_r) - T_v\, exp(-D/kT_v)}. \tag{53}$$

At $T_v \to 0, T_r \neq 0$ or $T_r \to 0, T_v \neq 0$ this formula reduces to

$$\Delta D = kT\frac{1 - (1 + D/kT)\; exp(-D/kT)}{1 - (1 + D/kT)}. \tag{54}$$

Here T stands for T_r or T_v, respectively. It stems from Eq. 54 that at low temperature the dissociation energy decreases by kT, that is by the average vibrational or end-over-end rotational energy. The limiting value $\langle D \rangle$ at $T \to \infty$ amounts to $\frac{1}{2}D$. This figure is obvious since, when all rotational (or vibrational) states up to dissociation limit are equally populated, the dissociation energy at the average should be halved. This differs considerably from the case, when the vibrational and rotational motions are at equilibrium and the average decrease of dissociation energy amounts to:

$$\Delta D = 2kT \, \frac{1 - \left[1 + D/kT + \frac{1}{2}(D/kT)^2\right] exp(-D/kT)}{1 - (1 + D/kT)\, exp(-D/kT)}. \tag{55}$$

In this case, at low temperature the decrease of dissociation energy amounts to $2kT$ corresponding to a sum of vibrational and two-dimensional rotational energies. Less obviously at $T \to \infty$ the effective dissociation energy reaches its terminal value $\langle D \rangle = \frac{1}{3} D$.

Fig. 6 demonstrates the temperature dependencies of $\langle D \rangle$ plotted according to relationships (54) and (55) for pure nitrogen and oxygen.

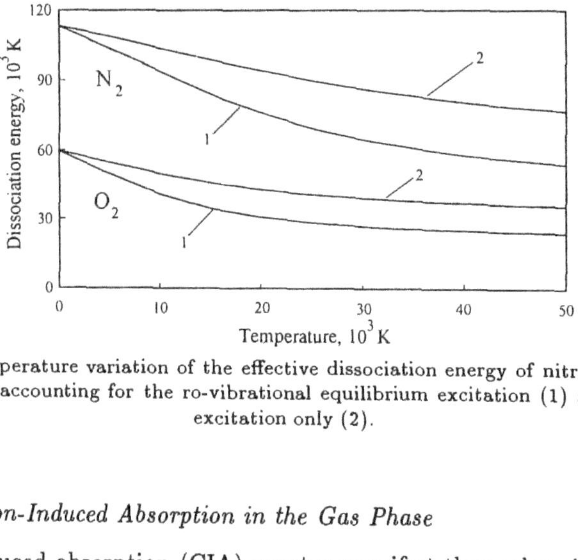

Figure 6: Temperature variation of the effective dissociation energy of nitrogen and oxygen molecules accounting for the ro-vibrational equilibrium excitation (1) and rotational excitation only (2).

4.2 *Collision-Induced Absorption in the Gas Phase*

Collision-induced absorption (CIA) spectra manifest themselves in the region of the homonuclear diatomic fundamental bands as weak, diffuse absorption bands, which appear as the gas density increases. Since their first detection by Welsh *et al.* in 1949, [19] these bands have attracted a good deal of attention from spectroscopists. CIA is known to affect the transmission of radiation in the Earth's atmosphere, [20,21] as well as to play a crucial role in the radiation budget of the giant planets, [22,23] and is significant in the atmospheres of cool stars. [24] An excellent reference book on collision-induced spectroscopy has been published recently by L. Frommhold. [25] An insight to the contemporary state of the art of collision-induced phenomena is provided by a collection of papers edited by G. Tabisz and M. Neuman. [26]

An important feature is that the study of CIA phenomenon enables one to achieve better insight into the nature of intermolecular interactions. During several decades, the study and interpretation of CIA band profiles served uniquely to characterize spectroscopically the van der Waals dimer formation in the gas phase. Nowadays, the use of the much more advanced molecular beam technique in conjunction with high-resolution spectroscopic tools results in obtaining a wealth of information about van der Waals species (see, e.g., [27,28]). Thorough investigation of collision-induced absorption phenomenon remains of great significance for both fundamental and applied molecular science.

It is a widely held belief that the mechanism of induced absorption in the region of homonuclear fundamentals can well be described by consideration of the transient dipole moment occuring during a binary collision. In the limit of low temperature some additional contribution arises due to formation of van der Waals dimers. The purely collisional nature of this phenomenon is emphasized even by its most frequently employed term 'collision-induced' absorption. Explicit analysis is needed, however, for the concrete systems to establish whether intermolecular interactions result in the formation of a manifold of bound and quasibound states or whether they really leave the interacting species unbound, thus keeping the purely elastic collisional nature of CIA intact.

Integrated Intensity

The integrated intensity of absorption I can be expressed in terms of the product of an average intensity of absorption per single molecular pair S_{ind} and the number of pairs N_p per unit volume in a gas at equilibrium:

$$I = \frac{1}{l}\int \alpha(\nu)\, d\nu = S_{ind}\frac{N_p}{V}. \tag{56}$$

Here $\alpha(\nu)$ is the absorption coefficient, l is the pathlength in the absorbing layer, V stands for the gas volume. The total number of pairs can be easily estimated as

$$N_p = \frac{N(N-1)}{2} \approx \frac{1}{2}N^2, \tag{57}$$

provided N stands for the total number of molecules in a gas. Note that only a fraction of the pairs, $x_{bm} = N_{bm}/N_p$, are in their bound or metastable states; the rest x_f experience elastic collision with a variety of impact parameters.

In general the dipole moment induced by one molecular unit due to interaction with another one is strongly coordinate dependent $\mu(r,\Omega)$. The average absorption intensity per single molecular pair S_{ind} can therefore be determined as the squared induced dipole moment μ^2, averaged over space with the appropriate Boltzmann distribution function:

$$S_{ind} = Q^{-1} \int \int_{0}^{\infty} \mu^2 \, exp(-U/kT) \, r^2 \, dr \, d\Omega, \tag{58}$$

where the partition function Q is given by

$$Q = \int \int_{0}^{\infty} exp(-U/kT) \, r^2 \, dr \, d\Omega.$$

At large intermolecular separations, the potential U is close to zero, so the integral Q is proportional to a gas volume V. The integral in the numerator of (58) does not depend on V and remains limited since the induced dipole moment drops rapidly with increasing r. Substitution of (58) and (57) into (56) results in the integrated absorption intensity being insensitive to whether the molecules are bound or not. In fact, the integrated intensity I equals:

$$I = c \left\langle \mu^2 \right\rangle \left(\frac{N}{V} \right)^2 . \tag{59}$$

Here c is constant and the angular brackets mean integration over the whole set of spatial variables. Despite the seeming CIA effect being insensitive to the formation of bound and quasibound states, the decomposition of integrated intensity or, more generally, CIA band profile into bound, metastable and free contributions is extremely valuable. This is because of at least two reasons. First, it is a widely accepted practice, which is, however, by no means justified, to find an average in (59) by using the simplest angular pre-averaged intermolecular potential and/or by making allowance for only linear or slightly curvilinear trajectories of colliding molecules. Thus, the contribution from bound plus metastable dimeric states is arbitrarily excluded from consideration, or at least diminished. Second, to build the CIA band envelope it is highly desirable to discriminate between bound, metastable and free states since each of these types of states can be characterized by its particular mechanism of line-broadening.

To reveal the true role of bound and quasibound dimeric contributions to the CIA integrated intensity, the integration over certain domains in phase space should be performed. Let an overall CIA intensity be represented as

$$I = \frac{N_p}{V} \left(S_{bm}^{ind} + S_f^{ind} \right) = I_{bm} + I_f . \tag{60}$$

Here subscripts bm and f refer to the average over the domains of bound plus metastable and free states, respectively. For the sake of simplicity, the induced dipole moment for dimers can be set coordinate independent, since the variations of intermolecular separation within the dimeric configuration are negligible. Consequently

$$S_{bm}^{ind} \sim \langle \mu^2 \rangle_{bm} \, x_{bm}$$

and the first term in (60) can be rewritten as follows:

$$I_{bm} = \frac{N_p}{V} S_{bm}^{ind} = \frac{N_p}{V} \langle \mu^2 \rangle_{bm} \, x_{bm}. \tag{61}$$

Here x_{bm} is a fraction of bound and metastable states relative to the total number of pair states $x_{bm} = N_{bm}/N_p$ and is average dipole moment $\langle \mu^2 \rangle_{bm}$ now considered to be constant. It seems worthwile to express the number of bound and metastable dimers via the mole fraction x_2 of dimers in equilibrium gas, so that $N_{bm} = x_2 N$, N being the total number of particles. Thus, instead of (61) the following relationship is valid:

$$I_{bm} = \frac{N}{V} \langle \mu^2 \rangle_{bm} \, x_2. \tag{62}$$

Making use of the mass-action law (9) together with the ideal gas equation of state $P = \frac{\rho}{\overline{m}} RT$, a portion of integrated CIA intensity which relates to the bound and metastable dimeric states assumes the following form:

$$I_{bm} = \langle \mu^2 \rangle_{bm} \frac{RT}{K_p} [\frac{x_1}{\overline{m}}]^2 \rho^2 \approx \langle \mu^2 \rangle_{bm} \frac{RT}{K_p} \frac{\rho^2}{m_1^2}. \tag{63}$$

Here R is gas constant, $\overline{m}$ and m_1 are mean and monomeric molecular masses, ρ is the gas density and it is assumed that the extent of dimerization is weak. It is seen that the term I_{bm} is proportional to the gas density squared. Formula (63) has been successfully employed in [29] to interpret the temperature dependence of CIA intensity in the region of the Fermi dyad of carbon dioxide. In that particular case, even complete neglect of the contribution due to elastic collisions did not give rise to substantial error.

The second term in (59) refers to free molecular pairs and can be represented as:

$$I_f = \frac{N_p}{V} S_f^{ind} = \frac{N^2}{2V} \langle \mu^2 \rangle_f \, . \tag{64}$$

It stems from the results discussed in the Sec. 3.3 that an accurate average of the dipole moment in (64) for interacting linear molecules can be performed using the following relationship:

$$\langle\mu^2\rangle_f = C(kT)^{7/2} \int [\int_0^{\sigma} \mu^2(r,\Omega)\, exp(-U(r,\Omega)/kT)\, r^2\, dr - \int_{\sigma}^{\infty} \mu^2(r,\Omega)\, exp(-G(r,\Omega)/kT)\, r^2\, dr]\, d\Omega. \tag{65}$$

To give an idea of how the temperature dependence of the integral of type (65) looks like, we shall consider below an example of calculations made with the use of a rather sophisticated intermolecular potential.

A Case Study of the CIA Intensity Subdivision for N_2, O_2, and CO_2

The intermolecular potential $U(r,\Omega)$ for symmetrical linear molecules can be chosen in the form of the site-site interaction model supplemented by electrostatic interactions:

$$U(r,\Omega) = \sum_{i,j} 4\epsilon_{ij} \left[\left(\frac{\sigma_{ij}}{r_{ij}}\right)^{12} - \left(\frac{\sigma_{ij}}{r_{ij}}\right)^{6} \right] + V_{QQ}. \tag{66}$$

Here ϵ_{ij} and σ_{ij} are Lennard-Jones parameters for the interaction of the ith atom of molecule 1 and the jth atom of molecule 2, r_{ij} is the distance between these atoms, term V_{QQ} is the energy of quadrupole-quadrupole interaction

$$V_{QQ} = -\frac{3Q}{4r^5}[1 + 2(5\, cos\theta_1\, cos\theta^2 - sin\theta_1\, sin\theta_2\, cos\phi + cos\theta_1\, cos\theta_2)^2 - 5(cos^2\theta_1 + 3cos^2\theta_1\, cos^2\theta_2 + cos^2\theta_2)]. \tag{67}$$

The coordinates notation is shown in Fig. 7, Q is the quadrupole moment of the monomer molecule. The parameters of the potential surface (66) used in the calculation [30] were taken from the work of Bouanich [31] and are shown in Table 1. The squared dipole moment $\mu^2(r,\Omega)$ induced in a sample molecule by a point quadrupole from the neighbouring molecule can be found through the following relationships [32]

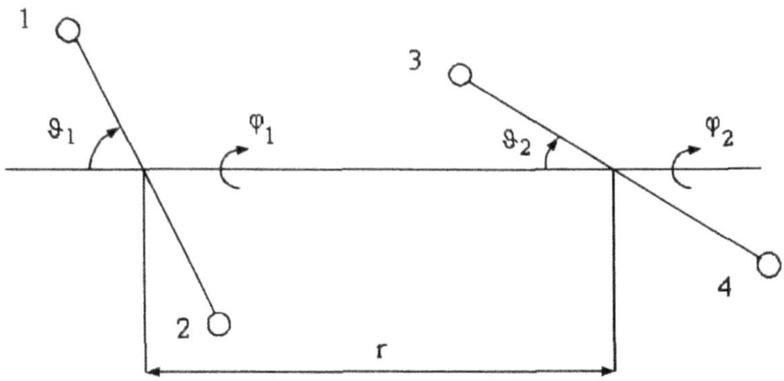

Figure 7: System of coordinates used in the calculations of induced dipole momentum.

$$\mu^2 = \mu_x^2 + \mu_y^2 + \mu_z^2,$$

where

$$\begin{aligned}
\mu_x &= F_x(\alpha_\perp + \Delta sin^2\theta_2\, cos^2\phi_2) + F_y\Delta sin^2\theta_2\, cos\phi_2\, sin\phi_2 \\
&\quad + F_z\Delta sin\theta_2\, cos\theta_2\, cos\phi_2 \\
\mu_y &= F_x\Delta sin^2\theta_2\, sin\phi_2\, cos\phi^2 + F_y(\alpha_\perp + \Delta sin^2\theta_2\, sin^2\phi_2) \\
&\quad + F_z\Delta sin\theta_2\, cos\theta_2\, sin\phi_2 \\
\mu_z &= F_x\Delta sin\theta_2\, cos\theta_2\, cos\phi^2 + F_y\Delta sin\theta_2\, cos\theta_2\, sin\phi_2 \\
&\quad + F_z(\alpha_\perp + \Delta cos^2\theta_2).
\end{aligned} \tag{68}$$

Here $\Delta = \alpha_\perp - \alpha_\|$, $\alpha_\perp$ and $\alpha_\|$ are the components of the polarizability tensor.

The components of the dipole field of molecule 2 at the position of molecule 1 are given by:

$$\begin{aligned}
F_x &= -\frac{3Q}{r^4} cos\theta_1\, sin\theta_1\, cos\phi_1 \\
F_y &= -\frac{3Q}{r^4} cos\theta_1\, sin\theta_1\, sin\phi_1 \\
F_z &= -\frac{3Q}{2r^4}(3cos^2\theta_1 - 1).
\end{aligned} \tag{69}$$

Table 1: Parameters of the site-site potential (66), taken from.[31]

	N_2	O_2	CO_2		
			$(ij) = (CC)$	$(ij) = (OO)$	$(ij) = (CO)$
$\epsilon_{ij}(K)$	36.3542	51.7349	70.7722	51.8037	60.5497
$\sigma_{ij}(\text{Å})$	3.35155	3.07045	2.99883	3.0058	3.0023
$Q(D\text{Å})$	1.4	0.4		4.3	

Knowing the potential surface (66), one can determine an auxiliary function $G(r, \Omega)$ by numerical approximation of the barrier height $\Phi_{eff}(r_h)$ dependence on coordinate r at any fixed orientation of the monomers. Then, the resulting $G(r, \Omega)$, together with the potential $U(r, \Omega)$ and the squared induced dipole moment $\mu^2(r, \Omega)$ given by formulae (68), and (69) are to be substituted into the formula (65) and analogous expressions for $\langle \mu_b^2 \rangle$ and $\langle \mu_m^2 \rangle$. To calculate all the integrals of type (65), an efficient numerical algorithm has been used.[30] The selection of results of these calculations for nitrogen is shown in Fig. 8 in terms of the temperature dependence of the ratios of $\langle \mu_f^2 \rangle$, $\langle \mu_b^2 \rangle$ and $\langle \mu_m^2 \rangle$ to $\langle \mu^2 \rangle$, which are denoted respectively as P_f, P_b and P_m.

It is seen from Fig. 8 that in the case of simple diatomics the role of tightly bound states is small near room temperatures. This makes it hardly possible to find distinct spectroscopic features of these dimers in equilibrium gases at ambient temperatures by the use of high resolution spectroscopy.

Note that for carbon dioxide, quasibound or metastable states dominate at normal temperatures.[30] Metastable states are thought to give rise to diffuse spectroscopic features due to their limited lifetime caused by fast intramolecular relaxation. That is why room temperature CIA spectra of carbon dioxide demonstrate no obvious fine structure of dimers, although the absorption of radiation seems to be mostly dimeric by its nature. The same may be true for the regions of continuous far-infrared absorption in water vapor, where the absorption can be associated with higher order intermolecular transitions in

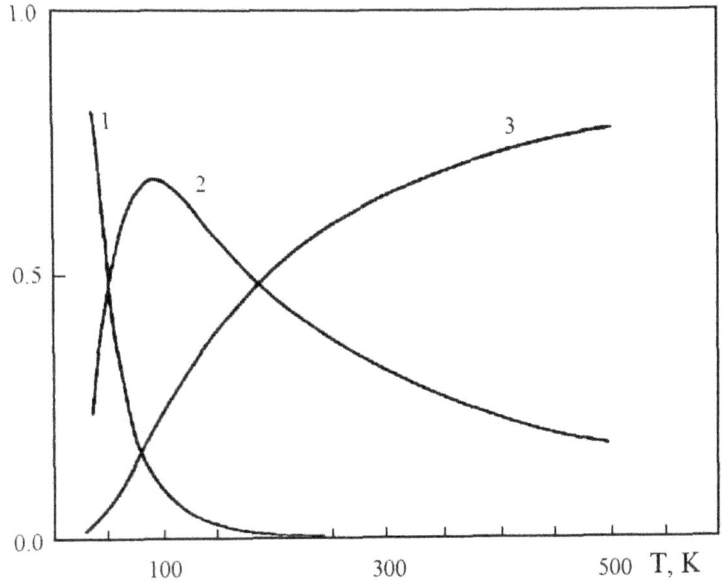

Figure 8: Temperature dependence of the normalized CIA intensities for the stable P_b (1), metastable P_m (2), and free P_f (3) states of the interacting nitrogen molecules.

a water dimer. In the phase space of interacting water molecules, metastable states were shown [33] to dominate over bound states above near room temperature. This may cause the formation of very broad and diffuse dimeric absorption instead of sharp peaks expected for rigid dimers, by analogy with the spectra of 'normal' molecular species.

In this Section we attempted to demonstrate that the effect of CIA can hardly be understood by accounting for the formation of free states only, that is by ignoring the formation of the substantial fraction of metastable and stable dimers. The absence of distinct spectroscopic features in the observed atmospheric spectra attributable to rigid dimers may not be sufficient to rule out of consideration their possible loosely bound modifications.

Acknowledgments

The author gratefully acknowledges valuable suggestions and assistance by Prof. L. M. Trafton in the course of proof-reading of the present Chapter.

References

1. R. Jaffe in *Thermophysical Aspects of Re-entry Flows*, eds J. N. Moss and C. D. Scott (AIAA, New York, 1986).
2. A. Vigasin, *Chem. Phys. Lett.* **242**, 33 (1995).
3. L. D. Landau and E. M. Lifshitz, *Statistical Physics* (Pergamon Press, Oxford, 1980).
4. *JANAF Thermochemical Tables, NSRDS-NBS 37*, eds D. R. Stull and H. Prophet (U.S.Department of Commerce, Washington, 1971).
5. *Thermodynamic Properties of Individual Substances*, eds L. V. Gurvich, I. V. Veyts, and C. B. Alcock (Hemisphere Publ. Co., 1991).
6. D. E. Stogryn and J. O. Hirshfelder, *J. Chem. Phys.* **31**, 1531 (1959).
7. J. B. Anderson, R. P. Andres, J. B. Fenn, and G. Maise in *Rarefied Gas Dynamics, Vol. 2*, ed. J. H. de Leeuw (Academic Press, N.Y., 1966), p.106.
8. J. M. Calo and J. H. Brown, *J. Chem. Phys.* **61**, 3931 (1974).
9. J. M. Calo, *J. Chem. Phys.* **62**, 4904 (1975).
10. G. Brocks, *J. Chem. Phys.* **88**, 578 (1988).
11. P. R. Certain and N. Moiseyev, *J. Phys.Chem.* **89**, 2974 (1985).
12. H. B. Levine, *J. Chem. Phys.* **56**, 2455 (1972).
13. B. L. Blaney and G. E. Ewing, *Ann. Rev. Phys. Chem.* **27**, 553 (1976).
14. A. Vigasin, *Chem. Phys. Lett.* **117**, 85 (1985).
15. W. Forst, *The Theory of Unimolecular Reactions* (Academic Press, New York, 1973).
16. T. Hill, *Statistical Mechanics* (McGraw-Hili, New York, 1956).
17. A. Vigasin, *Infrared Physics* **32**, 461 (1991).
18. I. Godnew, *Berechnung Thermodynamischer Funktionen aus Moleküldaten* (VEB Deutscher Verlag der Wissenschaften, Berlin, 1963).
19. M. F.Crawford, H. L. Welsh, and J. L. Locke, *Phys. Rev.* **75**, 1607 (1949).
20. C. Rinsland, M. A. H. Smith, J. M. Russel III, J. H. Park, and C. B. Farmer, *Applied Optics* **20**, 4167 (1981).
21. C. Rinsland, M. A. H. Smith, R. K. Seals, Jr., A. Goldman, F. J. Murcray, J. C. Larsen, and P. L. Rarig, *J. Geophys. Res.* **87**, 3119 (1982).
22. L. M. Trafton, see Chapter 6 of this book.
23. R. Tipping in *Phenomena Induced by Intermolecular Interactions*, ed. G. Birnbaum (Plenum Press, New York–London, 1985).
24. A. Borysow in *Collision and Interaction-Induced Spectroscopy*, ed. G. C. Tabisz and M. N. Neuman (Kluwer Acad. Press, Dordrecht, 1995).

25. L. Frommhold, *Collision-induced Absorption in Gases* (Cambridge University Press, Cambridge, 1994).
26. *Collision and Interaction-Induced Spectroscopy*, ed. G. C. Tabisz and M. N. Neuman (Kluwer Acad. Press, Dordrecht, 1995).
27. *Structure and Dynamics of Weakly Bound Molecular Complexes*, ed. A. Weber (Reidel, Dordrecht, 1987).
28. *Atomic and Molecular Clusters. Studies in Physical and Theoretical Chemistry*, ed. E. Bernstein (Elsevier, Amsterdam, 1990).
29. A. A. Vigasin, E. G. Tarakanova, and G. V. Tchlenova, *J. Quant. Spectrosc. Radiat. Transfer* **50**, 695 (1994).
30. S. Yu. Epifanov and A. A. Vigasin, *Chem. Phys. Lett.* **225**, 537 (1994).
31. J.-P.Bouanich, *J. Quant. Spectrosc. Radiat. Transfer* **47**, 243 (1992).
32. J. P. Colpa and J. A. A. Ketelaar, *Molecular Physics* **1**, 343 (1958).
33. S. Yu. Epifanov and A. A. Vigasin, *Molecular Physics* **90**, 101 (1997).

NON-RIGIDITY IN VAN DER WAALS MOLECULES: SOME CASE STUDIES

R. G. A. BONE[a]

Department of Chemistry, McMaster University, 1280 Main St. West, Hamilton, Ontario, L8S 4M1, Canada

Weakly bound complexes present significant challenges to both experimentalist and theoretician. In many cases, large-amplitude motions of the complexes make the deduction of structural parameters from spectroscopic data uncertain. When attempting to calculate those quantities, theoreticians are faced with the considerable cost and difficulty of locating stationary points on shallow potential energy surfaces. Often a useful measure of agreement between theory and experiment is the value of a tunneling splitting. But accurate calculation of barriers with *ab initio* methods is far from easy and the use of approximate functional forms to mimic the potential can be similarly fraught with difficulties. This chapter describes a number of complexes of acetylene and sulphur dioxide in order to show that non-rigidity may manifest itself in many different ways. In each case the consequence is a rich set of spectra whose interpretation caused controversy over a period of years. Whilst accompanying theory cannot yet offer quantitative agreement with experiment, its value is in identifying global minima and low-lying transition states from which likely interconversion pathways may be deduced. Accompanying group theory can often show that suggested mechanisms are consistent with spectroscopic observations.

1 Introduction

Van der Waals complexes deserve attention for many reasons. Primarily they open a window on the realm of intermolecular forces. Additionally, their existence might direct us to the intermediates and precursors to reactions that form the basis of chemistry in the cosmos and they may play a number of roles in atmospheric phenomena both in comets and small planets.[1] The conditions 'out there,' are condusive to the formation of weakly bound species: the low temperatures in regions far from stellar centres and the density of the species of interest make it highly probable that clusters form a measurable component of gas composition. Additionally, the small molecular species that have been observed in the cosmos are also known to form van der Waals (and/or hydrogen-bonded) complexes in the laboratory. The spectroscopic measurement of complexes has typically used micro-wave and infra-red radiation,[2,3,4] a wavelength region which has been successfully used in astronomical mea-

[a]Current address: Proteus Molecular Design Ltd., Lyme Green Business Park, Macclesfield, Cheshire, United Kingdom, SK11 0JL

surements. There is therefore a high expectation that the presence of van der Waals molecules in extraterrestrial settings may be readily confirmed.

Nevertheless, the elucidaton of their structures has never been straightforward, despite the fact that the experimental spectroscopist has a greater level of control than the astronomer over the conditions of observation and the species in question. For example, the use of isotopomers has played a critical role in structure-determination by leading to new sets of rotational constants, in the imposition of alternative selection rules on spectral transitions and in gaining insight into intermolecular exchange processes. Moreover, in the laboratory, the application of molecular beam methods to the formation and measurement of van der Waals molecules, whilst enabling a whole menagerie of species to be created and furnishing a wealth of high-precision information on their low-lying states, does not lead to easy comparison with atmospheric or remote measurements. Clusters are formed in molecular beams in non-equilibrium conditions and the pressures and temperatures involved are far lower than those typically found in the Solar System.

Despite such difficulties, evidence for the existence of van der Waals molecules in atmospheres has been gleaned from several sources. It is typical to assume a termolecular mechanism of formation and destruction [1] of a dimer, D, from monomers, M,

$$M + M + M \rightleftharpoons D + M. \tag{1}$$

In which case the third monomer serves to take away excess kinetic energy from the collision. Calculations of dimer mole fractions at equilibrium conditions suggest that dimers of nitrogen and oxygen may even be more abundant than 'trace' atmospheric constituents high in the Earth's atmosphere.[1] Furthermore, simple models show that the rate of dimer destruction according to Eq. 1 is far greater [1] than that due to photo-excitation or vibrational predissociation.[5] Indeed, continuum absorption portions of infrared spectra of the Earth's upper atmosphere, conventionally attributed to collision-induced transitions, have in fact been shown to be consistent with transitions amongst internal states of van der Waals dimers.[6]

Vibrationally averaged effects have always injected uncertainty into the derivation of basic structural parameters,[7,8] and in the case of fluxional behaviour, the spectra can be characteristic of the transition state for large-amplitude motion. Significant departures from a symmetric configuration can result in enhanced absorption intensity in the vibrational modes being observed. The most common result is that spectral lines are split by tunneling between equivalent configurations. It will normally require a high-resolution measurement to detect this effect but it manifests itself in almost all modes of vibration. Although the barriers to internal motion in van der Waals molecules are extremely small (compared with the binding energies of normal molecules), under the conditions at which they are found, kT is also sufficiently small that the lowest bound states are occupied and large-amplitude motions effectively mix different configurations. Of particular difficulty to interpret are the spectra of clusters for whom at least one rotational constant is approximately the same size as a tunneling splitting.

Traditionally, van der Waals molecules have been described with empirical methods deriving from the long range model of intermolecular forces. [9,10,11] Methods such as intermolecular perturbation theory [12] are attractive in that they offer an understanding of the forces involved but do not easily lead to accurate calculation. In the last two decades, as theoretical and computational chemistry have emerged as important fields in all areas of molecular study, it has become common to apply *ab initio* quantum mechanical methods to van der Waals systems. [14,15] Such methods seek to solve the electronic Schrödinger equation at a fixed set of nuclear coordinates, by expanding the wavefunction in a linear combination of atom-centred basis functions. [16,17] The least costly way of doing this (without resort to empirical parameterisations of integrals or energy functionals) is to obtain a converged set of molecular orbitals via a self-consistent field approximation and to add the electronic correlation energy via perturbation theory. [18] By now carrying out *ab initio* calculations of the structures of van der Waals molecules has successfully masqueraded as a scientific endeavour in its own right and will almost always accompany an experimental investigation. There are even examples where structural constants have been reliably computed *ab initio* without an unambiguous deduction from spectra alone; additionally, calculations may give information on transition state structures and energy barriers which are not always directly derivable from experiment. Often, merely an estimate of the overall binding energy can be valuable in thermochemical analysis. Once a wavefunction has been computed, a whole host of other useful properties can be computed. For example, from the electron density, bonding interactions [19] may be studied using the theory of Atoms in Molecules. [20]

But *ab initio* calculations remain difficult for a number of reasons. Technically, it is hard to locate stationary points (minima in transition states) on shallow potential energy surfaces. The weakness of the binding forces and the often isotropic nature of the intermolecular pair potentials leads to enormous difficulties in using gradient-based methods to find equilibrium structures. Nevertheless, with advances in computer technology, the CPU requirements of such searches have become less demanding. Indeed for the species described here, fully optimized geometries were obtainable at adequate levels of theory. Despite this, one still has to consider costly methods of calculation. For these molecules, whether dominated by electrostatic forces or not, correlation energy is a necessary component of the calculations, because estimates of the dispersion forces are not possible without it. Dispersion forces achieve an overall contraction in optimum intermolecular distances regardless of orientation, even when electrostatic effects are not favourable. The need for a correlated level of theory leads to the requirement that large basis sets are used and thus the cost of calculation is driven up. A benefit to the use of a large basis set is that the basic electrostatic properties of the individual monomers are likely to be well represented. Errors in describing monomer charge distributions can lead to very poor descriptions of complexes. A side-effect of finite basis set methods is the need for some correction for the so-called "Basis Set Superposition Error" (BSSE).[21] This artifact of calculation implies that energy differences between bound complexes and separated monomers are over-estimated. The basis set used for the dimer exceeds that used for each separate monomer implying that they do not share the same 'zero' of energy in the calculation. There is still vociferous debate about the most rigorous way to correct for this problem[22,23,24,25] and basis set quality as well as size influences the effect.[26] In the data presented here, BSSE has been taken into account when calculating energy differences but its influence on optimized geometries has not been considered.

This chapter outlines a number of case-studies of van der Waals molecules which exhibit non-rigid behaviour. The examples chosen are from the chemistry of SO_2 and C_2H_2. Whilst these are not the most common species either in the interstellar medium or the Earth's atmosphere, they each have a rich variety of cluster formation and exhibit a representative selection of the non-rigid motion that has been observed in the growing catalogue of small van der Waals clusters. Consequently they serve as useful prototypes for future investigations. Their study is prefaced by a discussion of interconversion tunneling, symmetry and non-rigidity.

2 Tunneling Phenomena and Symmetry

When a molecular structure possesses geometric symmetry,[27] there may be more than one equivalent 'version' of it on the potential energy surface.[28] In essence, an element of symmetry relates two or more identical nuclei (unless we are considering reflections in the plane containing them).[29] Different versions of a structure arise when distinct labels are attached to atoms of the same element so that more than one unique (non-superimposable) labelling of the structure becomes possible. The existence of several versions of a structure (although physically indistinguishable from one another) becomes apparent when motions between them occur on the timescale of observation.[27] Proper symmetry operations of the system now include those permutations which effectively convert one version into another.[30] In practice, the lowest barrier on the potential energy surface must not be high enough to quench the effect; a temperature dependence can therefore be expected. Examples of equilibrium structures which visit several versions of themselves are widespread in chemistry. For example, the NMR spectrum of PF_5 at room temperature shows that all five fluorine atoms are equivalent whereas at low temperatures a pair of 'apical' fluorines distinguishable from the three 'equatorial' fluorines[27] of a trigonal bipyramidal minimum energy structure. With sufficient energy, a pseudorotation mechanism allows fluorine atoms in each of the two geometrically distinct positions to interchange with one another. Similarly, carbocations can be 'scrambled'. Bullvalene easily undergoes an internal rearrangement which permutes the carbon atoms of its skeleton.[31] In inorganic and organic chemistry, such molecules are referred to as 'floppy' or 'fluxional', terms which should be considered synonymous with 'non-rigid' in physical and theoretical chemistry.[29] Indeed, in the realm of molecular spectroscopy a 'non-rigid' molecule is one for which interconversion tunneling can be measured.[30]

Whilst it is usually straightforward to think of a fluxional process in mechanistic terms, the quantum mechanical picture is of tunneling between versions. Wavefunctions for the different versions penetrate the barriers between them, mix and cause splittings of the levels which are localised in each well. Transitions between such sets of split levels now themselves appear to be split or perturbed. So-called interconversion tunneling has been observed in the vibrational and rotational spectra of many molecules and, amongst high-resolution spectroscopists, is a primary indicator of non-rigidity. (It should be understood that not all spectroscopic methods will yield direct indication of tunneling in the form of a quantifiable splitting.)

The textbook case of tunneling phenomena is the 'umbrella' inversion of ammonia, NH_3.[33] When the three protons have been labelled, there are just two versions of the pyramidal (C_{3v}) minimum energy structure: they differ only in the cyclic order of the three hydrogens when viewed in the same sense down the principal axis. The potential curve for the motion is a classic double-minimum. The transition state for interconversion is the planar (D_{3h}) structure (for which there is just one version).

In cases where tunneling is observed, the classification and assignment of the spectra has been aided by the Molecular Symmetry (MS) group formalism, introduced by Longuet-Higgins[30] and brought to the spectroscopist by Hougen[34] and Bunker.[29] For molecules undergoing interconversion tunneling, the perturbed vibrational levels have symmetry labels derived from the MS group, not the more familiar molecular point group. Point group symmetry operations are either 'proper' (pure rotations) or 'improper' (comprising inversion, reflection or rotation-reflections).[42] It can be shown that proper symmetry operations may be represented by permutations of the labels (and hence coordinates) of identical nuclei whereas improper operations require the product of a space-fixed inversion operation with the appropriate permutations. The MS group expands the set of operations which represent the point group symmetries, to include those permutations and permutation-inversions which represent the interconversion of those versions which are accessible to one another through tunneling.[34] Recent work in this area has demonstrated the close relationship between transition state symmetries and the MS group for the tunneling.[28,36] In short, the collection of symmetry operations is pertinent to any set of accessible structures on the potential energy surface of the motion.

For complexes of small molecules, we frequently find that there are several possible versions of their equilibrium structures. Since it is invariably true that barriers on their intermolecular potential energy surfaces are low, tunneling splittings can be a signature of their spectra. In many cases, one can hardly dare to assume that a single equilibrium structure has any meaning under the conditions of observation. There is by now a long precedent of observing interconversion tunneling in hydrogen-bonded molecules in terrestrial laboratories. The HF dimer is the archetypal example[37] and demonstrates facile exchange of positions of the protons without bond-breaking. Similarly, both the water dimer,[38,39] and ammonia dimer,[40,41] exhibit complicated interconversion tunneling mechanisms. Hydrogen exchange is of acute interest to those studying organic and aqueous systems but the role of tunneling in the positional exchange of heavier elements, whilst often overlooked, is important to the study of other chemistries. By now, it has become accepted that interconversion tunneling is not exclusive to complexes containing the lighter elements.

3 Occurrence of Argon, C_2H_2 and SO_2 in Atmospheres and the Cosmos

3.1 Argon

Argon is minor component of the Earth's atmosphere,[42] as well as being the third and fourth most abundant constituent of the atmospheres of Mars and Venus,[43] respectively (where CO_2 is the dominant species). There are suggestions that it is also a measurable component of Titan's atmosphere.[44]

3.2 C_2H_2

Acetylene has not been widely identified in planetary atmospheres, although the presence of small organic moieties in inter-stellar space (both in giant clouds and the grains which go to make up comets) has suggested that it plays a discernible role. The majority of astrophysical searches for C_2H_2 have focused on the outer planets of the Solar System and their respective moons (particularly Titan) which all have dense organic hazes surrounding them. Colussi *et al.*, suggest that dimers of acetylene may be almost non-existent in the upper atmosphere of Titan because of the local conditions: the partial pressure of acetylene is close to its sublimation pressure and the temperature is close to its sublimation point.[45] (Additionally, entropic factors mitigate against the formation of larger clusters — trimers, tetramers, etc.). In fact, what clusters of acetylene there are, are likely to be present as ultrafine particles[45] or aerosols.[46,47] (This may be a general result: the typical models of dimer formation in equilibrium conditions assume that the temperature is well above that of sublimation.[48] Other non-polar species, *e.g.*, CO_2 form such particles.[49]) On the other hand, the presence of methylacetylene and diacetylene in Titan's stratosphere has also been confirmed[50,51] and the appreciation that species trapped in solid particles may undergo photochemical reactions justifies the study of C_2H_2 complexes. Indirect evidence of this is the abundance of methane (a likely photodecomposition product of C_2H_2) in both Titan's atmosphere and that of Neptune.[47]

3.3 SO_2

Present in trace quantities in the Earth's atmosphere[42] (at ppb levels on average), SO_2 is certain to participate in acid rain formation. Its complexes may make a contribution to chemistry in the higher reaches of the atmosphere either propelled there from volcanic emissions or by upward diffusion from man's industrial processes. SO_2 is a reactive moiety and can be photolytically activated, undergo radical reactions and carry out chemistry in water droplets, as well as form aerosols by reaction with organic species.

Though a small component of the interstellar medium, SO_2 is most likely to be found in the more oxidising planetary atmospheres, in particular the moons of Jupiter and Saturn. Ozone, rather than SO_2 has been identified on Rhea and Dione[52] whereas SO_2 itself has been observed on Europa[53,54] and Callisto.[55,56] The first interpretation of the spectra of Europa[53] was that of sulphur ions from the Jovian magnetosphere implanted in water-ice.

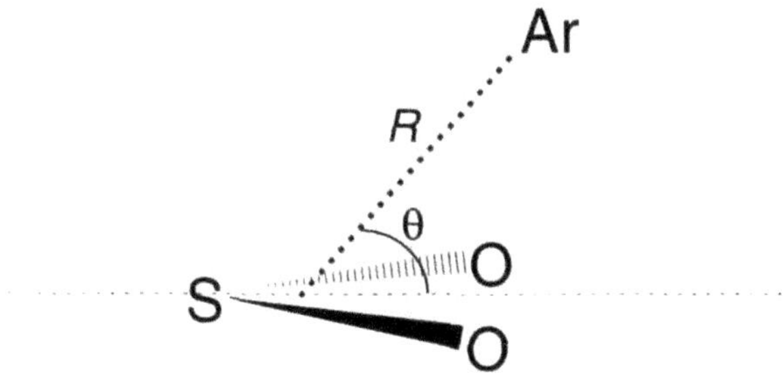

Figure 1: $Ar-SO_2$: C_s global minimum.

Nevertheless, subsequent study [54] showed that the spectral data pointed to SO_2 ice itself. Similarly, spectra on Callisto can be well fit by models of SO_2 ice absorption, [55,56] allowing for the possibility of mixed water-SO_2 solids. [55] Photo-oxidation of SO_2 in low-temperature matrices showed that $(SO_2)_2$ was a crucial species in SO_3 formation. Venus's highly oxidising atmosphere has a high proportion of H_2SO_4 and can therefore be expected to contain a rich sulphur chemistry.

4 Examples of Non-Rigidity in van der Waals Molecules

4.1 $Ar-SO_2$

The $Ar-SO_2$ complex exhibits the simplest case of tunneling: a symmetric double-well potential. Spectroscopic studies (both in the microwave [58,59] and the infrared [60]) affirmed its structure to be a near-prolate asymmetric top with C_s symmetry, in which the argon atom sits above the plane of the SO_2 molecule (Fig. 1). Early ambiguities in deducing the angle, θ, between the vector joining the argon to the SO_2 center-of-mass and the SO_2 principal axis[58] were resolved in later work, [59,60] finding a value of around 60°. But, with fittings to a rigid-rotor model being imperfect, [58] the spectrum was shown to be consistent with an interconversion tunneling mechanism. It was suggested that the SO_2 monomer rotated about its a-inertial axis, (*i.e.*, maintaining the plane of symmetry which relates the two oxygens to one another throughout its motion), because of the low associated reduced mass. [58]

Because the overall center-of-mass of the complex stays fixed during the internal rotation, the argon atom effectively travels around the SO_2 molecule, sampling two equivalent local minima. The tunneling coordinate is therefore

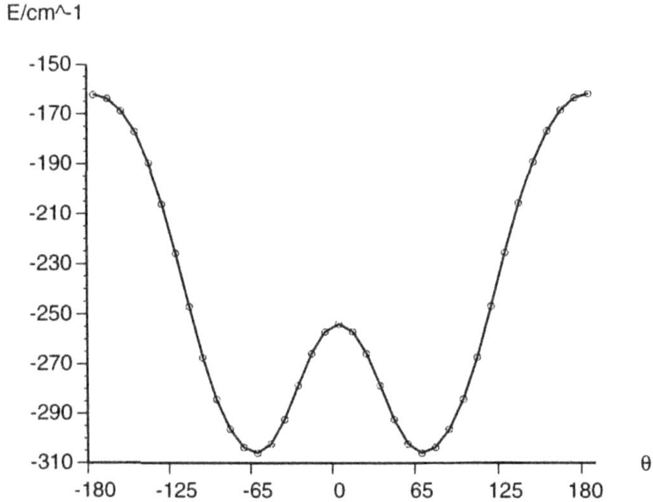

Figure 2: Interconversion tunneling potential of Ar$-SO_2$.

the angle θ, in Figure 1. The ground state tunneling splitting is 980 MHz which, in a heavy-atom system, was considered to be unusually high, implying that the barrier to inversion is extremely small.[58] Indeed, using a WJKB approximation and assuming that the argon atom passes through the 'V' of the SO_2 rather than around its apex, DeLeon *et al.*[58] deduced a barrier height of just 10 cm^{-1}. More recent theoretical work,[60] fitting the observed tunneling splittings with a two-dimensional hindering potential (at fixed intermolecular separation) suggested that the barrier is 80 cm^{-1} and that an alternative barrier, to orthogonal motion, *i.e.*, with the argon atom in the plane of the SO_2 molecule, but alongside the S$-$O bond, is above 150 cm^{-1}. A second calculation,[60] in which electrostatic and induction forces were modelled by atom-sited multipoles and dispersion forces with atom-atom Lennard-Jones coefficients, gave poor quantitative agreement finding that the minimum energy structure was in fact at the C_{2v} structure, the geometry of the experimental minimum being just 8 cm^{-1} higher. Nevertheless, this empirical model confirmed the supposedly high barrier to internal rotation of the SO_2 monomer about its C_2-axis. Consequently there remained considerable uncertainty over the barrier height and the extent of the motion of the argon atom both from experiment and from low-level calculation.

Extensive *ab initio* calculations[61] using a correlated method, second order Møller-Plesset perturbation (MP2) theory,[18] established that even a simple four-atom system such as Ar$-$$SO_2$ has a complicated potential energy surface. The global minimum was found to be very similar to the structure deduced from experiment with $\theta = 68.5°$. In addition, four further geometrically distinct stationary points were identified and motions between them suggested. The tunneling coordinate was computed, pointwise using a basis set in which monomer properties are well represented. At each value of θ, the angle was kept constant and all other geometric parameters were minimized to ensure that the energy obtained would be truly that of the actual motion. The relative simplicity of the motion makes this an ideal test case for the accuracy of *ab initio* calculations. The potential is represented in Figure 2. One unusual observation was a point of bifurcation at 110° on the tunneling coordinate (*i.e.*, between the global minimum and a planar C_{2v} structure in which the argon atom 'bonds' to the sulphur atom.[61]) The implication is that the argon atom is unlikely to circulate fully around the SO_2 molecule along the tunneling coordinate. It will either remain trapped in the double well inside the 'V' of the SO_2 or will skirt off around the 'side' of the SO_2 molecule if it has sufficient energy.

The energy barrier between the C_{2v} transition state and the global minimum was calculated to be 59 cm^{-1} (17 cm^{-1} when corrected for BSSE and zero-point energy effects[61]). This is in fair agreement with experiment.

4.2 $(SO_2)_2$

The dimer of SO_2 proved to be a very hard case for spectroscopists. For several years after the first recorded measurement,[62] there was silence in the literature concerning the structure of this species. Prediction of the gas-phase structure was made difficult by the realisation that the two monomers were symmetrically inequivalent, *i.e.*, that the orientation of the monomers did not correspond to that found in the crystal structure.[63] (Had such an alignment been adopted, the resulting complex would have been nonpolar.) It was deduced from the early microwave measurement[62] that the two monomers changed positions via a not-well understood interconversion tunneling mechanism whose frequency was 70 kHz. Only when Taleb-Bendiab *et al.*[64] obtained spectra of many isotopically substituted species did a clearly-defined structure emerge. Their model is of a complex with the C_s point group, in which the plane of symmetry is the *ac*-plane with one monomer lying in it (Fig. 3). One oxygen atom of the in-plane monomer species acts as an (electron) donor towards the sulphur atom of the other. The angle of inclination of the monomer principal axes is 66°. Systematic absences in the spectra precluded a accurate estimate of the interconversion tunneling barrier and transition states for the motion could

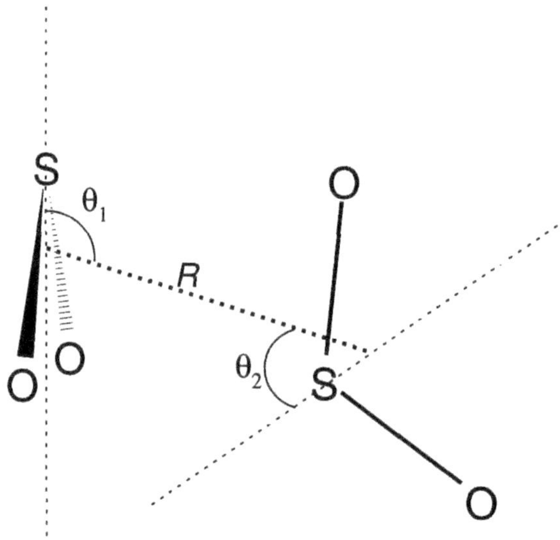

Figure 3: $(SO_2)_2$: C_s global minimum.

not be reasonably inferred. Whilst this system was undoubtedly harder to unravel than Ar$-SO_2$, there was a chance that calculations could both confirm the global minimum structure and deduce the likely transition states which interconverted versions of it.

The theoretical study by this author and coworkers [65] assumed that there would be strong parallels with the water dimer system, for which many stationary points were already known. [66] Even so, the cost of calculations on $(SO_2)_2$ and the extreme shallowness of the potential energy surface blighted a quick solution to the problem. Although six geometrically distinct stationary points were located, only 3 low-lying ones were deemed significant. Interest settled on structures with C_s, C_i and C_2 symmetry. (A higher-lying C_{2h} symmetry structure could well serve as a precursor to oxygen exchange, as observed in SO_2 trapped in solid matrices. [67]) The characterisation of these species as minima or transition states was sensitive to the level of theory employed and the energy differences between them were so small that serious interpretability was open to question.[65] Even so, a picture emerged in which the C_s structure seemed to be the likely global minimum, (though with the monomer axes inclined at an angle differing considerably from the experimental interpretation), and the C_i structure would be the lowest lying transition state.

Visualisation of normal mode motions at both of these structures led to the proposition of a cyclic periodic potential utilising four versions of each structure, (Fig. 4). The C_i symmetry structures allow for the successive interchange of positions of 'donor' and 'acceptor' monomers. It should be noted, though, that there are 8 possible versions of each of the C_s and C_i forms. This potential is closed and only allows for communication between 4 in each case. The potential energy surface of $(SO_2)_2$ is therefore segmented in such a way that groups of 4 versions are separated from one another by a different and probably higher barrier. A given version may only interconvert to 3 others using the C_i symmetry transition states.

It is possible to model this cyclic potential with a simple empirical sinusoidal function (following an example set by Fraser *et al.*[68]):

$$V = \frac{1}{2}\sum_k V_{4k}(1 - \cos 4k\alpha) \tag{2}$$

where α is an internal angle such that successive versions of the C_s minimum are at $\alpha = 0, \pi/2, \pi$ and $3\pi/2$.

We add the rotational kinetic energy Hamiltonian,

$$\hat{T} = -\frac{\hbar^2}{4I}\frac{d^2}{d\alpha^2} \tag{3}$$

where I is the b moment of inertia of SO_2 and expand solutions in a free rotor basis:

$$|m\rangle = \frac{1}{\sqrt{2\pi}}e^{im\alpha} \tag{4}$$

Matrix elements, h, of the Hamiltonian with this basis are related to the magnitude of the tunneling splitting. For the 4 versions **a**, **b**, **c** and **d** of $(SO_2)_2$ in Figure 4, the tunneling matrix is:

$$\begin{array}{c} \\ \mathbf{a} \\ \mathbf{b} \\ \mathbf{c} \\ \mathbf{d} \end{array} \begin{array}{c} \begin{array}{cccc} \mathbf{a} & \mathbf{b} & \mathbf{c} & \mathbf{d} \end{array} \\ \begin{pmatrix} 1 & 0 & h & h \\ 0 & 1 & h & h \\ h & h & 1 & 0 \\ h & h & 0 & 1 \end{pmatrix} \end{array}$$

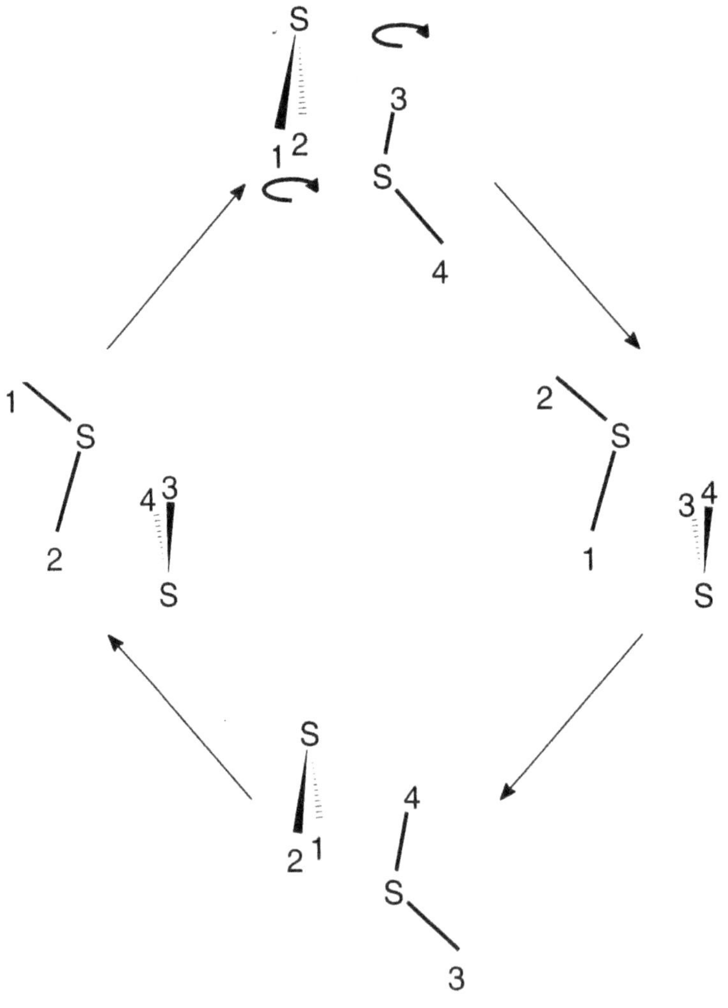

Figure 4: $(SO_2)_2$: Interconversion tunneling pathway.

and the MS group[29] for the interconversion tunneling is C_{4v}(M). The tunneling splits energy levels into 2 levels at $\pm 2h$ and an unperturbed degenerate pair. Trial and error leads us to find that, in this model, the observed 70 kHz splitting is consistent with a barrier of 56 cm^{-1}. This is within the range of the *ab initio* calculations. We should note that the splitting is not only dependent on the barrier height, but also the moment of inertia for the motion. In this case, the latter quantity is important in dampening the tunneling. The small size of the tunneling splitting originally led spectroscopists to think that the barrier was high. In fact it is remarkably low and it is the high inertia of the motion that mitigates the overall effect.

The whole SO_2–SO_2 interaction potential is surprisingly isotropic in the region of the global minimum. This may be a result of deficiencies in the level of theory — we have noted that the equilibrium angular inclination is wrong — but it remains consistent with the interpretation of a low barrier process. In fact, it could be a high degree of vibrational averaging which causes the observed structure to differ from the calculated one.

The case of $(SO_2)_2$ has highlighted the power of theory to provide insights into interconversion processes, not only by identifying transition states, but by showing that more than one interpretation may be consistent with the available spectra.

4.3 $(C_2H_2)_2$

The $(C_2H_2)_2$ system has now been well-studied and understood. Early investigations of the complex attempted to resolve a dilemma about the structure: whether it was 'T'-shaped, C_{2v} symmetry or nonpolar, C_{2h} symmetry (Fig. 5). Microwave spectra by Fraser *et al.*,[68] and infrared measurements by Ohshima *et al.*,[69] appeared to resolve this debate by explaining the complicated spectrum by a 'T'-shaped dimer undergoing a cyclic interconversion tunneling process. The mechanism is reminiscent of that for $(SO_2)_2$, above, but, in this case, the four equivalent transition states have C_{2h} symmetry, and all possible versions of both global minimum and the transition state are accessed along the potential. This analysis is consistent with a group theoretical rule for cyclic potentials of this sort, which is that the point groups of the minimum and transition states must have the same order as each other.[28]

As with $(SO_2)_2$, the tunneling motion can be interpreted in terms of motion along a low frequency normal mode from the global minimum to the transition state; the two monomers execute a gear-wheel type motion.[68]

The tunneling splitting, 2.1 GHz, is very close to the B rotational constant of the complex, 1.9 GHz, a fact which hampered easy interpretability of the

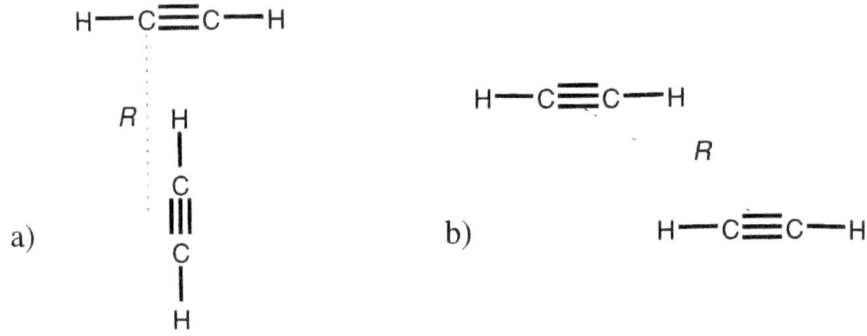

Figure 5: $(C_2H_2)_2$: a) C_{2h} transition state; b) C_{2v} global minimum.

spectrum. Whilst this splitting is many times larger than that observed in $(SO_2)_2$, the moment of inertia of $(C_2H_2)_2$ is considerably smaller than that of $(SO_2)_2$. Consequently, a small barrier to interconversion was also consistent with the observations for $(C_2H_2)_2$. The cyclic potential of the form of Eq. 2 used by Fraser *et al.*, required a barrier of 33 cm^{-1} to reproduce the observed splitting.

Early *ab initio* calculations [70] found a barrier in surprisingly good agreement with the experimental figure. Subsequent analysis, though, suggested that the effect of BSSE may well have vitiated those results: calculations with larger basis sets gave a barrier height of around twice the experimental value. [71] Whilst such a number is sufficient to effectively quench tunneling in such a light system, it was not an unreasonable assessment at the time. The success of these early results spawned a host of calculations on other acetylenic systems. [72,73,71]

An important feature of homogeneous acetylenic clusters is that the centre of mass to center of mass separation between adjacent monomers is almost independent of orientation. The structures and potential energy surfaces of these systems are dominated by two orientations of interaction, the two stationary points in $(C_2H_2)_2$. Both the 'T-shaped' (sometimes called 'hydrogen-bonded' [74]) and the slipped parallel forms correspond to stationary points in the interaction function of two quadrupoles. The lowest non-vanishing multipole moment of C_2H_2 is the quadrupole and the electrostatic model of van der Waals molecules [9,10,11] expands the interaction energy in a power series of multipole-multipole terms. The quadrupole-quadrupole term dominates the C_2H_2–C_2H_2 interaction and at fixed centre-of-mass to centre-of-mass separation, the 'T'-shaped structure is the lower in energy of the two. The various *ab initio* calculations[70,71] confirm this result with the later contribution[71] obtaining a centre of mass separation very close to the experimentally determined value.[68]

4.4 $(C_2H_2)_3$

When splittings were observed in high energy rotational transitions[75] in an otherwise text-book symmetric top spectrum for $(C_2H_2)_3$, there was speculation on another tunneling interpretation.

The global minimum of $(C_2H_2)_3$ is a planar cyclic structure of C_{3h} symmetry (Fig. 5).[70] Whilst the ideal perpendicularity of adjacent monomers is distorted in this system, compensation comes from the fact that a cyclic structure is the only conformation in which three pairwise interactions are attainable.

Extensive *ab initio* calculations on the system[72] revealed many stationary points and subsequent group theoretical analysis played a part in unravelling their respective relationships.[28] Using a rule of thumb that each pairwise interaction between C_2H_2 monomers would contribute an approximately fixed amount to the overall binding energy, it was clear that no other possible conformation of the three monomers could approach the energy of the global minimum. Inevitably, then, the barrier to interconversion would be substantially higher than that found in $(C_2H_2)_2$ and would be similar to the magnitude of one pairwise interaction. This would not be inconsistent with the minute nature of the observed splittings, but made the task of rationalising the motion somewhat harder. (Neither is such a barrier unprecedented: for example, in $(H_2O)_2$ the calculated barrier[66] to tunneling is 207 cm^{-1} and in $(HF)_2$ it is 428 $cm^{-1}$by the same methods.[66]) Ultimately, attention focused on the two lowest-lying transition states, a planar, C_{2v} symmetry structure and a C_s symmetry structure in which one monomer is rotated perpendicular to the plane of the other two (Fig. 5).

Group theory played a role again[28] and it was straightforwardly shown that there are 16 versions of the $(C_2H_2)_3$ global minimum, partitioned into two groups of eight. The two groups differ from one another in the ordering of the monomers (once differentiated by labels) when the cluster is viewed down a fixed axis. Within each group of eight, it is possible for a given monomer to have either end pointing in or out of the ring.

Analysis of normal modes at both the global minimum and the transition states suggests that the C_{2v} structure facilitates a rearrangement in which the cyclic order of the monomers is reversed, and the C_s structure enables any given monomer to exchange the positions of its internal and external hydrogens. The latter structure, the lowest energy transition state, lies at 408 cm^{-1} above the

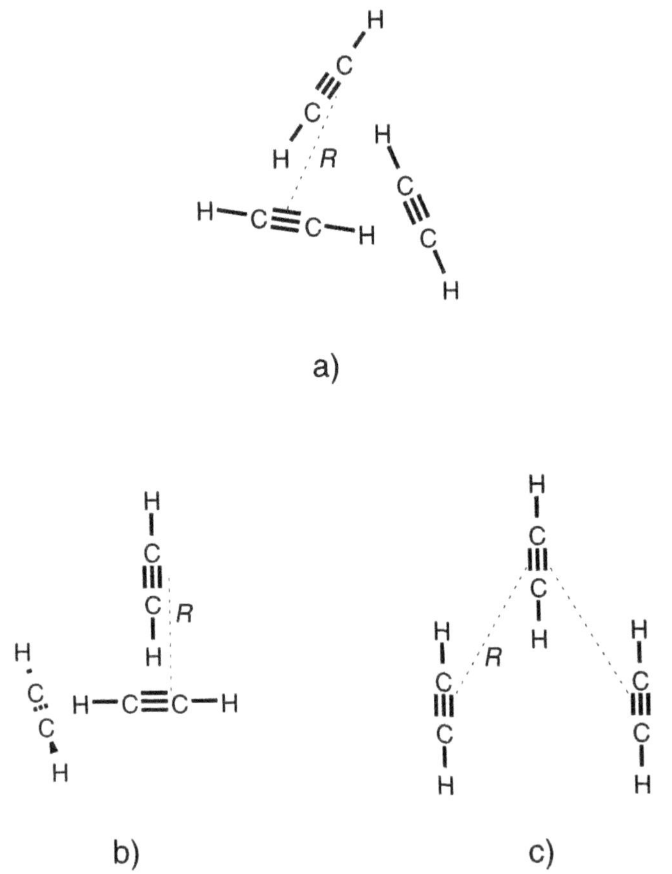

Figure 6: $(C_2H_2)_3$: a) C_{3h} global minimum; b) C_s transition state; c) C_{2v} transition state.

global minimum according to the best level of theory.[28] Whilst 8 versions of the global minimum can be interconnected by 24 versions of a transition state, leading to a complex splitting pattern, the resolution of the spectrum does not permit such a detailed analysis. The path between any adjacent pair of versions of the global minimum (differing only in the orientation of a single monomer) can be modelled as a 2-fold cyclic potential. (Two 'mirror-image' forms of the C_s transition state can effect the same net interconversion.)

For the calculated tunneling splitting to correspond reasonably with that observed, the barrier height needs to be as low as 75 cm^{-1}. Part of the reason here is that the barrier is 'wider' than in, say, $(C_2H_2)_2$, due to being a 2-fold cyclic potential instead of a 4-fold potential. Even so, the most reasonable conclusion is that, although more than one type of interconversion tunneling may be possible in $(C_2H_2)_3$, it has not conclusively been observed at this time. Further reasons should be explored to explain the splittings in the vibration-rotation spectrum of this species.

Such a result cannot be said to be a failure for theory. On the contrary, theory has successfully demonstrated that interconversion tunneling in this cyclic trimer is a rather unlikely process. The three factors of reduced mass (for the in-plane motion, which would involve the concerted movement of all three monomers), barrier height and barrier width all conspire to quench tunneling. The theoretical exploration of tunneling processes has been an exciting foray into the field of trimers, showing that their potential energy surfaces are rich and complex. By contrast, water trimer has been similarly extensively studied and its spectrum is rich, yielding many possible tunneling mechanisms.[76,77]

4.5 $(C_2H_2)_4$ and $(C_2H_2)_5$

Analysis of the infrared C–H stretch band of molecular beams of C_2H_2 found evidence of larger clusters than the trimer, though only the tetramer has been analysed in detail.[78] Its spectrum was consistent with a puckered cyclic structure of S_4 symmetry (Fig. 6).

In a similar manner to the dimer and trimer, the structure is composed of perpendicular interactions between adjacent monomers. But there appears to be no reason for the tetramer to be non-planar. Calculations support this viewpoint and, even at the correlated level, find that the global minimum has C_{4h} symmetry.[73] Although there is no evidence of interconversion tunneling from the spectrum, there is scope for investigation of large-amplitude motion because of the large deviation of the calculated minimum from the experimentally observed structure.

The answer lies in the vibrational frequencies. The lowest energy frequency

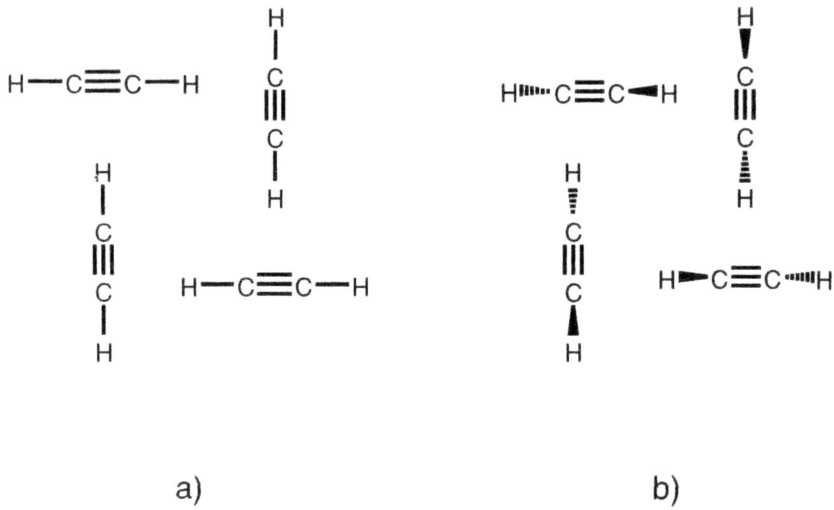

Figure 7: $(C_2H_2)_4$: a) C_{4h} global minimum; b) S_4 puckered form.

at the global minimum is effectively zero and corresponds to an out-of-plane antisymmetric ring-puckering motion of B_u symmetry. The reduced mass for such a motion is negligible and no intermolecular quadrupole-quadrupole interactions are disrupted in the process. The structures of the species obtained by the ring-puckering have S_4 symmetry. Investigation of structures formed by displacing along this coordinate found that the energy rises only very slowly. The ring-puckering potential is thus best described as the limit of a double minimum problem in which there is no barrier to inversion (*i.e.*, the planar structure is a critical point of fourth order).[42] The simplest mathematical description of the coordinate would be a simple quartic in the distance of a monomer centre of mass to the plane containing the cluster centre of mass, perpendicular to the S_4 symmetry axis.[73]

There are other small intermolecular frequencies, so it is likely that other distortions are simultaneously taking place so that the cluster flexes in orthogonal directions also. Nevertheless, the out-of-plane motion is the dominant one when it comes to evaluating overall fluxionality.

In order to be consistent with the observation of a time-averaged displaced form instead of a planar C_{4h} form, it can easily be assumed that, even under the conditions of observation, levels higher than the ground state mode of the ring-puckering vibration will be populated. Consequently the vibrational wavefunction will have large amplitude away from its midpoint (the planar structure).

The spectrum of the next larger cluster, $(C_2H_2)_5$, had been identified and tentatively assigned to another cyclic structure, because of the symmetric top pattern of transitions.[78] Such a cluster would be bound by 5 adjacent monomer interactions. Whilst calculation shows that a lower energy structure of C_{2h} symmetry may exist (containing 6 adjacent monomer interactions)[73] it has not been conclusively observed. The mechanisms and pathways of formation of these species have yet to be understood, so the cyclic moiety may be favoured for kinetic rather than thermodynamic reasons. Even prior to the conclusive elucidation of the pentamer structure, we may speculate that the cyclic form undergoes another type of fluxional process: a kind-of 'pseudo-rotation' typical of 5-membered-ring molecules. Such a prediction is consistent with the presence of a pair of degenerate intermolecular vibrational modes at low-frequency.

4.6 $Ar-C_2H_2$

We conclude with a study of a mixed-monomer ('hetero-molecular') cluster, $Ar-C_2H_2$. Due to the use of argon carrier gas in many of the molecular beam experiments in which C_2H_2 clusters were measured, the spectrum of $Ar-C_2H_2$ has been observed on many occasions and extensively studied.[80,81,82,83,84] The cluster has been prototypical in the characterisation of weak interactions because it is a very simple example of attraction between two nonpolar species. Its properties, (geometry, induced dipole moment, etc.) would provide useful support for theories of intermolecular forces.[85]

A 'T'-shaped equilibrium form for the complex was consistent with the spectroscopy,[80,82] but for a long time the value of the intermolecular separation, R, was in doubt, with values ranging from 3.2 Å[80] to 4.04 Å.[82] The uncertainty in the structural parameters stemmed from a high-amplitude in-plane vibrational motion.[81] The angle θ (Fig. 8) was placed at $\geq 60°$ from values of rotational constants[81] and $\geq 66°$ from hyperfine analysis.[80] But values of the equilibrium angular displacement are dependent on an assumed value of the intermolecular separation (and vice versa).

Results of *ab initio* calculations[86] with a high quality basis set on both constituent monomers showed that the potential energy surface was extremely flat and highly anharmonic in both the stretching and angular degrees of freedom. Even so, there is a fine-scale structure. The C_{2v} structure is a transition state and the global minimum has C_s symmetry with $\theta = 65°$. (Concurrent calculations by other workers[87] at a different level of theory found $\theta = 66°$.)

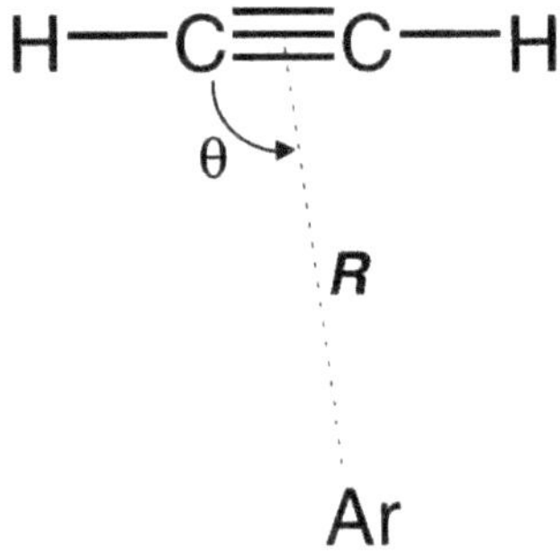

Figure 8: Ar–C_2H_2: C_s global minimum.

There are two equivalent C_s structures on either side of the C_2 symmetry axis. The barrier between the two structures is so insignificant that it is unlikely that vibrational levels are trapped within the two equivalent basins. It is encouraging to find that the angular displacement corresponding to the *ab initio* global minimum is close to that found experimentally. The highest barrier on the surface is at $\theta = 22°$ corresponding to a transition state. This barrier is probably sufficient to prevent the argon atom from reaching a position collinear with the C_2H_2 molecule under the conditions of observation.

The *ab initio* calculations found equilibrium intermolecular separations of 3.79 Å in the C_{2v} symmetry structure and 3.86 Å in the planar minimum. Both these values are within the range found experimentally and both are close to the sum of van der Waals radii on argon and carbon. This laid to rest some confusion over how the two molecules could hold together either at an extended or a compressed separation.

Numerical fitting was carried out in order to determine more accurate values for the harmonic vibrational frequencies and anharmonic corrections. Finite difference calculations were found to be unreliable on such flat surfaces, whereas least-squares fitting to Morse and quartic functional forms were in close agreement with each other.[86]

For this species, as with $(C_2H_2)_4$ it was found that non-rigidity, though not leading to tunneling splittings, had a profound effect on the interpretation of vibration-rotation spectra. It can barely be expected that a single structure provides a good representation of the form of these molecules. It is much fairer to think of a dynamic picture in which the constituent monomers are undergoing large amplitude librations.

5 Conclusions

Detailed consideration of six van der Waals clusters has revealed a number of distinct types of fluxional behaviour, each of which have observable effects on the vibration-rotation spectra of the species. Experimentally it is hard to interpret such spectra. Often, isotopic substitution must be employed. For species in the cosmos, which we are observing at long-range, we will not have the opportunity to carry out the controlled, extensive measurement and analysis that a laboratory scientist is able to. Consequently an awareness and appreciation of fluxional behaviour will be an asset to the interstellar chemist when thinking of these molecules.

We can see that large-amplitude motions may either give rise to vibrational averaging effects that lead to difficulties in deducing structural constants, or they may lead to interconversion tunneling and cause spectra to be complicated by splittings of spectral lines. The latter behaviour is both the most complex and the most interesting and many factors play a part in the magnitude of the effect. They are, principally, the barrier height to the process, the moment of inertia of the species involved and the number of versions of the structures participating in the motion.

Theory plays an invaluable part in the analysis and characterisation of these species. Often only a very small portion of a spectrum is available from astronomical observations. A correct identification of the species responsible can therefore rely heavily upon an accurate assignment from quite limited data. Of the many factors contributing to the complexity of a spectrum, non-rigidity and interconversion tunneling are of great significance for van der Waals molecules. Large amplitude motions have a dramatic effect on the rotational constants which are deduced from spectra. (In many instances rigid-rotor models are barely applicable.) Whilst calculations of tunneling splittings directly from *ab initio* computed potential energy surfaces is rarely of high accuracy except for the simplest of molecules, a mere estimate can in most cases be sufficient to distinguish between a choice of mechanisms.

As more and more complexes are isolated and studied in laboratories on Earth, we will have greater opportunities to confirm the identity of such species in the atmospheres of distant worlds and be better placed to understand the roles they play in foreign atmospheres.

Acknowledgments

EPSRC (U.K.) is acknowledged for a 1992-4 NATO postdoctoral Fellowship. I am grateful to Professor Richard Bader for the use of his laboratory. Computer facilities at McMaster University were employed for calculations on the $Ar-SO_2$ and $Ar-C_2H_2$ molecules. All reported calculations took place with the quantum chemistry program CADPAC.[88]

References

1. J. M. Calo and R. S. Narcisi, *Geophys. Res. Lett.* **7**, 289 (1980).
2. R. E. Miller, *J. Phys. Chem.* **90**, 3301 (1986).
3. D. J. Nesbitt, *Chem. Rev.* **88**, 843 (1988).
4. K. R. Leopold, G. T. Fraser, S. E. Novick, and W. Klemperer, *Chem. Rev.* **94**, 1807 (1994).
5. G. E. Ewing, *Chem. Phys.* **29**, 253 (1978).
6. L. S. Bernstein, D. C. Robertson, J. A. Conant, and B. P. Sandford, *Appl. Opt.* **18**, 2454 (1979).
7. D. J. Nesbitt and R. Namaan, *J. Chem. Phys.* **91**, 3801 (1989).
8. A. Ernesti and J. M. Hutson, *Chem. Phys. Lett.* **222**, 257 (1994).
9. A. D. Buckingham, *Adv. Chem. Phys.* **12**, 107 (1967).
10. A. D. Buckingham and P. W. Fowler, *Can. J. Chem.* **63**, 2018 (1985).
11. A. D. Buckingham, P. W. Fowler, and J. M. Hutson, *Chem. Rev.* **88**, 963 (1988).
12. A. J. Stone, "The Calculation of Intermolecular Potential Energy Surfaces", in *Dynamics of Polyatomic Van der Waals Complexes* (Plenum Press, New York, 1990), p. 329.
13. G. C. Maitland, M. Rigby, E. B. Smith, and W. A. Wakeham, *Intermolecular Forces* (Oxford University Press, 1981).
14. J. H. van Lenthe, J. G. C. M. van Duijneveldt-van de Rijdt, and F. B. van Duijneveldt, *Adv. Chem. Phys.* **69**, 521 (1987).
15. G. Chalasinski and M. M. Szczesniak, *Chem. Rev.* **94**, 1721 (1994).
16. A. T. Szabo and N. S. Ostlund, *Modern Quantum Chemistry: An Introduction to Advanced Electronic Structure Theory* (MacMillan, 1987).
17. R. McWeeny, *Methods of Molecular Quantum Mechanics, (2nd Ed.)* (Academic Press, 1989).
18. C. Møller, M. S. Plesset, *Phys. Rev.* **46**, 618 (1934).
19. R. G. A. Bone and R. F. W. Bader, *J. Phys. Chem.* **100**, 10892 (1996).
20. R. F. W. Bader, *Atoms in Molecules—A Quantum Theory*, International Series of Monographs on Chemistry, No. 22 (Oxford University Press, 1990).
21. S. F. Boys and F. Bernardi, *Molec. Phys.* **19**, 553 (1970).
22. M. Gutowski, J. G. C. M. van Duijneveldt-van der Rijdt, J. H. van Lenthe, and F. B. van Duijneveldt, *J. Chem. Phys.* **98**, 4728 (1993).

23. M. Gutowski and G. Chalasinski, *J. Chem. Phys.* **98**, 5540 (1993).
24. M. M. Szczesniak and S. Schiener, *J. Chem. Phys.* **84**, 6328 (1986).
25. E. R. Davidson and S. J. Chakravorty, *Chem. Phys. Lett.* **217**, 48 (1994).
26. D. Feller, *J. Chem. Phys.* **96**, 6104 (1992).
27. R. S. Berry, *Rev. Mod. Phys.* **32**, 447 (1960).
28. R. G. A. Bone, T. W. Rowlands, N. C. Handy, and A. J. Stone, *Molec. Phys.* **72**, 33 (1991).
29. P. R. Bunker, *Molecular Symmetry and Spectroscopy* (Academic Press, 1979).
30. H. C. Longuet-Higgins, *Molec. Phys.* **6**, 445 (1963).
31. K. Bergmann, S. Gortler, J. Manz, and H. Quasi, *J. Am. Chem. Soc.* **115**, 1490 (1993).
32. R. P. Bell, *The Tunnel Effect in Chemistry*, (Chapman and Hall, 1978).
33. G. Herzberg, *Infrared and Raman Spectra of Polyatomic Molecules* (van Nostrand, 1945).
34. J. T. Hougen, *J. Phys. Chem.* **90**, 562 (1986).
35. M. Hammermesh, *Group Theory and its Application to Physical Problems* (Pergamon, New York, 1962).
36. R. G. A. Bone, "Generalising the Molecular Symmetry Group of Longuet-Higgins to Asymmetric Tunneling Problems", in *Structure and Dynamics of Non-Rigid Molecular Systems* (Kluwer, 1995), p. 67.
37. T. R. Dyke, B. J. Howard, and W. Klemperer, *J. Chem. Phys.* **56**, 2442 (1972).
38. T. R. Dyke, *J. Chem. Phys.* **66**, 492 (1977).
39. G. T. Fraser, *Int. Rev. Phys. Chem.* **10**, 189 (1991).
40. D. Nelson Jr., G. T. Fraser, and W. Klemperer, *J. Chem. Phys.* **83**, 6201 (1985).
41. D. Nelson Jr. and W. Klemperer, *J. Chem. Phys.* **87**, 139 (1987).
42. J. Heicklen, *Atmospheric Chemistry* (Academic Press, 1976).
43. K. Fox and S. J. Kim, *J. Quant. Spect. Radiat. Trans.* **40**, 177 (1988).
44. R. E. Samuelson, R. A. Hanel, V. G. Kunde, and W. C. Maguire, *Nature* **292**, 688 (1981).
45. A. J. Colussi, S. P. Sander, and R. R. Friedl, *Chem. Phys. Lett.* **178**, 497 (1991).
46. T. Dunder and R. E. Miller, *J. Chem. Phys.* **93**, 3693 (1990).
47. P. N. Romani and S. K. Atreya, *Geophys. Res. Lett.* **16**, 941 (1989).
48. B. L. Blaney and G. E. Ewing, *Ann. Rev. Phys. Chem.* **27**, 553 (1976).
49. G. E. Ewing and De T. Sheng, *J. Phys. Chem.* **92**, 4063 (1988).

50. W. C. Maguire, R. A. Hanel, D. E. Jennings, V. G. Kunde, and R. E. Samuelson, *Nature* **292**, 683 (1981).
51. V. G. Kunde, A. C. Aikin, R. A. Hanel, D. E. Jennings, W. C. Maguire, and R. E. Samuelson, *Nature* **292**, 686 (1981).
52. K. S. Noll, T. L. Roush, D. P. Cruikshank, R. E. Johnson, and Y. J. Pendleton, *Nature* **388**, 45 (,)1997.
53. A. L. Lane, R. M. Nelson and D. L. Matson, *Nature* **292**, 38 (1981).
54. K. S. Noll, H. A. Weaver, and A. M. Gonnella, *J. Geophys. Res.* **100**, 19057 (1995).
55. K. S.Noll, R. E. Johnson, M. A. McGrath, and J. J. Caldwell, *Geophys. Res. Lett.* **24**, 1139 (1997).
56. A. L. Lane and D. L. Domingue, *Geophys. Res. Lett.* **24**, 1143 (1997).
57. J. R. Sodeau and E. K. C. Lee, *J. Phys. Chem.* **84**, 3358 (1980).
58. R. L. DeLeon, A. Yokozeki, and J. S. Muenter, *J. Chem. Phys.* **73**, 2044 (1980).
59. L. H. Coudert, K. Matsumura, and F. J. Lovas, *J. Molec. Spectr.* **147**, 46 (1991).
60. A. K. Lewin, M. A. Walsh, and T. R. Dyke, *J. Molec. Spectr.* **151**, 334 (1992).
61. R. G. A. Bone, *Chem. Phys.* **178**, 255 (1993).
62. D. Nelson Jr., G. T. Fraser, and W. Klemperer, *J. Chem. Phys.* **83**, 945 (1985).
63. B. Post, R.S. Schwartz, and I. Fankuchen, *Acta Crystallogr.* **5**, 372 (1952).
64. A. Taleb-Bendiab, K. W. Hillig II, and R. L. Kuczkowski, *J. Chem. Phys.* **94**, 6956 (1991).
65. R. G. A. Bone, C. R. Le Sueur, R. D. Amos, and A. J. Stone, *J. Chem. Phys.* **96**, 8390 (1992).
66. B. J. Smith, D. J. Swanton, J. A. Pople, H. F. Schaefer III, and L. Radom, *J. Chem. Phys.* **92**, 1240 (1990).
67. N. N. Lichtin, I. Laulicht, and S. Pinchas, *Inorg. Chem.* **3**, 537 (1964).
68. G.T Fraser, R.D. Suenram, F.J. Lovas, A.S.Pine, J.T. Hougen. W.J. Lafferty, and J.S. Muenter, *J. Chem. Phys.* **89**, 6028 (1988).
69. Y. Ohshima, Y. Matsumoto, and M. Takami, *Chem. Phys. Lett.* **147**, 1 (1988).
70. I. L. Alberts, T. W. Rowlands, and N. C. Handy, *J. Chem. Phys.* **88**, 3811 (1988).
71. R. G. A. Bone and N. C. Handy, *Theor. Chem. Acta* **78**, 133 (1990).
72. R. G. A. Bone, C. W. Murray, R. D. Amos, and N. C. Handy, *Chem. Phys. Lett.* **161**, 166 (1989).

73. R. G. A. Bone, R. D. Amos, and N. C. Handy, *J. Chem. Soc. Far.* **86**, 1931 (1990).
74. A. C. Legon and D. J. Millen, *Chem. Rev.* **86**, 635 (1986).
75. G.W. Bryant, D.F.Eggers, and R.O.Watts, *J. Chem. Soc. Far. 2* **84**, 1443 (1988).
76. K. Liu, J. G. Loeser, M. J. Elrod, B. C. Host, J. A. Rzepiela, N. Pugliano, and R. J. Saykally, *J. Am. Chem. Soc.* **116**, 3507 (1994).
77. D. J. Wales, *Science* **271**, 925 (1996).
78. G.W. Bryant, D.F.Eggers, and R.O.Watts, *Chem. Phys. Lett.* **151**, 309 (1988).
79. P. G. Mezey, *Potential Energy Hypersurfaces*, Studies in Physical and Theoretical Chemistry, **53** (Elsevier, New York, 1987).
80. R. L. DeLeon and J. S. Muenter, *J. Chem. Phys.* **72**, 6020 (1980).
81. T. A. Hu, D. G. Prichard, L. H. Sun, and J. S. Muenter, *J. Molec. Spectr.* **153**, 486 (1992).
82. Y. Ohshima, M. Iida, and Y. Endo, *Chem. Phys. Lett.* **161**, 202 (1989).
83. M. Takami, T. Ohshima, S. Yamamoto, and Y. Matsumoto, *Faraday Discuss. Chem. Soc.* **86**, 1 (1988).
84. Y. Ohshima, Y. Matsumoto, M. Takami, and K. Kuchitsu, *J. Chem. Phys.* **99**, 8385 (1993).
85. C. R. Le Sueur, A. J. Stone, and P. W. Fowler, *J. Phys. Chem.* **95**, 3519 (1991).
86. R. G. A. Bone, *J. Phys. Chem.* **98**, 3126 (1994).
87. R. Bemish, P. A. Block, L. G. Pederson, W. Yang, and R. E. Miller, *J. Chem. Phys.* **99**, 8585 (1993).
88. CADPAC5: The Cambridge Analytic Derivatives Package, Issue 5, Cambridge 1992. A suite of quantum chemistry programs developed by R. D. Amos with contributions from I. L. Alberts, S. M. Colwell, N. C. Handy, D. Jayatilaka, P. J. Knowles, R. Kobayashi, N. Koga, K. E. Laidig, P. E. Maslen, C. W. Murray, J. E. Rice, J. Sanz, E. D. Simandiras, A. J. Stone and M.-D. Su.

DIMERIC ABSORPTION IN THE ATMOSPHERE

A.A. VIGASIN
Obukhov Institute of Atmospheric Physics,
Russian Academy of Sciences,
Pyzhevsky per. 3, Moscow, 109017 Russia

Possible spectroscopic manifestations of van der Waals and hydrogen-bonded molecules in planetary atmospheres are reviewed. Special attention is paid to existing experimental spectroscopic patterns which can be interpreted as weakly bound species in the atmosphere. In most cases the spectroscopic signatures of dimers stand on the background formed by collision-induced diffuse absorption. Possible contribution of the weakly bound dimers of hydrogen, nitrogen, oxygen and carbon dioxide to collision-induced absorption phenomena in the atmosphere is considered. The problem of the water vapor continuum absorption is also considered in its relation to the problem of dimeric absorption.

1 Introduction

In this Chapter we focus attention on the possible manifestations of weakly bound species in the absorption and emission spectra of planetary atmospheres. Our primary goal is to gather scarce data which may relate to weakly bound atmospheric species. The main source for our data is spectroscopic material taken *in situ*, but data taken in the laboratory are also considered in this Chapter. To reveal the true role played by weakly bound complexes in the absorption and emission of atmospheric radiation, one has to elucidate initially a set of related problems. The question may be raised how many dimers are there? Simple estimates show that neutral dimers and bimolecular complexes are as abundant in the atmosphere as conventional gaseous species involved in various atmospheric processes. Apparently, the very low content of van der Waals species in the atmosphere does not allow them to be considered as important as known radiatively significant minor or trace atmospheric constituents. One of the principal distinctions of weakly bound species from 'normal' gaseous components of the atmosphere is the ability of the former to exist naturally in the atmosphere only with their more abundant monomeric progenitors. Because of this, the spectra of dimers are often masked by the spectra of their parent monomers. Hence, it is very problematic in most cases to make the spectral signature of dimers to stand out of a diffuse or non-resolved background. In the following pages several examples are given, however, of the spectral manifestation of a dimer, which may be distinguished unequivocally in atmospheric spectra. Dimeric absorption manifests itself in most cases either as a slight modification of the spectra of parent monomers, as absorption which occur in the additional regions of forbidden monomeric transitions, or in the regions where monomeric absorption is not possible at all. Moreover, in the regions of forbidden monomeric transitions, purely dimeric features are often masked by collision-induced spectroscopic patterns, which, by their nature, are also due to intermolecular interaction in a pair of monomers exhibiting a collision. Probably the clearest illustration of the overlap of dimeric and collision-induced spectroscopic manifestations is provided by the so-called 'simultaneous' transitions in collisional pairs. These are identical in nature to combination tones excited in different monomeric units within a dimer and as such are intrinsically unavailable in the isolated monomers. Due attention is paid to the problem of interpreting the nature of the continuum absorption, the particular case of water vapor being the best known and most important example. Detailed insight into the nature of this phenomenon is useful for understanding the general causes and consequencies of bimolecular absorption.

2 Weakly Bound Species in Planetary Atmospheres

Molecular absorption is known to differ among the planets as a result of varying composition and properties of the atmospheres. All celestial bodies in the solar system possessing a substantial atmosphere may be classified into three groups (see Table 1). The Earth and Titan — a satellite of Saturn — are the only bodies in the solar system having massive nitrogen atmospheres. The atmospheres of Venus and Mars are known to be composed mainly from carbon dioxide. Molecular hydrogen dominates in the atmospheres of the outer giant planets — Jupiter, Saturn, Uranus, and Neptune. The hydrogen dimer, being one of the weakest dimers in nature, is so far the only weakly bound neutral molecular system detected unequivocally in an astronomical object. Our account below begins with this very spectacular dimer example.

2.1 Jupiter, Saturn and Titan

The far-infrared atmospheric spectra of the giant planets are known to be formed mainly due to collision-induced transitions in molecular hydrogen.[2] The main portion of the atmospheric emission occurs in the infrared and the fraction of outgoing radiation in the infrared is about two times the absorbed solar radiation.[3] That is why even minor features in the hydrogen collision-induced spectrum may turn out to be of significance for the comprehension of the total radiative budget of the giant planets. Specifically, these features are of interest for retrieval of the height profiles of gaseous constituents based on spectroscopic

Table 1: Gaseous composition (in per cents) of mainly nitrogen (1), carbon dioxide (2) and hydrogen (3) atmospheres of some celestial bodies in the solar system.[1]

		N_2	O_2	CO_2	Ar
1	Earth	78.08	20.95	$3.4 \cdot 10^{-4}$	0.93
	Titan	~ 89			10

		CO_2	N_2	O_2	Ar
2	Mars	95.3	2.7	0.13	$1.6 \cdot 10^{-2}$
	Venus	95	3.5	< 0.001	$0.5 - 1.2 \cdot 10^{-4}$

		H_2	He	CH_4	NH_3
3	Jupiter	89	11	$6.5 \cdot 10^{-4}$	$1.8 \cdot 10^{-4}$
	Saturn	94	6	$8 \cdot 10^{-4}$	$2 \cdot 10^{-4}$
	Uranus	90	10	*traces*	
	Neptune	90	10	*traces*	

observations made from space vehicles. Spectra of the atmospheres of Jupiter, Saturn, and Titan were recorded in the course of Voyager–1 and Voyager–2 space missions in 1979–1981. The spectral resolution of the on-board InfraRed Interferometric Spectrometer (IRIS) used was rather low ($\Delta\nu \approx 4$ cm^{-1}). Very weak but sharp features in these spectra have attracted much attention since Frommhold *et al.* [4] and McKellar [5] attributed them to spectral lines of the hydrogen dimers $(H_2)_2$. Fig. 1 clearly demonstrates the similarity of the *in*

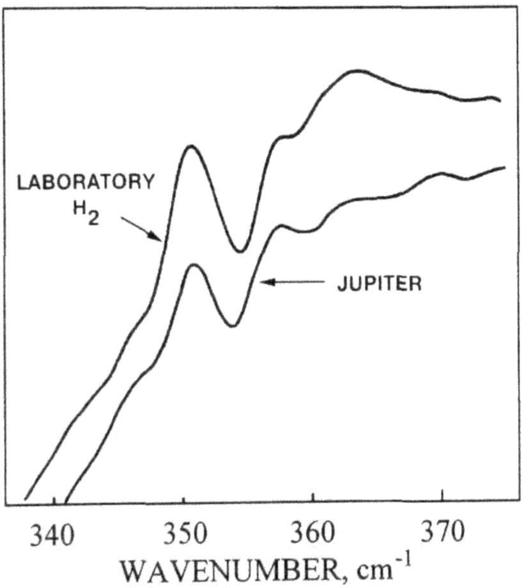

Figure 1: A portion of experimental hydrogen spectrum in comparison with *in situ* spectrum of the Jovian atmosphere as given by McKellar. Reprinted with permission from Ref. 6. ©1988 The American Astronomical Society.

situ spectrum of the atmosphere of Jupiter with a laboratory spectrum taken by McKellar [6] at low temperature. The resolution of the laboratory spectrum was artificially degraded to 4.3 cm^{-1} spectral resolution appropriate to the IRIS spectrometer. Note that the laboratory far-infrared spectrum was first taken [6] at T $\approx$ 77 K, whereas the average ambient temperature of giant planet atmospheres is thought to be of the order of 100 K or even more. At low resolution, two peaks in the atmospheric spectrum, situated at $\nu = 351$ cm^{-1} and $\nu = 584$ cm^{-1}, were assigned to the $(H_2)_2$ dimer. The position of one of these peaks in the laboratory hydrogen spectrum, taken at higher resolution and very low temperature is shown in Fig. 2. New absorption features of

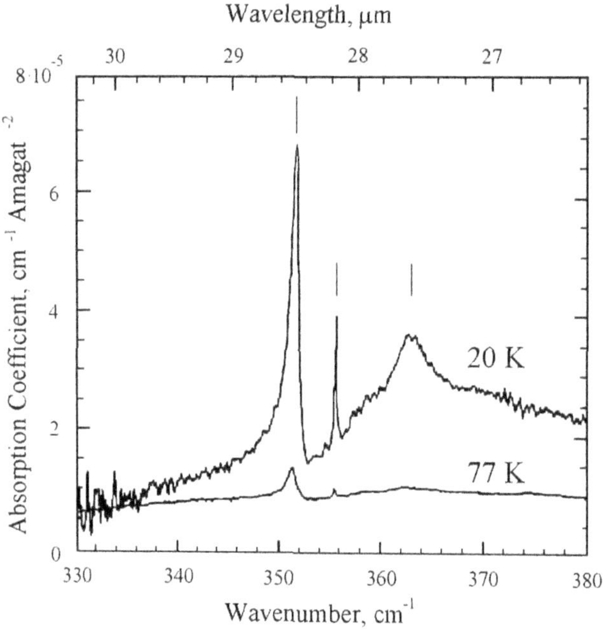

Figure 2: Far infrared hydrogen absorption spectra at different temperatures. Vertical bars show the position of dimeric absorptions. Reprinted with permission from Ref. 7. ©1991 The American Institute of Physics.

the hydrogen dimer have been found recently by Kim *et al.*[8] and Trafton *et al.*[9] in the equatorial spectrum of Jupiter, Saturn and Neptune. The $(H_2)_2$ dimer emission is tentatively proposed by Trafton *et al.*[10,11] to explain the features observed in the Jovian atmosphere emission spectrum near the H_2 fundamental. To reveal the nature of the observed features in the spectrum of hydrogen it is worthwhile to characterize briefly the structure and molecular properties of the $(H_2)_2$ dimer.

The first *ab initio* quantum-chemical calculation relevant to the equilibrium structure of the $(H_2)_2$ dimer was performed in 1972.[12,13] The most rigorous calculations seem to belong to Burton and Senf.[14] It turned out that the equilibrium structure of the $(H_2)_2$ dimer refers to the T–shape configuration (C_{2v} point group) and is characterized by the intermolecular bond energy $D_e \approx 600\ J/mole$ or $D_e/R \approx 70K$. Bearing in mind that the anisotropy of interaction of two hydrogen molecules is very small, intermolecular potential can be effectively averaged over the orientations of the monomeric units. The dissociation energy of the resulting quasidiatomic hydrogen dimer is, according to,[15] merely about 3.5–4 K. This means that only the ground state of the translational vibration mode is stable in the intermolecular potential well. The rotational excitation of the monomeric constituents of a dimer transforms it into the Feshbach resonance state, which is capable of spontaneously decaying. It has been emphasized, however, in, [7] that the coupling of intra- and intermolecular motions within a dimer is very inefficient and this may result in a rather small (<0.05 cm^{-1}) contribution of the predissociation effect to the overall broadening of the observed spectral lines.

Watanabe and Welsh [16] first detected the fundamental transition of the hydrogen dimer with rotational resolution using compressed hydrogen gas in 1964. Later spectroscopic study of the $(H_2)_2$ dimer is mainly due to the work of McKellar. [5,6,7,17,18] The spectra were recorded using a cooled long-path optical cell with a base length of 3.50 m providing multiple traversals giving total optical path from 13 m to 110 m. The cell was cooled to temperatures of about 20 K [7,17,18] or 77 K [6] and mixtures of isotopic species H_2 and D_2 were employed to ensure the appearance of $(H_2)_2$, $(D_2)_2$, $H_2 \ldots D_2$, $(HD)_2$ and $HD \ldots D_2$ molecular complexes. Spectra of the hydrogen constituent of $H_2 \ldots N_2$ and $H_2 \ldots CO$ complexes were taken in the mid and far infrared regions. [19] In most cases the well-resolved spectral features of the dimers were assigned assuming free internal rotation of the individual hydrogen molecules within the dimer. This means that vibrational v_1 and v_2 and rotational j_1 and j_2 quantum numbers remain good labels. Let the end-over-end rotation of a dimer be characterized by rotational quantum number l. In the intermolecular potential well of the $(H_2)_2$ dimer, only the states with $l = 0$ and $l = 1$ are bound. At $l = 2$ and $l = 3$, the two hydrogen molecules are captured in rotationally quasibound or shape resonance state while at higher l, only free states are allowed.

Table 2 reproduces the wave numbers and identifications of the far infrared transitions which belong to the $(H_2)_2$ dimer, as observed and assigned by McKellar and Schaefer. [18]

It is seen from Fig. 2 that these transitions appear near the maxima of the collision-induced rotational transitions of the H_2 monomers in their ground vibrational state (subscript zero is an indication of the ground state). It is assumed that the transitions occur with $\Delta j = 2$ (S–transitions) from initial level $j'' = 0$ $(S_0(0))$ or $j'' = 1$ $(S_0(1))$ and various quantum numbers l. Selection rules for this latter demand that $\Delta l = \pm 1, \pm 3$. Because of this, the centers of the transitions in a dimer do not coincide with those of the $S_0(0)$ and $S_0(1)$ transitions in the free H_2 molecule.

Fox and Kim [20] have estimated the mole fraction of the hydrogen dimers in the atmospheres of Jupiter and Saturn to be of the order of 10^{-4}. In the far infrared spectrum of Titan very weak features have been found, which

Table 2: Wave numbers and assignment of observed $(H_2)_2$ transitions in the far infrared.[18]

	ν, cm^{-1}	$l' \leftarrow l''$
$S_0(0)$	351.295(20)	$1 \leftarrow 2$
	355.425(5)	$1 \leftarrow 0$
	362.78(30)	$3 \leftarrow 0$
$S_0(1)$	584.12(5)	$1 \leftarrow 2$
	588.300(10)	$1 \leftarrow 0$
	588.503(10)	$1 \leftarrow 0$
	588.803(10)	$2 \leftarrow 1$
	589.51(4)	$2 \leftarrow 1$
	595.6(3)	$3 \leftarrow 0$

look similar to those of the $(H_2)_2$ dimer. The suggestion has been put forward by Borysow and Frommhold,[21] Fox and Kim,[20] and Bachet *et al.*[22] to associate these spectral features with the $H_2 \ldots N_2$ complexes. An estimate of the content of these complexes in the atmosphere of Titan[20] is that the $H_2 \ldots N_2$ mole fraction is of the order of $4 \cdot 10^{-6}$. Detailed analysis of the laboratory spectra of $H_2 \ldots N_2$, $H_2 \ldots CO$ and $H_2 \ldots Ar$ complexes has been carried out by McKellar.[19,23] He concluded that because of the low resolution of the Voyager spectra, the $S_0(0)$ and $S_0(1)$ peaks of $H_2 \ldots Ar$, $H_2 \ldots CO$ and even of $H_2 \ldots CH_4$ should be indistinguishable from those of $H_2 \ldots N_2$ complexes. Consequently, the up-to-date spectral data do not permit one to ascribe the features in Titan's atmosphere spectra to $H_2 \ldots N_2$ complexes unequivocally, albeit this attribution seems quite likely and the presence of a significant amount of these complexes in the atmosphere of Titan is almost certainly correct. At least recent model calculations[24] of the thermal emission spectrum of Titan's atmosphere properly account for the fine structure of $H_2 \ldots N_2$ and $H_2 \ldots CH_4$ dimers in the far infrared region.

2.2 Mars and Venus

Carbon dioxide molecules which dominate the atmospheres of Mars and Venus are known to be able to form rather strong van der Waals bonds. The $(CO_2)_2$ dimers, as well as a number of binary and even ternary complexes involving CO_2 molecules, are characterized spectroscopically in much detail using laboratory controlled infrared and Raman techniques. The details of these investigations are beyond the scope of the present review due to space limitations. Our comments below extend only to inclusion of those comprehensive features which seem to be important for atmospheric applications.

Mannik and Stryland[25] were the first to have interpreted collision-induced spectrum of compressed carbon dioxide taken in the region of the Fermi doublet in terms of a superposition of free–free and dimeric transitions. Based on this analysis, it was assumed for a long time in the literature that the most stable configuration of the $(CO_2)_2$ dimer is the T–shaped structure, belonging to C_{2v} point group (see Fig. 3). This conclusion seemed to be in line with the detection

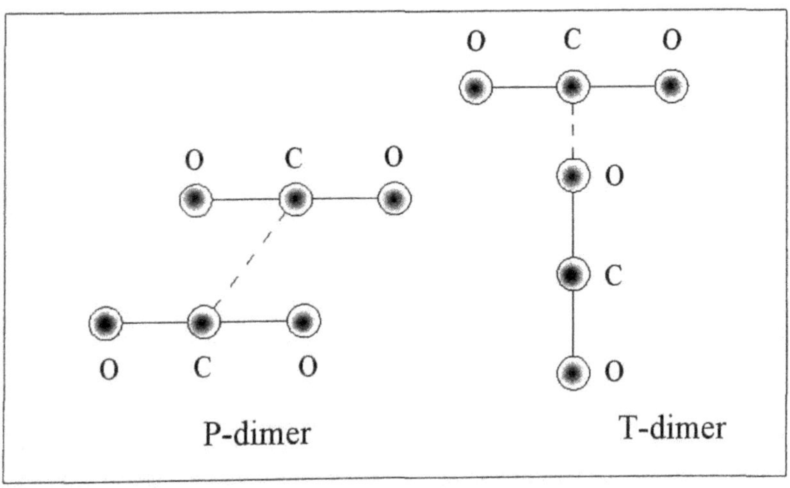

Figure 3: The structure of the carbon dioxide dimers.

of a dipole moment for the $(CO_2)_2$ dimer in the early electric deflection study by Lobue *et al.*[26] Quantum-chemical calculations (see e.g. [27–30]), however, favor a centrosymmetric slipped parallel configuration, having C_{2h} symmetry, which is often called P–structure (see Fig. 3). The most accurate *ab initio* data (see e.g.[28,29]) show that the T–structure is a transition state, which refers to a saddle point on the potential energy surface, and can be achieved in the course of geared motion of the monomeric units between two equilibria points.

In 1985 Nibler *et al.*[31] have reported on the first detection of the $(CO_2)_2$ dimer band in CARS (Coherent AntiStokes Raman Scattering) spectra taken in free jet expansion. The main conclusion of this work is that the $(CO_2)_2$ dimer should have an inversion center compatible with P–structure. In a subsequent publication from the same group[32] it has been shown, however, that an assignment was an error.[31] The most reliable observation of the $(CO_2)_2$ dimer feature in the Raman spectra belongs to Huisken *et al.*[33] who also used the CARS technique with a slit nozzle expansion of the neat and Ar seeded CO_2 gas. At that time, the structure of the $(CO_2)_2$ dimer was unequivocally established and it has been proven that the carbon dioxide dimer forms a centrosymmetric P–structure. This was done using a diode laser[34] or difference frequency laser[35,36] techniques in conjunction with the gas-dynamic generation of dimers in free expansion into vacuum. The high-resolution infrared absorption spectra obtained in these studies made it possible to determine a set of precise rotational constants for the $(CO_2)_2$ dimer. Moreover some force field constants, characterizing the curvature of intermolecular potential energy surface near the equilibrium configuration, have been estimated via centrifugal distortion coefficients. In a similar way the complexes of $Ar\ldots CO_2$, $N_2\ldots CO_2$ and $H_2O\ldots CO_2$ have been characterized[37,38,39] following a thorough examination of their infrared absorption spectra taken at high resolution (see Table 3). Pine and Fraser[51] have studied the broadening due to predissociation and the shifts of absorption band centers caused by the formation of dimers and trimers of the carbon dioxide molecule or the complexes of CO_2 with noble gas Ar, Kr and Ne atoms. Analysis of the microwave spectra of a water admixture containing carbon dioxide in free expanding jets enabled Peterson *et al.*[52,53] to determine the equilibrium structures of triple complexes $H_2O\ldots(CO_2)_2$ and $(H_2O)_2\ldots CO_2$.

Detailed quantum-chemical studies of the equilibrium structure, dissociation energy and vibrational spectrum of the $(CO_2)_2$ dimer open the possibility of direct calculation of the equilibrium constant and thermodynamical properties of the $(CO_2)_2$ dimer formation. Bearing in mind that physical properties of the atmospheres of Mars and Venus are relatively well known due to numerous spacecraft missions, one might suppose that the height profiles of the $(CO_2)_2$ dimers in these atmospheres can be predicted rather accurately. First esti-

Table 3: Molecular characteristics of some van der Waals complexes.

Molecular pair		Point group	R_c, Å	D_e, K	References
H_2	H_2	C_{2v}	3.44	≤ 60	[14]
	Ar	$C_{\infty,v}$	3.557	83.5	[40]
	N_2	–	3.4	≥ 65	[41]
N_2	N_2	D_{2d}	3.50	183	[42]
	Ar	C_{2v}	3.9 ± 0.2	≥ 138	[43]
O_2	$O_2(^5A_g)$	D_{2h}	3.18	221	[44]
	$O_2(^3B_{1u})$	D_{2h}	3.28	201	[44]
	$O_2(^1A_g)$	D_{2d}	3.28	181	[44]
H_2O	N_2	C_s	3.37(4)	600	[45-48]
	CO_2	C_{2v}	2.836	1500	[39,49,50]
CO_2	CO_2	C_{2h}	3.599(7)	~ 700	[27,35]
	N_2	C_{2v}	3.73	650	[38]
	Ar	C_{2v}	3.493	~ 300	[37]

mates for the altitude profiles of the $(CO_2)_2$ dimer in planetary atmospheres have been published by Fox and Kim.[20] More accurate profiles have been generated independently by Vigasin and Chlenova.[54]

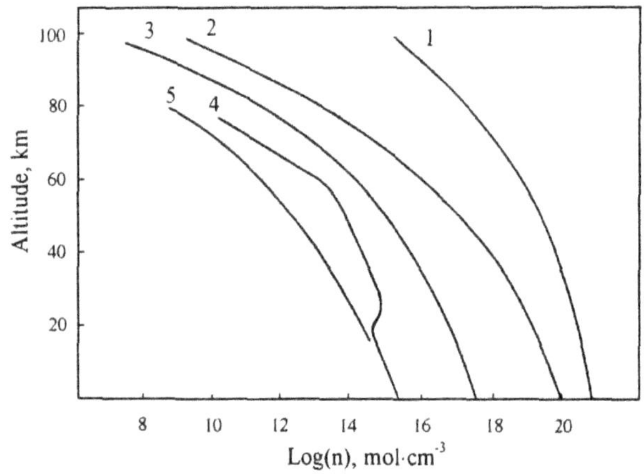

Figure 4: Equilibrium number densities for the van der Waals complexes in the atmosphere of Venus. 1–total number density, 2–$(CO_2)_2$, 3–$CO_2 \ldots N_2$, 4–$CO_2 \ldots H_2O$, 5–$CO_2 \ldots Ar$.[54]

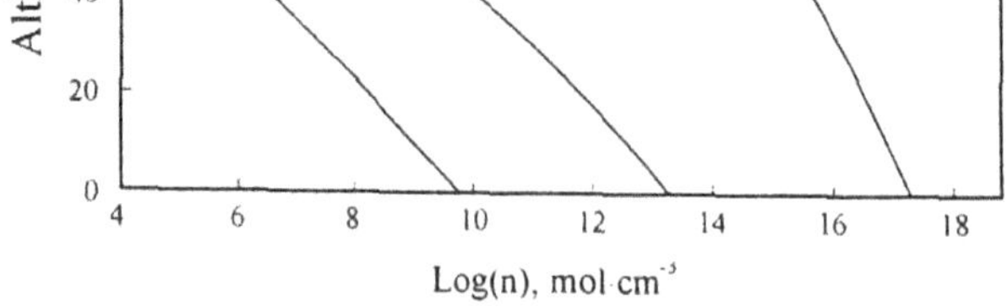

Figure 5: Equilibrium number densities for the van der Waals complexes in the atmosphere of Mars. 1–total number density, 2–$(CO_2)_2$, 3–$CO_2 \ldots H_2O$.[54]

Figs. 4, 5 demonstrate some results of these calculations. For the Martian atmosphere the model of Moroz *et al.*[55] has been employed in our calculations.[54] The structure and composition of the Venusian atmosphere referred to[54] use the data of.[56,57] The equilibrium constants for the $(CO_2)_2$

and $CO_2 \ldots H_2O$ dimers were calculated using available experimental (structure) and calculated (vibrational frequencies) data. The dimer mole fraction is governed by the mass-action law as described in Chapter 1. For $CO_2 \ldots Ar$ and $CO_2 \ldots N_2$ complexes, the corresponding mole fractions were calculated through the use of the approximate formula, suggested by Anderson *et al.* [58] after numerical calculations by Stogryn and Hirshfelder [59] and also reported in Chapter 1. In a series of works by Slanina *et al.* [60,61,62] the sensitivity of the $(CO_2)_2$ dimer height profile to details of the intermolecular potential has been investigated with the use of Stogryn and Hirshfelder [59] formalism (see also Chapter 4). The data of Slanina *et al.* [61] fit well those of [54] relevant to Venusian atmosphere.

The low temperature of the Martian troposphere ($\approx$ 200 K) and low near surface density of the carbon dioxide ($\rho \approx 1.5 \cdot 10^{-5}$ g/cm^3) provide favorable conditions for the observation of well-resolved spectroscopic features of the CO_2 complexes. Our estimates show, however, that even near surface number density of $(CO_2)_2$ dimers does not exceed ca. 10^{-13} mol/cm^3. This makes it hardly probable that these dimers play a significant role in the transmission of radiation through the atmosphere of Mars. The possible weak manifestation of dimers in the total visible atmospheric column are masked by spectroscopic features from the carbon dioxide absorption band wings and, probably, the absorption by the Martian dust. It is assumed that the difference between the model carbon dioxide atmospheric transparency and that of the real Martian atmosphere can be totally ascribed to the aerosol absorption.

In the atmosphere of Venus the extent of association of CO_2 molecules may reach about 10 percent. [54,60] This means that the $(CO_2)_2$ dimer can be formally considered as the most abundant constituent of the Venusian atmosphere after the monomeric carbon dioxide gas itself. The statistical physics partitioning of phase space for the interacting CO_2 pairs, which has been discussed in the Chapter 1, shows that about one half of the CO_2–CO_2 molecular pairs are firmly bound or metastable at $T \leq 700$ K, i.e. at the temperature typical of the lower atmosphere of Venus. The impact of this on the radiative modelling of the Venus atmosphere is, however, almost negligible since the dimer absorption seems to be effectively masked by the broad absorptions due to the CO_2 monomers. Because of this, it seems worthwhile to elucidate the role played by the carbon dioxide dimers in the overall spectroscopic properties of compressed carbon dioxide gas.

It has been shown in [63,64] that the variation of bandshape of the CO_2 infrared absorption bands at 1.434, 1.606 and 2.06 μm with increasing density can be described within experimental accuracy by consideration of monomer–dimer equilibrium. All investigators agree that intermolecular interactions cause the variation of spectral bandshape with density. However, in a number of works (see e.g. [65]) it is claimed that the effect of association can be completely neglected in comparison with the effects of line-mixing and the duration of molecular collisions. All available present day concepts of the infrared absorption bandshapes are empirical to some extent, that is why there currently exists no recipe for the rigorous solution of this particular problem. The author's belief is that purely spectroscopic evidence should not be treated independently from thermodynamic considerations, but that they should conform to each other. This viewpoint makes very probable an involvement of dimeric states in shaping the contours of IR-permitted absorption bands in compressed carbon dioxide. Additional support for this belief comes from the consideration of the dipole forbidden transitions. It has been shown in [66] (see also Sec. 4.2 of Chapter 1) that the temperature dependence of the absorption band intensity in the region of the Fermi doublet follows that, which stems from the mass-action law. This means that a large set of the spectroscopic events embracing the formation of dipole permitted and forbidden transitions in pressurized carbon dioxide can be successfully described in terms of monomer–dimer equilibrium, which comply with independent thermodynamic considerations.

2.3 Earth

The transparency of the Earth's atmosphere in the infrared is determined mainly by absorption bands of water vapor, carbon dioxide, ozone and several other minor constituents. The nitrogen and oxygen molecules, which dominate in the atmosphere, do not possess dipole moments and the relevant pure rotational and vibro-rotational transitions are dipole forbidden. Note that weak quadrupole moments permit these homonuclear diatomics to have quadrupole absorption spectra. As a matter of fact, very weak quadrupole lines have been observed recently in the region of the N_2 and O_2 fundamentals in atmospheric spectra.[67] It seems worthy of mention that magnetic-dipole transitions are allowed in the oxygen molecule so a number of dipole transitions from low-lying excited electronic states to the ground triplet state fall in the infrared, visual and ultraviolet spectral range.

Besides the line spectrum, which is typical for most infrared active atmospheric species, a so-called continuum spectrum is often referred to in considerations of atmospheric absorption. Apparently, the nature of this continuum absorption is due to intermolecular interactions. Because of this, in the following pages we shall pay special attention to this particular spectral phenomenon. The more important contributions to the continuum background in atmospheric spectra include:

— Diffuse electronic bands of molecular oxygen pairs, which can be found over the wide spectral range from 0.3 to 1.26 μm,

— The water vapor continuum, which appears in the spectral range from the microwave to middle infrared, the most important wavelength range being between 8 and 13 μm,

— The continuum absorption in the middle infrared from 4.0 to 6.0 μm, which is due to absorption in the CO_2 4.3 μm transition band head with some contribution from the dipole-forbidden transitions in CO_2, N_2 and O_2.

There currently exists no widely recognized unified and reliable model to mimic the behavior of the continuum absorption over the wavelength ranges given above. The absence of resolved spectroscopic lines, which characterizes these continua absorptions, makes it possible to model the observed absorptions, starting from essentially different principles. Very roughly, one of two postulates are invoked. One says that all events may be reduced to an effect on the distant lineshape wings, the centers of which are characteristic of the isolated monomeric molecules. The other holds that all observed effects can be well explained in terms of absorption by dimers, formed from the conventional monomeric gaseous constituents of the atmosphere. Below, we shall concentrate on a treatment of the observations, which could be considered as supporting this latter postulate.

Prior to the subsequent discussion of the nature of diffuse and continuum absorption in atmospheric spectra, the abundance of weakly bound complexes in the Earth's atmosphere deserves a brief discussion. There exists only a handful of calculations devoted to this problem. A short communication by Calo and Narcisi[68] remains probably the only serious attempt to estimate the abundance of a number of van der Waals species in the atmosphere and to indicate their possible role in nucleation processes, photochemistry, infrared absorption, and emission in the natural atmosphere. It has been demonstrated that the equilibrium state of some dimers makes them as abundant as minor rather than as trace atmospheric constituents. The variety of dimers is rich, since N different gaseous components of the atmosphere give rise approximately to $N^2/2$ various complexes (more exactly $N(N-1)/2$ intermolecular complexes plus N self-associated dimers). The height distribution of complexes is governed by the mass-action law and normally the abundance of complexes drops more rapidly with altitude than that of monomers due to the quadratic dependence on the monomer's density in the corresponding atmospheric layers. It should be emphasized that for the parent species distributed inhomogeneously in the atmosphere (such as ozone, SO_2, HNO_3, water vapor etc.), the character of the vertical distribution of the dimers may be significantly more complicated. Among these associated species, particular attention is paid to water dimers due to their possible involvement in attenuation of radiation in the relatively transparent atmospheric windows in the near and far infrared. The ground-level content and the vertical profile of the water dimer have been topics of concern in a number of investigations.[68,54,70] We shall comment on the results of such studies in the Section below, while considering the problem of the water vapor continuum.

The middle infrared wavelength range situated in the 4.3 μm CO_2 ($10^00 \leftarrow 00^00$) absorption band head is of particular interest. The rotational constant B' of the upper vibrational state in the ν_3 mode is lower than that (B'') of the CO_2 molecule ground state. Because of this the ν_3 absorption band has a sharp band head at $\nu_b = 2397$ cm^{-1}. At $\nu > \nu_b$, up to 2600 cm^{-1}, the atmospheric spectrum can be considered as a superposition of the combination bands of N_2O (2462 cm^{-1} and 2600 cm^{-1}), CO_2 (2429 cm^{-1}), CH_4 (2612 cm^{-1}), and vibro-rotational lines of H_2O with an admixture of weak quadrupole N_2 transitions.[71] Wide ν_3 CO_2, ozone and nitrous oxide absorption bands dominate for $\nu < 2397$ cm^{-1} as well as the band of isotopic $^{13}CO_2$ at $\lambda \approx 4.4$ μm. In the neighborhood of the 4.3 μm band and in the CO_2 band head, continuum absorption is present which may be associated with absorption mechanisms mainly due to interactions of homo– and heteromolecular nitrogen, oxygen, carbon dioxide, and water pairs. Despite the very low intensity of this absorption, it is firmly detected in the solar radiation spectrum transmitted either through the total atmospheric column[72] or in the lower stratosphere,[71] including upper troposphere.[73] Four factors are thought to determine this continuum absorption. These are:

1. Collision-induced absorption in the neighborhood of the nitrogen fundamental centered at 2329.91 cm^{-1}.

2. Absorptions due to the most abundant van der Waals complexes $(N_2)_2$, N_2–O_2 and N_2–CO_2.

3. Water dimer $(H_2O)_2$ absorption.

4. H_2O vibro-rotational lines far-wings, extending to the neighborhood of the CO_2 4.3 μm band head.

Bernstein *et al.*[73] have attempted to associate absorptions they observed at altitudes higher than 5 km with N_2–N_2 and N_2–CO_2 complexes. Their main reason was that the perturbation of N_2 or CO_2 molecules is most effective in collisions of long duration, which occur at closest intermolecular separations.

These conditions are the most favorable for the formation of firmly bound and metastable dimers. Dianov-Klokov and Malkov [72] claimed that the nature of continuum absorption in the 4.3 μm neighborhood should be associated with absorption due to short-living N_2–N_2 and N_2–O_2 complexes.

Controversial opinions may be found in the literature with respect to the role played by oxygen dimers or free collisional pairs in the absorption of mid-infrared radiation by the gas phase. For instance McKellar *et al.* [74] believe that it is generally accepted that the diffuse absorption bands of oxygen have nothing in common with the quasistable oxygen dimers. By contrast Gijseman *et al.* [75] claim that the contributions from tightly bound and free collisional pairs to the intensity of collision-induced absorption at room temperature are of the same order of magnitude. Dianov-Klokov [76] concluded from the absence of rotational structure in the observed oxygen pair spectra that the life-time of oxygen couples in the atmosphere does not exceed $\sim 2 \cdot 10^{-11}$ sec. This approximate estimate gives us an idea of the oxygen pair, which decays under atmospheric conditions before undergoing successive collisions.

In a series of works [77,78,79] Shardanand and associates attributed the concentration dependence of the oxygen absorption in the Herzberg continuum to absorption by $(O_2)_2$, O_2–N_2, and O_2–Ar complexes. Blake and McCoy [80] and Greenblat and associates [81] claim that the contribution of O_2–O_2 van der Waals complexes to the observed oxygen pair absorption can be ruled out, since the expected temperature dependence of the dimeric pressure coefficient should be much stronger than that observed experimentally. They believe that collision-induced absorption of unbound pairs provides a good account of the temperature dependence shown by experimental data.

Estimates (see Chapter 1) show that the temperature dependence of the overall integrated induced intensity does not allow one to distinguish the bound molecular pair contribution from that of free pairs. However, statistical physics partitioning of this intensity (see Chapter 1) shows that the contribution of nitrogen or oxygen dimeric (mainly metastable) states to pure rotational collision-induced intensity of absorption does not exceed about 20% of the total absorption at T $\sim$ 200–300 K. It can be thought therefore that in the neighborhood of the rotational induced spectrum, the contribution of these dimers is not negligible at ambient temperatures; however, it plays a secondary role after free-free transitions, occurring in the course of elastic collisions of N_2 and/or O_2 molecules.

One should infer that the same is true for collision-induced absorption in the range of the oxygen fundamental at ν=1556 cm^{-1}. Timofeyev and Tonkov [82] have shown that absorption of the molecular pairs O_2–O_2 and O_2–N_2 should be accounted for in modeling of the radiative transfer in the atmosphere especially at slant paths with tangent altitudes below 10 km. Stratospheric measurements of the oxygen absorption band performed by Rinsland *et al.*[83] at high resolution ($\Delta\nu \approx 0.02$ cm^{-1}) demonstrated that the Timofeyev and Tonkov's[82] absorption coefficients at lower stratospheric temperatures have an absolute accuracy better than 20 % in the wavelength range 1435–1725 cm^{-1}. The neglect of the oxygen continuum absorption in this range may result in significant errors while interpreting quantitatively the results of remote sensing experiments in the atmosphere.

Collision-induced spectra in the region of the oxygen and nitrogen fundamental transitions were recorded for the first time in 1949 by Crawford *et al.*[84] Later, these absorption bands were the subject of numerous experimental studies at low temperature as well as at a variety of near-room temperatures. Long *et al.*[85] were the first to discover structural patterns in the vicinity of the band center superimposed on the smooth collision-induced background of the nitrogen fundamental band, which might be ascribed to the nitrogen dimers. According to McKellar,[86] the experimental spectrum of $(N_2)_2$ dimers differs considerably from that calculated by Brocks and van der Avoird,[87] who used an intermolecular potential suggested by Berns *et al.*[42] This potential leads to the crossed ground state configuration of monomers in a dimer, having D_{2d} point group symmetry (see Table 2). Barriers to internal rotations of the monomers within a dimer amount to merely $\sim$ 30 K, therefore the monomers are practically free to rotate even at low temperatures. The perturbation of rotational movement due to the intermolecular interaction results in the appearance of weak regular ripples on the diffuse collision-induced contour, the minima of which coincide in a surprisingly accurate way with the position of vibro-rotational transitions in N_2 monomers, calculated using selection rules appropriate to collision-induced spectra. These ripples have been detected by Henderson *et al.*[88,89] for the O_2–Ar and N_2–Ar systems and by McKellar[86] and Lafferty *et al.*[90] for pure nitrogen. Qualitatively, these ripples may be successfully interpreted either as combinations of intramolecular vibration with hindered rotations within a dimer[88] or as an evidence of line-mixing effects.[91] Lafferty *et al.*[90] point out that the amplitudes of these weak ripples are practically independent of temperature and at near room temperature are of the same order of magnitude as at 77 K. This seems to disfavor the dimeric nature of the observed weak features, whereas the line-mixing mechanism suggested in[91] looks more convincing.

Despite the apparent importance of the N_2 and O_2 complexes for atmospheric applications, there is only a handful of laboratory data, concerning these atmospheric species. No detailed laboratory studies of oxygen or nitrogen dimers at high resolution have been reported hitherto. For the mixed

N_2–O_2 complexes, only the early electric deflection study by Novick *et al.*[92] deserves to be mentioned, in which the absence of the dipole moment for this system has been demonstrated. Among other relevant publications, the work by Dianov-Klokov and Malkov[72] can be cited in which collision-induced absorption in a mixture of N_2 and O_2 has been investigated at pressures up to 20 atm. The temperature dependence of the absorption in the neighborhood of the oxygen fundamental has been studied by Orlando *et al.*[93] at $T = 225$–356 K to account for the formation of $(O_2)_2$ and N_2–O_2. In a recent paper Menoux *et al.*[94] corroborated the conclusion made in,[72] that the efficiency of collisional induction for the N_2–O_2 system is about a factor of 0.8 that for the N_2–N_2 system at $T = 297$ K and varies slightly with temperature. Shardanand[78] notes that the presence of N_2 enhances the absorption of O_2 in the Herzberg continuum.

A high-resolution infrared absorption spectrum for the N_2–CO_2 complexes has been recorded by Walsh *et al.*[34] using the method of multiple-traversals diode laser spectroscopy for the free supersonic expansion of a mixture, containing 0.5% CO_2 and 5% N_2 in Ne as a carrier gas. The N_2–CO_2 complex has been shown to have a T–shaped structure (see Table 3). Fredin *et al.*[95] have detected weak absorption at 2328.1 cm^{-1} in the spectrum of CO_2 trapped in a nitrogen matrix, which can be assigned to the slightly shifted stretching fundamental of nitrogen due to the N_2–CO_2 interaction.

The study of the 'ozone hole' phenomenon stimulated particular interest to the oxygen continuum absorption in the atmosphere. The spectroscopic identification and retrieval of the column abundance of such species as OClO, BrO, NO_2 require exact knowledge of the reference spectra, including underlying weak and diffuse oxygen absorption features.

Pioneering studies of oxygen spectra over a wide range of external conditions, including the natural atmosphere, have been carried out by Dianov-Klokov (see e.g. [72,76,96,97,98]). He believed that absorption bands of oxygen pairs, seen in the natural atmosphere are due to short-living dimers, which decay before collisions. The formation of firmly bound dimers of diatomics is hardly possible at atmospheric conditions. The most probable and abundant metastable dimers are capable of spontaneous decay caused by intramolecular energy redistribution.

The oxygen dimers are a subject of concern in a good wealth of experimental and theoretical speculations. It is only very recently, however, that Bussery and Wormer[44] have succeeded in building a complete *ab initio* intermolecular potential for the ground state of $(O_2)_2$, which is a considerable advancement over the pre-existing truncated O_2–O_2 potentials.[99,100] The oxygen molecule is known to have a triplet ground state $O_2(^3\Sigma_g^-)$, a fact which makes this molecule a rare exception. The formation of an oxygen dimer results in the $(O_2)_2$ ground state splitting into singlet, triplet and quintet states (see Table 3). According to [44] these states are energetically fairly close. This makes it exceptionally difficult to model the $(O_2)_2$ spectrum. Goodman and Brus [101] have found that ground state singlet-triplet splitting for the $(O_2)_2$ dimer trapped in a Ne matrix amounts to 55 cm^{-1}. This value is somewhat in excess of the splitting predicted in; [44] however, both estimates are of the same order of magnitude as the values expected for the intermolecular vibrational frequencies. Model calculations show (see e.g. [102]) that intermolecular frequencies are bracketed within a range of wave numbers from about 10 to 50 cm^{-1}. Vibrational movements of the monomeric units within a dimer should be determined by the spin properties of the potential energy surface. Van der Avoird and Brocks [103] have shown that the spin-rotational fine structure of the $(O_2)_2$ spectrum is very sensitive to the dimer geometry, the nature of internal movements and the magnetic interaction coupling. Because of this, it may be thought that the spectrum of $(O_2)_2$ dimers is very dense and difficult to interpret. The oxygen dimers themselves represent a challenge both to theoreticians and experimentalists. In laboratory controlled conditions they could be easily formed in adiabatically cooled flows and they may be probed using a variety of absorption (in the microwave, infrared, visual and ultraviolet wavelength ranges) and Raman techniques.

The history of spectroscopic observations relevant to oxygen pairs in the atmosphere ascends from the works of Janssen [104,105] published more than a century ago. In the thirties the spectra of gaseous compressed and liquified oxygen became a subject of very intensive investigations. For a number of bands in the visual and near-UV spectral ranges, Salow and Steiner [106,107] have found a quadratic intensity growth as a function of the oxygen partial pressure varying from 40 to 160 atm. The independence of the intensity on the buffer gas pressure was established in [106,107] as well as the band shape insensitivity to the overall pressure in a mixture. Salow and Steiner have supposed that the observed spectral bands are due to short-living oxygen pairs. In works by Herman, [108] Vassy and Vassy, [109] and later Dufay [110] attempts were made to detect the bands, already observed in the spectrum of compressed oxygen, in the spectrum of the setting sun (the ground-based paths employed in [109] have a transmission path varying from 2.5 to 25 km using artificial or extraterrestrial light as an optical source).

The main absorptions of oxygen pairs in the atmosphere fall in the range 3000 to 12600 Å. Following Ellis and Kneser, [111] it is convenient to classify these transitions according to the sum electronic states of the isolated O_2 molecules to account for the vibrational repetitions (see Table 4). Some transitions shown

Table 4: Characteristics of the electronic transitions in the oxygen dimer from its ground-state $^3\Sigma_g^- + ^3\Sigma_g^-$.

Upper $(O_2)_2$ *state*		*Wavelength* $\lambda, \mu m$		*Absorption coefficient* K_{max}, $km^{-1}\ atm^{-2}$		
Electronic	*Vibrational*	[98]	[114]	[98]	[106]	[108]
I group	0	1.26		0.051		
$^3\Sigma_g^- + ^1\Delta_g$	1	1.06		0.033		
II group	0	0.764		(0.004)		
$^3\Sigma_g^- + ^1\Sigma_g^+$	1	0.688				
III group	0	0.629		0.020	0.0205	0.017
$^1\Delta_g + ^1\Delta_g$	1	0.5769	0.5750	0.028	0.0315	0.035
	2	0.5325		(0.004)	0.0019	
IV group	0	0.4773	0.4769	0.016	0.020	0.021
$^1\Delta_g + ^1\Sigma_g^+$	1	0.4470	0.4440	(0.002)	0.0014	
V group	0	0.3808	0.3797	(0.006)	0.0086	
	1	0.3612	0.3612	(0.012)	0.014	
$^1\Sigma_g^+ + ^1\Sigma_g^+$	2	0.3439	0.3439		0.0041	
	3	0.3282			0.0006	

in Table 4 are fairly close to analogous transitions appropriate to the isolated monomers. For instance, the emission of singlet oxygen from the $^1\Delta_g$ and $^1\Sigma_g^+$ states situated at 1.26 μm and 0.76 μm are well known from observations of the upper atmosphere and are called 'ir–atmospheric' and 'atmospheric' emissions respectively, the latter being observed for the first time in compressed oxygen by Egoroff [112] over a century ago. The states $^1\Delta_g$ and $^1\Sigma_g^+$ in the oxygen molecule are metastable with a radiative lifetime of the order of 65 min and 7 sec, respectively. [113] Intermolecular interactions partially eliminate the magnetic dipole restriction to dipole transitions and make a number of new transitions allowed, which may not be present in the monomeric molecule. For instance the so-called 'simultaneous' transitions occur if the wave functions in both monomers are strongly mixed thus making possible an electronic transition in molecule 1 to be accompanied by a vibrational transition in molecule 2. Absorption coefficients in the maxima of the observed bands are given in Table 4 according to measurements by Dianov-Klokov, [98] Salow, [106] and Herman, [108] as they are quoted in. [98] The bracketed values are measured less accurately comparative to other data in the same column. Some of these data entries were slightly reconsidered in the latest publications, namely by Greenblat *et al.*, [81] Newnham *et al.*, [115] and Hermans *et al.* [116] Table 4 accumulates, however, more representative list of transitions.

Based on the results of his own investigation of the oxygen spectrum in the pressure range from 2 to 35 atm, Dianov-Klokov [98] has concluded that diffuse bands of the first group (see Table 4) should manifest themselves more intensely in the atmosphere than the bands at shorter wavelengths. Despite the 1.26 μm and 1.06 μm $(O_2)_2$ bands being masked by the atmospheric O_2 bands at 12684 Å and 10684 Å , respectively, the contribution of the O_2 monomeric line spectrum may be separated from the diffuse spectrum, which belongs to O_2–O_2 pairs in the high resolution spectrum. Note that in the monomer spectrum the band $^3\Sigma_g^+ \rightarrow {}^1\Delta_g(v = 0)$ ($\lambda \approx 1.26\mu$m) is at least ten times more intense than that at $^3\Sigma_g^+ \rightarrow {}^1\Delta_g(v = 1)$ ($\lambda \approx 1.06$ μm). [117] According to Blickensderfer and Ewing, [118] based on the calculations of the Franck-Condon factor for the oxygen couple, [119] the ratio of intensities at 1.26 μm to 1.06 μm bands should be about 200. The same ratio measured in the spectrum of oxygen pairs is surprisingly low; both transitions are of the same order of intensity. The reason for such a drastic change in intensity is not yet clear. It looks like the deformation of the upper electronic-vibrational state upon formation of the complex results in an increase of the Franck-Condon factor for the 1.06 μm transition. Due to this increase in intensity, the prospects for the observation of the 1.06 μm band in the atmosphere are rather favorable. Dianov-Klokov [98] notes that in the vicinity of 1.06 μm the intensities of the monomer and bimolecular transitions become equal to each other at ca. 2 atm oxygen pressure. Since the pair density is at least about 10^{-3} of the monomer density, it may be concluded that the line-strength of the 1.06 μm transition is enhanced three order of magnitude upon formation of a dimer.

Perner and Platt [114] have observed a number of weak absorption bands (see Table 4) using differential optical absorption at atmospheric optical path lengths up to 10 km. According to an estimate made in, [114] oxygen dimers absorb about $1.5 \cdot 10^{-3}$ of the total solar energy coming into the atmosphere. The excited states of $(O_2)_2$ dimer may be effectively involved, as is noticed in [114] for oxidation reactions involving trace gases , such as

$$CO + (O_2)_2^* \rightarrow CO_2 + O_3.$$

Here the asterisk denotes an oxygen dimer in the excited $2^1\Delta_g$ or $2^1\Sigma_g^+$ singlet state. Slanger *et al.* [120] have shown that ozone molecules are generated in the course of direct photoionization of oxygen dimers matrix-isolated at low-temperatures under the action of radiation at $\lambda = 2480$ Å. Recently Clary *et al.* [121] speculated about an involvement of the oxygen dimer in the anomalies observed for the vibrational relaxation rates for highly vibrationally excited oxygen molecules. It would be extremely interesting to include oxygen dimers, especially those in their excited singlet states, in the modeling of ozone chemical transformations in the atmosphere. As a concluding note, the chemical generation of singlet oxygen is known to be very efficient in a mixture containing such familiar substances for the ozone destruction cycle as atomic chlorine and the ClO radical. [122] It is these reactions which enabled singlet oxygen to reach a concentration of up to 40 % yielding an efficient chemical pumping of the first iodine vapor laser. [123]

3 Water Vapor Continuum Absorption

Water vapor continuum absorption in the atmosphere is reported to cover a wide spectral range extending from several wave numbers to several thousand wave numbers. This weak absorption is reliably determined in the intervals between roto-vibrational lines in the transparent atmospheric windows. The spectral range from 8 to 13 μm is of particular interest, since the losses within this window mainly determine the radiative cooling of the Earth's atmosphere underlying surface. Substantial efforts have also been made to satellite-borne remote temperature sensing of the surface. In addition, this atmospheric window is of special concern for the problem of transmittance of powerful lasers through the atmosphere, especially that of the CO_2–laser generated in the vicinity of $\lambda = 10.6$ μm.

The nature of continuous absorption is still controversial. In 1938 Elsasser [124] suggested that the spectral continuum absorption is determined by the sum of distant line wings from isolated water molecules in the atmosphere. Before ca. 1960 the optical thickness τ in the 8–13 μm window was assumed to be a linear function of total humidity w. Attempts to fit experimental data under this assumption failed, however, since they led to negative values of τ. This made Goody and Bignell *et al.* [125,126] to conclude that the $\tau(w)$ dependence is nonlinear.

Attenuation of infrared radiation in the natural atmosphere is a function of the variation of temperature, density, and humidity with altitude. It depends also on atmospheric aerosols which are very unstable in their composition and abundance. To reveal the explicit characteristics of the water vapor continuum absorption and to exclude the influence of the aerosols, thorough laboratory experiments have been performed. The more important of these are those made by Bignell, [127] Burch *et al.*, [128] Aref'ev *et al.*, [129,130] and Hinderling *et al.* [131] A good number of reviews are dedicated to various aspects of the problem of the water vapor continuum. The papers by Burch *et al.* [128] and Bohlander *et al.* [132] summarize the state of the art of this problem up to the early 1980s. Detailed journal publications by Adiks *et al.*, [133] Hinderling *et al.*, [131] Aref'ev, [134] Grant, [135] and Thomas [136] contain extensive bibliographies, concerning in particular the propagation of laser radiation through atmospheric water vapor.

The excess continuum absorption by water vapor is determined as the difference between the measured and calculated absorption in the microwindows between spectral lines. This absorption is a specific function of wavelength, gas temperature, and pressure, which affects the self-broadening and the broadening by foreign gases, if any. These characteristic dependencies, which hold in the infrared and near-microwave, may be briefly summarized from [128] as follows:

— there are no spectral features narrower than 40 cm^{-1},

— at a given frequency and temperature, the absorption is proportional to the product of the partial pressure of water squared $P^2_{H_2O}$ and the thickness of the absorbing layer l,

— absorption is decreasing with increasing temperature the most rapidly far from the strongest water monomer lines,

— the effect is much more strongly pronounced for self-broadening than for foreign gas broadening, being proportional in this latter case to the product of the partial pressures of the water vapor and the broadening gas.

The attenuation of radiation may be characterized by the spectral absorption coefficient:

$$\sigma(\nu) = \sigma_s(\nu) + \sigma_f(\nu) = w\left[C_s(\nu,T)P_{H_2O} + C_f(\nu,T)P_f\right]. \tag{1}$$

Here subscripts s and f denote parameters for self- and foreign-broadening, respectively.

In a series of works by Bauer *et al.* [134–138] the temperature and pressure dependencies of the water vapor continuum absorption are investigated at selected wavelengths in the millimeter wave range 150–240 GHz. The gaps between atmospheric water vapor lines in this spectral range are particularly important for remote sensing of minor gaseous constituents and temperature in the Earth's and planetary atmospheres. Besides pure water vapor, the mixtures of water with N_2, CO_2, and Ar have been taken for these experiments. It has been established that continuum absorption in H_2O–CO_2 mixture is higher than that observed for N_2 or Ar as foreign gases. Also the temperature dependence of absorption slightly differs among various mixtures. In all cases the temperature dependence of absorption was fit to the power law $(T_0/T)^n$. It is worth noting that none of considered in these works theoretical line-shape models were found adequate to extensive experimental observations in the millimeter wave range.

Generally the same is true for the laboratory and field experimental data taken in the infrared. A recent paper by Tobin *et al.* [142] is devoted to experimental study of self- and N_2-broadened continuum within the water vapor ν_2 band. The measured continuum absorption was found in rough agreement with that of Burch *et al.* [128] Despite minor discrepancies, the data of [142] agree also with experimental observations of Thériault *et al.* [143] in which the atmospheric transmission spectra were recorded over a horizontal path of 5.7 km. Tobin *et al.* point out that an accurate near-wing line shape theory (valid roughly within 5–25 cm^{-1} of the line center) does not exist at the present time and the applicability of existing far-wing theories to the in-band continuum is dubious.

For the sake of applications a portion of experimental data taken in [128] was reduced by Roberts *et al.* [144] to the following approximate expression for the coefficient $C_s(\nu,T)$:

$$C_s(\nu,T) = C_s(\nu, 296K)exp\left[T_0\left(\frac{1}{T} - \frac{1}{296}\right)\right]. \tag{2}$$

Here T_0 is an empirical parameter which is a function of the wavelength and temperature. In the spectral range λ = 8–13 μm, and in the temperature interval from 280 K to 350 K, the parameter T_0 equals 1800 K according to, [144] or 3.58 kcal/mole.

In case the water vapor is broadened by nitrogen it has been established that the relative contribution of the $C_f(\nu, T)$ term is small; for instance Burch and Gryvnak [128] point out that at $\lambda \leq 14\mu m$ the magnitude of $C_f(\nu, T)$ for H_2O-N_2 mixture is comparable to the experimental data error. The temperature dependence of $C_f(\nu, T)$ is weak in case of nitrogen broadening.

The continuum absorption in pure water vapor and in a mixture with a foreign gas can be associated either with the non-lorentzian shape of the monomer water lines or with binary complexes of the water molecule. Basic arguments which were put forward in the literature in favor or against the involvement of van der Waals species in continuum absorption are briefly discussed below.

3.1 Water Complexes and Continuum Absorption

In the work by Varanasi *et al.*[145] it was suggested for the first time that the continuum absorption of water vapor might be associated with the absorption due to the hydrogen-bonded water dimers. Pioneering works of Victorova and Zhevakin [146,147] and Braun and Leidecker [148] played an important role in drawing attention to the structure and spectra of the water dimer in connection with the problem of the continuum absorption. This hypothesis was further advanced in a number of subsequent publications (see, e.g. Refs. [69,130,132,144,149,150]). Actually, the density and temperature dependencies of continuum absorption may be easily explained in terms of dimeric absorption, bearing in mind that the equilibrium number density of dimers and intermolecular complexes are yielded by the mass-action law.

In a mixture of water vapor with a foreign gas a series of hydrogen-bonded (or van der Waals) species can be formed. Until a gas is strongly pressurized, the principal corrections to isolated monomers are due to the water dimer and water–foreign gas 1:1 complexes. Therefore the spectral absorption coefficient far from monomer lines can be represented in the following form:

$$\sigma(\nu) = S_d(\nu)n_d + S_c(\nu)n_c. \tag{3}$$

Here S_d, n_d, and S_c, n_c are the absorption cross-sections and number densities of dimers and complexes, respectively. It has been shown in Chapter 1 that at low densities n_d is proportional to the water vapor partial pressure squared, while n_c is proportional to the product of partial pressures $P_{H_2O} \cdot P_f$. Hence the dependence of the spectral absorption coefficient on pressure given by formula (3) coincides with what is detected in observations (cf. Eq. 1).

Interpretation of the temperature dependence of absorption is less trivial, though, as we shall see in forthcoming paragraphs, formula (3) enables one to suggest a reasonable model of the measured continuum. Not going into much detail we can invoke the following expressions which stem directly from the mass-action law formulated in Chapter 1:

$$\begin{aligned} n_d &= \frac{1}{kT} P_{H_2O}^2 K_d \sim P_{H_2O}^2 T^{-n} \exp(D_0/kT), \\ n_c &= \frac{1}{kT} P_{H_2O} P_f K_c \sim P_{H_2O} T^{-m} \exp(D_0/kT). \end{aligned} \tag{4}$$

Here K_d and K_c are the equilibrium constants for the water dimer and water–foreign gas complexes, respectively, D_0 stands for dissociation energy. An inclusion of pre-exponential temperature dependence is very important here since, especially for very weak complexes with $D_0 \leq kT$, the number density of complexes (4) may turn out to be almost insensitive to the value of dissociation energy D_0. Instead, the temperature dependence of continuum absorption is mainly determined in this case by the pre-exponential power law, which, in turn, is determined by the temperature dependence of truncated partition functions for the coupled bimolecular states (see Chapter 1). The effective dissociation energy decrease due to thermal excitation of a complex should be also accounted for, e.g. through the use of thermally averaged dissociation energy as it is described in detail in Chapter 1, Sec. 4.1.

For relatively strong water dimer the enthalpy ΔH_d^0 of formation can be roughly estimated using exponential factor (4) from the intensity of continuum absorption measured as a function of temperature, that is by employing the simplified formula:

$$\sigma \sim \exp(-\Delta H_d^0/kT). \tag{5}$$

For instance, the estimate by Aref'ev *et al.*,[129] based on measurements of CO_2-laser radiation transmission in the atmosphere yielded $\Delta H^0 = -4.98\ kcal/mole$. This value is to be compared to the value derived from the best experimental data for the dimer, retrieved from the thermal conductivity measurements by Curtiss *et al.*, [155] which yield $\Delta H^0 = -3.59 \pm 0.5\ kcal/mole$ and from the accurate ir intensity data of Gorbaty *et al.*, [152] which lead to $\Delta H^0 = -3.98 \pm 0.9\ kcal/mole$.

These values are fairly close to quantum-chemical estimates of the water dimer dissociation energy, which amount to 3.5–4 kcal/mole (see e.g. [153–156]). This coincidence has been regarded for a long time as very strong evidence in favor of the dimeric nature of continuum absorption.

Even more convincing arguments can be retrieved upon consideration of the temperature dependence of the millimeter-wave continuum absorption for a series of the water vapor binary mixtures with foreign gases. The data of Bauer *et al.* [134–138] on the temperature and pressure dependences of absorption

Table 5: Intermolecular well depths D_e and dissociation energies D_0 for several water complexes.

Molecular pair		D_e, K	D_0, K	Basis set	References
H_2O	H_2O	2860	1700	VTZ/MP4	[156]
	CO_2	1512	1140	6-31+G(2d,2p)/MP2	[50]
	N_2	630	350	MP4/Sdf(b-ext)	[47]
	Ar	190[a]			

[a]see text for explanations

at 190 GHz (for pure H_2O) and 230 GHz (for pure H_2O and its mixtures with Ar, N_2, and CO_2) were fit using the formulae (3) and (4).[157] In all cases the quality of a fit is very high.

Figures 6,7 demonstrate that deviations of the calculated and experimental data are almost negligible. Same is true for the rest of calculated data for H_2O–N_2 and H_2O–Ar, not displayed here.

Three parameters are retrieved from this fit for each considered gas mixture. The model suggests to interpret them as absorption cross-section of a selected complex (or dimer) at chosen wavelength, the power exponent (denoted by n and m in (4)), and dissociation energy D_0 of the corresponding van der Waals molecule. Little can be said at the present time about the spectra of van der Waals species in the mm-wave range, although the order of magnitude of obtained cross-sections seems reasonable. True values of n and m parameters are determined by a complicated interplay of hindered internal rotations and intermolecular vibrations within a complex. It can be shown from the first principles, however, that for all mixtures the numbers n, m have to be bracketed between 1.5 and 5. The fitting procedure returned the numbers which fall in this interval: $n = 3.5$ for H_2O–H_2O, $m = 5$ for Ar–H_2O, 2.8 for N_2–H_2O, and 2.9 for CO_2–H_2O.

Table 5 accumulates computed intermolecular potential energy well characteristics for selected water complexes. Upper rows contain data gained by sophisticated *ab initio* quantum-chemical calculations. For H_2O–Ar complex

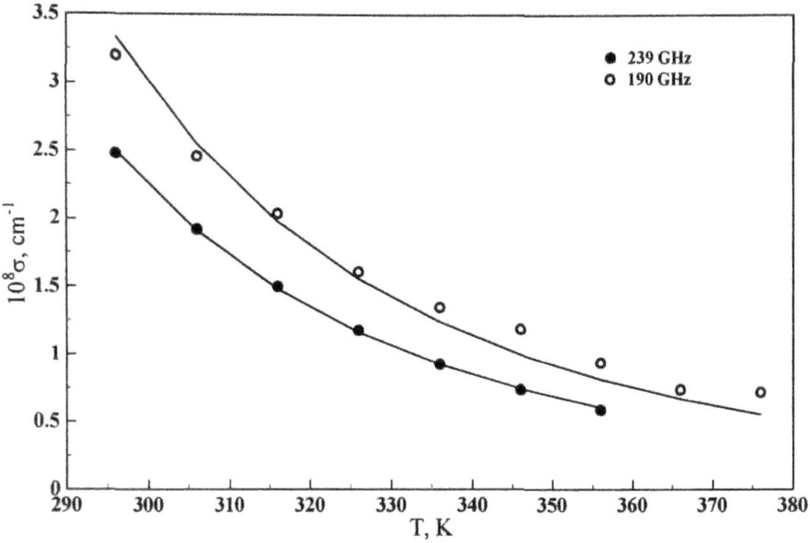

Figure 6: Comparison of the experimental continuum absorption in pure water vapor at 190.31 GHz [137] and 239.370 GHz [139] with model calculations.[157] $P_{H_2O} = 1$ Torr.

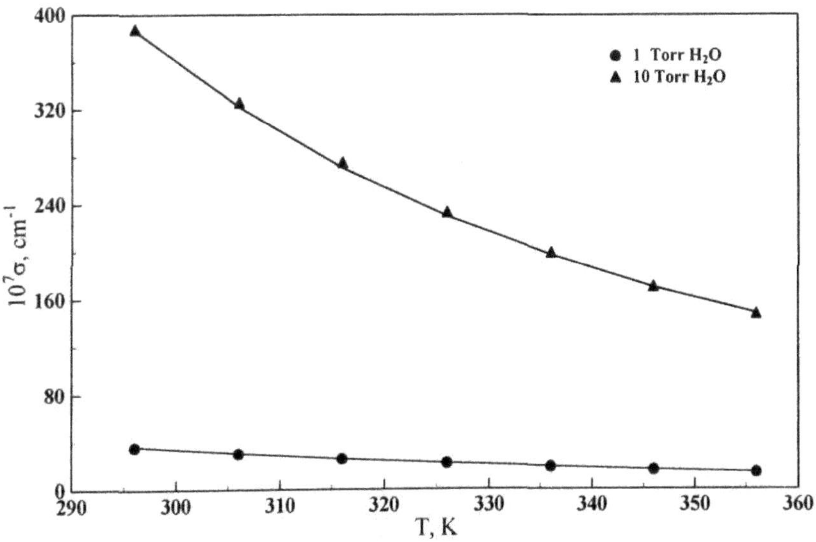

Figure 7: Comparison of the experimental continuum absorption of a mixture H_2O–CO_2 at 239.370 GHz [140] with model calculations.[157] $P_{CO_2} = 750$ Torr, $P_{H_2O} = 1$ Torr (circles), $P_{H_2O} = 10$ Torr (triangles).

the well depth D_e is roughly estimated through the use of approximate relationship valid for the Lennard-Jones potential

$$D_e = \frac{1}{72} k_s R_{c.m.}^2, \tag{6}$$

taking experimentally determined in [158] the center-of-mass separation ($R_{c.m.} = 3.75$ Å) and stretching force constant ($k_s = 0.0134$ mdyn/Å).

Figure 8 displays the correlation between model parameter D_0 and literature estimates of dissociation energies taken from Table 5. The straight line shows the correlation coefficient equal to unity. The error bars depicted in this figure represent 20 % error of each value. This uncertainty very roughly characterises the inaccuracy of both the electronic energy calculations and the estimates of the ground state level.

Figure 8 demonstrates such a strong correlation of the fitted D_0 parameter with dissociation energy, that there remains little doubt in the physical meaning ascribed to D_0 in our model.[157] Consequently the suggested associative model of continuum absorption seems to be approved.

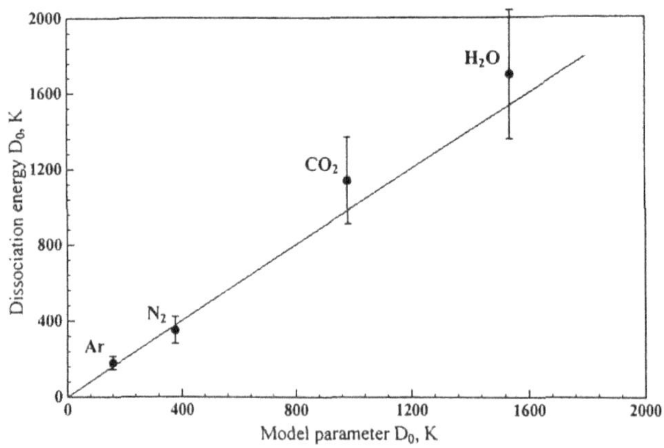

Figure 8: Correlation of D_0, the parameter obtained in the course of fitting of the mm-wave continuum absorption with D_0, the dissociation energy of bimolecular complexes.

3.2 About Controversies of Dimeric Hypothesis

It has been shown by Montgomery [159] that an excess absorption measured in the vicinity of 1200 cm^{-1} increases with temperature within the interval from ca. 400 K to 470 K, while at lower temperatures 300–400 K it decreases, as might be expected from relationship (4), appropriate to dimers. True origin of the anomalous temperature dependence obtained by Montgomery [159] is still unclear, partially since no more experimental data on the water vapor continuum absorption at elevated temperatures are available up-to-now.

Hinderling *et al.* [131] have examined the differences in the recorded height profiles of the line and continuum absorptions in the diffusion chamber at various CO_2-laser transitions. On the basis of these measurements, an upper limit of the dimer absorption cross section has been derived. They concluded that the dimer cross section may not exceed ca. $9.5 \cdot 10^{-22}$ cm^2/mole. According to [131] this figure is by a factor of 27 less than that necessary to fit the observed continuum absorption coefficients. This made it possible to conclude that $(H_2O)_2$ dimers are not the main contributors to the attenuation of ir-radiation in the atmosphere. [131]

Similar viewpoints have become very popular in recent years mainly due to the extensive advancement of theories aimed to model the shape of the absorption in the near and distant line wings. The models suggested by Thomas and Nordstrom [160,161] and especially by Ma and Tipping *et al.* [155–160] are capable of reproducing many of the observed features of the continuum absorption without any reference to water dimer absorption. Moreover these theoretical models attempt to represent the line shape by using the intermolecular potential with no adjustable parameters. The accuracy of such models remain, however, insufficient for some practical applications. Also they seem not convincing enough from the physical viewpoint, since the intermolecular interactions are often taken into account in an inconsistent way. For instance all researchers agree that finite duration of collisions is crucial for the line shape theory. Rigorous partitioning in phase space should, however, precede the statistical average over the domain of elastic collisions. Therefore, the existing line shape theories are justified only when the role of bound and quasibound intermolecular states is really negligible. In case of the water molecules forming rather strong complexes such situation may arise more probably as an exception.

The so-called CKD models of continuum absorption suggested by Clough, Kneizys, Davies *et al.* [168,169] are relatively successful in reproducing either the spectral shape or the temperature dependence of absorption in the millimeter wave range and in the infrared and are, therefore, widely used in atmospheric applications. These models, however, are empirical by their nature and contain a good number of adjustable parameters. Therefore they do not allow one to make a judgement on the true nature of continuum absorption.

One of the arguments which is usually put forward against the involvement of dimers in far infrared water vapor absorption is that there are no fundamental dimer transitions which fall near 10 μm. The short-wavelength wing of the intermolecular fundamentals is obviously unable to explain the observed absorption. It was suggested that this difficulty could be overcome by accounting for the higher order intermolecular vibrational transitions in the water dimer.[170] Subsequent calculations performed in harmonic[149,171] and anharmonic[172] approximations have demonstrated that the absorption intensity in the region of the overtone and combination transitions is sufficient to explain the magnitude and spectral behavior of the observed continuous absorption (see Fig. 9).

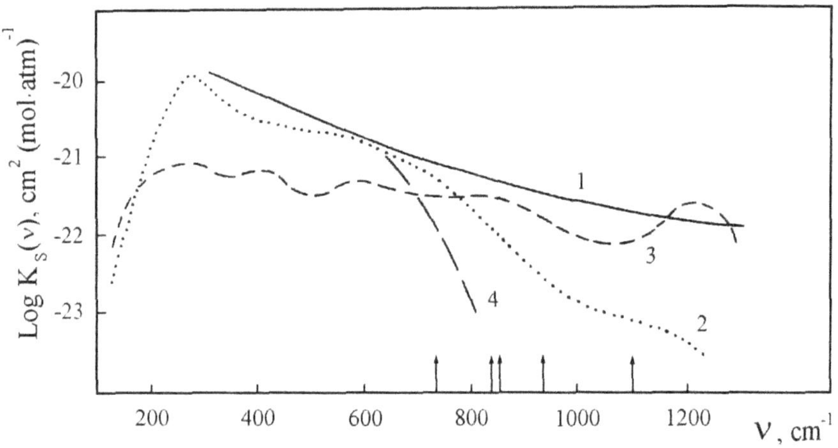

Figure 9: Water vapor continuum absorption. 1–experimental,[128] 2, 3–water dimer spectrum calculated in harmonic[149] and anharmonic[172] approximations, 4–the wing of intermolecular fundamental bands. Arrows in the bottom show the position of some overtones and combination intermolecular transitions in a dimer (as calculated in[149]).

It is a widely held belief that number density of the water dimers in the atmosphere is insufficient to explain the observed continuum absorption. A series of independent calculations shows that the mole fraction of water dimers near room temperature amounts to 10^{-3}–10^{-4} at $P_{H_2O} \sim 10$ Torr. Note that it is usual practice to account for minor and trace gas phase atmospheric constituents at the level of 10^{-4}–10^{-9} while considering the transmission of radiation through the atmosphere.

Another argument often put forward against the involvement of dimers invoke the absence of distinct spectroscopic features in the recorded spectra which might be attributed to the water dimers. This seems especially surprising for the millimeter wave range where the spectra are better resolved than in the infrared. The statistical physics partitioning of phase space for interacting water molecules made by Epifanov and Vigasin [173] shows that the fraction of tightly bound water dimers at atmospheric conditions does not exceed one half of the bound plus metastable dimeric states. The significant contribution to the overall absorption arising from the floppy metastable dimers enables one to infer the diffuse character of the dimeric absorption, which agrees with laboratory observations of continuum absorption at near room temperature and above.

3.3 Concluding Remarks

Despite the controversies mentioned above, there exist no experimental data nowadays which conflict dramatically with the idea of the dimeric nature of continuous absorption.

Following conclusions may be drawn from the above consideration.

— The true nature of the water vapor continuum absorption is often not crucial for the applications. The empirically determined characteristics of the continuum absorption (see e.g. [144]) and the existing parametrized line-shape models (e.g. CKD models) are capable to roughly predict the atmospheric phenomena which depend on the water vapor continuum.

— In order to reveal the physical nature of the continuum, the use of only spectroscopic data seems insufficient, because the absence of distinct spectral features favors the creation of models which may turn out to be groundless despite an impressive level of their complexity and beauty.

— A thorough investigation of the thermodynamics and statistical physics of interacting water molecules must accompany the purely spectroscopic models constructed for spectral regions including continuum absorption.

— There is no experimental or theoretical evidence, that water dimers and water complexes may not be responsible for the observed continuous absorption. It seems probable, however, that the absorbing pairs are in their excited floppy states, which make them resemble scattering resonances more than rigid ground state dimers.

Acknowledgments

It is with a great deal of pleasure that I extend my thanks to Professor L. M. Trafton for his valuable suggestions and assistance in the course of proofreading of this Chapter. The author's thanks are also due to Profs. and/or Drs. L. Frommhold, F. Huisken, Z. Slanina, W. J. Lafferty, A. R. W. McKellar, and A. Bauer for providing their published materials, many useful discussions and encouragement. The help of Drs. T. G. Adiks, G. V. Chlenova, E. G. Tarakanova, and S. Yu. Epifanov in carrying out a number of investigations is gratefully acknowledged.

References

1. P. Brimblecombe *Air, Composition and Chemistry* (Cambridge University Press, Cambridge, 1986).
2. L. Frommhold *Collision Induced Absorption in Gases* (Cambridge University Press, Cambridge, 1993).
3. R. Tipping in *Phenomena Induced by Intermolecular Interactions* ed. G. Birnbaum (Plenum Press, N.Y., 1985), p. 727.
4. L. Frommhold, R. Samuelson, and G. Birnbaum, *Astrophys. J.* **283**, L79 (1984).
5. A. R. W. McKellar, *Canad. J. Phys.* **62**, 760 (1984)
6. A. R. W. McKellar, *Astrophys. J.* **326**, L75 (1988).
7. A. R. W.McKellar, *J. Chem. Phys.* **92**, 3261 (1990).
8. S. J. Kim, L. M. Trafton, T. R. Geballe, and Z. Slanina, *Icarus* **113**, 217 (1995).
9. L. M. Trafton, S. Kim, T. R. Geballe, and S. Miller, *Icarus* 1997, submitted.
10. L. M. Trafton and J. K. G. Watson, *Astrophys. J.* **385**, 320 (1992).
11. L. M. Trafton, *J. Geophys. Res.* 1997, submitted.
12. O. Tapia and G. Bessis, *Theor. Chimica Acta* **25**, 130 (1972).
13. O. F. Bender and H. F. Schaefer III, *J. Chem. Phys.* **57**, 217 (1972).
14. P. G. Burton and U. E. Senf, *J. Chem. Phys.* **79**, 526 (1983).
15. R. G. Gordon and J. K. Cashion, *J. Chem. Phys.* **44**, 1190 (1966).
16. A. Watanabe and H. L. Welsh, *Phys. Rev. Lett.* **13**, 810 (1964).
17. A. R. W. McKellar and H. L. Welsh, *Canad. J. Phys.* **52**, 1082 (1974).
18. A. R. W. McKellar and J. Schaefer, *J. Chem. Phys.* **95**, 3081 (1991).
19. A. R. W. McKellar, *J. Chem. Phys.* **93**, 18 (1990).
20. K. Fox and S. J. Kim, *J. Quant. Spectrosc. Radiat. Transfer* **40**, 177 (1988).
21. A. Borysow and L. Frommhold, *Astrophys. J.* **303**, 495 (1986).
22. G. Bachet, D. Gautier, and A. Coustenis, *Compt. Rend. Acad. Sci.* **307**, 133 (1988).
23. A. R. W. McKellar, *J. Chem. Phys.* **105**, 2628 (1996).
24. R. Courtin, D. Gautier, and C. P. McKay, *Icarus* **114**, 144 (1995).

25. L. Mannik, J. C. Stryland, and H. L. Welsh, *Canad. J. Phys.* **49**, 3056 (1971).
26. J. M. Lobue, J. K. Rice, and S. E. Novick, *Chem. Phys. Lett.* **112**, 376 (1984).
27. N. Brigot, S. Odiot, S. H. Walmsley, and J. L. Witten, *Chem. Phys. Lett.* **49**, 157 (1977).
28. A. J. Illies, M. L. McKee, and H. B. Schlegel, *J. Phys.Chem.* **91**, 3489 (1987).
29. R. Eggenberger, S. Gerber, and H. Huber, *Molecular Physics* **72**, 433 (1991).
30. K. Domansky, O. Kitao, and K. Nakanishi, *Chem. Phys. Lett.* **199**, 525 (1992).
31. G. A. Pubanz, M. Maroncelli, and J. W. Nibler, *Chem. Phys. Lett.* **120**, 313 (1985).
32. K. W. Brown, A. D. Richardson, N. H. Rich, and J. W. Nibler in *Proc. XIIIth Intern. Conf. Raman Spectroscopy*, ed. W. Kiefer *et al.*, (John Wiley, Chichester, 1992).
33. F. Huisken, L. Ramonat, J. Santos, V. V. Smirnov, O. M. Stelmakh, and A. A. Vigasin, *J. Molec. Struct.* **410-411**, 47 (1997).
34. M. A. Walsh, T. H. England, and T. R. Dyke, *Chem. Phys. Lett.* **142**, 265 (1987).
35. K. W. Jucks, Z. S. Huang, R. E. Miller, G. T. Fraser, A.S.Pine, and W. J. Lafferty, *J. Chem. Phys.* **88**, 2185 (1988).
36. K. W. Jucks, Z. S. Huang, D. Dayton, R. E. Miller, and W. J. Lafferty, *J. Chem. Phys.* **86**, 4341 (1987).
37. S. W. Sharpe, R. Sheeks, C. Wittig, and R. A. Beaudet, *Chem. Phys. Lett.* **151**, 267 (1988).
38. M. A. Walsh, T. R. Dyke, and B. J. Howard, *J. Molec. Struct.* **189**, 111 (1988).
39. K. I. Peterson and W. Klemperer, *J. Chem. Phys.* **80**, 2439 (1984).
40. W. Klemperer in *Adv. Laser Chem.*, ed. A. H. Zevail (Springer, Berlin, 1978), p. 398.
41. B. L. Blaney and G. E. Ewing, *Ann. Rev. Phys. Chem.* **27**, 553 (1976).
42. R. M. Berns and A. van der Avoird, *J. Chem. Phys.* **72**, 6107 (1980).
43. L. Beneventi, P. Casavecchia, G. G. Volpi, C. C. K. Wong, and F. R. W. McCourt, *J. Chem. Phys.* **98**, 7926 (1993).
44. B. Bussery and P. E. S. Wormer, *J. Chem. Phys.* **99**, 1230 (1993).
45. H. O. Leung, M. D. Marshall, R. D. Suenram, and F. J. Lovas, *J. Chem. Phys.* **90**, 700 (1989).

46. D. B. Kitchen, T. P. Ruane, and L. C. Allen, *Chem. Phys. Lett.* **141**, 525 (1987).
47. P. Sandler, J. oh Jung, M. M. Szczęśniak, and V. Buch, *J. Chem. Phys.* **101**, 1378 (1994).
48. J. Sadlej, B. Rowland, J. P. Devlin, and V. Buch, *J. Chem. Phys.* **102**, 4804 (1995).
49. B. Jönsson, G. Karlström, and H. Wennerström, *Chem. Phys. Lett.* **30**, 58 (1975).
50. P. A. Block, M. D. Marshall, L. G. Pedersen, and R. E. Miller, *J. Chem. Phys.* **96**, 7321 (1992).
51. A. S. Pine and G. T. Fraser, *J. Chem. Phys.* **89**, 100 (1988).
52. K. I. Peterson, R. D. Suenram, and F. J. Lovas, *J. Chem. Phys.* **90**, 5964 (1989).
53. K. I. Peterson, R. D. Suenram, and F. J. Lovas, *J. Chem. Phys.* **94**, 106 (1991).
54. A. A. Vigasin and G. V. Chlenova, *Solar Syst. Res.* **24**, 129 (1990).
55. V. I. Moroz, M. N. Izakov, and V. M. Linkin, *Engeneer model for the atmosphere of Mars (variant MA-87)*, Preprint N 1449 (Space Research Institute, Moscow, 1988).
56. V. I. Moroz, V. I. Istomin, and L. M. Mukhin, *Proposal for VIRA (Venus International Reference Atmosphere): Chemical composition of Atmosphere below 100 km*, Preprint N 830 (Space Research Institute, Moscow, 1983).
57. V. V. Kerzhanovich, M. Ya. Marov, and V. I. Moroz, *Proposal for VIRA (Venus International Reference Atmosphere): Dynamics and Structure below 100 km*, Preprint N 831 (Space Research Institute, Moscow, 1983).
58. J. Andreson, R. P. Anders, J. B. Fenn, and G. Maise in *Rarefied Gas Dynamics*, ed. J. H. de Leeuw (Academic Press, N. Y., 1966).
59. D. E. Stogryn and J. O. Hirshfelder, *J. Chem. Phys.* **31**, 1531 (1959).
60. Z. Slanina, *Theor. Chimica Acta* **195**, 45 (1992).
61. Z. Slanina, K. Fox, and S. J. Kim, *Theor. Chimica Acta* **200**, 33 (1992).
62. Z. Slanina, K. Fox, and S. J. Kim, *J. Quant. Spectrosc. Radiat. Transfer* **47**, 91 (1992).
63. T. G. Adiks, G. V. Tchlenova, and A. A. Vigasin, *J. Quant. Spectrosc. Radiat. Transfer* **29**, 575 (1989).
64. A. A. Vigasin, N. N. Filippov, and G. V. Chlenova, *Optics and Spectroscopy* **72**, 56 (1992).
65. J.-M. Hartmann and C. Boulet, *J. Chem. Phys.* **94**, 6406 (1991).
66. A. A. Vigasin, E. G. Tarakanova, and G. V. Tchlenova, *J. Quant. Spectrosc. Radiat. Transfer* **50**, 695 (1994).

67. D. Reuter, D. E. Jennings, and J. W. Brault, *J. Mol. Spectrosc.* **115**, 294 (1986).
68. J. M. Calo and R. S. Narcisi, *Geophys. Res. Lett.* **7**, 289 (1980).
69. A. A. Vigasin and G. V. Chlenova, *Izvestiya, Atmos. and Ocean Phys.* **19**, 527 (1983).
70. Z. Slanina, *J. Atmospheric Chem.* **6**, 185 (1988).
71. C. Rinsland, M. A. H. Smith, J. M. Russel III, J. H. Park, and C. B. Farmer, *Applied Optics* **20**, 4167 (1981).
72. V. I. Dianov-Klokov and I. P. Malkov, *Izvestiya, Atmos. and Ocean Phys.* **9**, 411 (1973).
73. L. S. Bernstein, D. C. Robertson, J. A. Conant, and B. P. Sandford, *Applied Optics* **18**, 2454 (1979).
74. A. R. W. McKellar, N. A. Rich, and H. L. Welsh, *Canad. J. Phys.* **50**, 1 (1972).
75. O. L. J. Gijzeman, E. A. Ogryzlo, and R. P. H. Rettschnick, *J. Chem. Phys.* **52**, 3718 (1970).
76. V. I. Dianov-Klokov and O. A. Matveyeva, *Izvestiya, Atmos. and Ocean Phys.* **4**, 235 (1968).
77. Shardanand, *J. Quant. Spectrosc. Radiat. Transfer* **20**, 265 (1978).
78. Shardanand, *J. Quant. Spectrosc. Radiat. Transfer* **18**, 525 (1977).
79. Shardanand and Prasad Rao, *J. Quant. Spectrosc. Radiat. Transfer* **17**, 433 (1977).
80. A. J. Blake and D. C. McCoy, *J. Quant. Spectrosc. Radiat. Transfer* **38**, 113 (1987).
81. G. D. Greenblat, J. J. Orlando, J. B. Burkholder, and A. R. Ravishankara, *J. Geophys. Res.* **95**, 18.577 (1990).
82. Yu. M. Timofeyev and M. V. Tonkov, *Izvestiya, Atmos. and Ocean Phys.* **14**, 437 (1978) (in Russian).
83. C. Rinsland, M. A. H. Smith, R. K. Seals, Jr., A. Goldman, F. J. Murcray, J. C. Larsen, and P. L. Rarig, *J. Geophys. Res.* **87**, 3119 (1982).
84. M. F. Crawford, H. L. Welsh, and J. L. Locke, *Phys. Rev.* **75**, 1607 (1949).
85. C. A. Long, G. Henderson, and G. E. Ewing, *Chem. Phys.* **2**, 485 (1973).
86. A. R. W. McKellar, *J. Chem. Phys.* **88**, 4190 (1988).
87. G. Brocks and A. van der Avoird, *Molecular Physics* **55**, 11 (1985).
88. G. Henderson and G. E. Ewing, *J. Chem. Phys.* **59**, 2280 (1973).
89. G. Henderson and G. E. Ewing, *Molecular Physics* **27**, 903 (1974).
90. W. J. Lafferty, A. M. Solodov, A. Weber, W. B. Olson, and J.-M. Hartmann, *Applied Optics* **35**, 5911 (1996).
91. A. A. Vigasin, *J. Quant. Spectrosc. Radiat. Transfer* **53**, (1996).

92. S. E. Novick, P. B. Davies, and T. R. Dyke, *J. Amer. Chem. Soc.* **95**, 8547 (1973).
93. J. J. Orlando, G. S. Tyndall, K. E. Nickerson and J. G. Calvert, *J. Geophys. Res.* **96**, 20755 (1991).
94. V. Menoux, R. Le Doucen, C. Boulet, A. Roblin, and A. M. Bouchardy, *Applied Optics* **32**, 263 (1993).
95. L. Fredin, B. Nelander, and G. Ribbegård, *J. Mol. Spectrosc.* **53**, 410 (1974).
96. V. I. Dianov-Klokov, *Optics and Spectroscopy* **16**, 409 (1964) (in Russian).
97. V. I. Dianov-Klokov, *Optics and Spectroscopy* **21**, 233 (1966) (in Russian).
98. V. I. Dianov-Klokov, *Thesis Dr. Sci.*, (Inst. Atmosph. Phys. USSR, Moscow, 1964.) (in Russian)
99. M. C. van Hemert, P. E. S. Wormer, and A. van der Avoird, *J. Chem. Phys.* **51**, 1167 (1983).
100. P. E. S. Wormer and A. van der Avoird, *J. Chem. Phys.* **81**, 1929 (1984).
101. J. Goodman and L. E. Brus, *J. Chem. Phys.* **67**, 4398 (1977).
102. Z. Slanina, P. Hobza, and R. Zahradnik, *Coll. Czechoslovak Chem. Commun.* **39**, 228 (1974).
103. A. van der Avoird and G. Brocks, *J. Chem. Phys.* **87**, 5346 (1987).
104. J. Janssen, *Compt. Rend. Acad. Sci.* **102**, 1352 (1886).
105. J. Janssen, *Compt. Rend. Acad. Sci.* **106**, 1118 (1888).
106. H. Salow and W. Steiner, *Z. Phys.* **99**, 137 (1936).
107. H. Salow and W. Steiner, *Nature* **134**, 463 (1934).
108. L. Herman, *Ann. de Physique (Paris)* **11**, 548 (1939).
109. A. Vassy and M. E. Vassy, *J. de Phys. et Radium (Paris)* **10**, 403 (1939).
110. J. Dufay, *Ann. Astrophys. (Paris)* **5**, 93 (1942).
111. J. W. Ellis and H. O. Kneser, *Z. Phys.* **86**, 583 (1933).
112. M. Egoroff, *Compt. Rend. Acad. Sci.* **101**, 1143 (1885).
113. H. Okabe, *Photochemistry of Small Molecules* (Wiley, N.-Y., 1978).
114. D. Perner and U. Platt, *Geophys. Res. Lett.* **7**, 1053 (1980).
115. D. A. Newnham, J. Ballard, and M. S. Page in *Proc. Atmospheric Spectroscopy and Applications Workshop* (Reims, September 1996) p. 31.
116. C. Hermans, A. C. Vandaele, M. Van Roozendaele, P. C. Simon, M. Carleer, S. Fally, R. Colin, B. Coquart, A. Jenouvrier, and M. F. Merienne in *Proc. Atmospheric Spectroscopy and Applications Workshop* (Reims, September 1996) p. 83.
117. L. Herzberg and G. Herzberg, *Astrophys. J.* **105**, 353 (1947).
118. R. P. Blickensderfer and G. E. Ewing, *J. Chem. Phys.* **51**, 5284 (1969).

119. R. M. Badger, A. C. Wright, and R. F. Whitlock, *J. Chem. Phys.* **43**, 4345 (1965).
120. T. G. Slanger, L. E. Jusinsky, G. Black, and G. E. Gadd, *Science* **241**, 945 (1988).
121. R. Hernandez, R. Toumi, and D. C. Clary, *J. Chem. Phys.* **102**, 9544 (1995).
122. *Singlet O_2. Physico-Chemical Aspects*, ed. A. A. Friman, V.1, (Boca Raton, 1985).
123. W. E. McDermot, N. R. Pchelkin, D. I. Bernard, and R. R. Bousek, *Appl. Phys. Lett.* **32**, 469 (1978).
124. W. M. Elsasser, *Phys. Rev.* **53**, 768 (1938).
125. W. T. Roach and R. M. Goody, *Quart. J. Roy. Meteor. Soc.* **84**, 319 (1958).
126. K. J. Bignell, F. Saiedy, and P. A. Sheppard, *J. Opt. Soc. Amer.* **53**, 466 (1963).
127. K. J. Bignell, *Quart. J. Roy. Meteor. Soc.* **96**, 390 (1970).
128. D. E. Burch and D. A. Gryvnak in *Atmospheric Water Vapor*, ed. A. Deepak, T. D. Wilkerson, and L. H. Ruhnke (Academic Press, N.-Y., 1980), p. 47.
129. V. I. Dianov-Klokov, V. M. Ivanov *et al.*, *J. Quant. Spectrosc. Radiat. Transfer* **25**, 83 (1981).
130. V. N. Aref'ev and V. I. Dianov-Klokov, *Optics and Spectroscopy* **42**, 488 (1977) (in Russian).
131. J. Hinderling, M. W. Sigrist, and F. K. Kneubühl, *Infrared Physics* **27**, 63 (1987).
132. R. A. Bohlander, R. J. Emery, D. T. Llewelyn-Jones *et al.* in *Atmospheric Water Vapor*, ed. A. Deepak, T. D. Wilkerson, and L. H. Ruhnke (Academic Press, N.-Y., 1980) p. 241.
133. T. G. Adiks, V. N. Aref'ev and V. I. Dianov-Klokov, *Sov. J. Quantum Electron.* **5**, 481 (1975).
134. V. N. Aref'ev, *Atmosph. Opt.* **2**, 1034 (1989) (in Russian).
135. W. B. Grant, *Applied Optics* **29**, 451 (1990).
136. M. E. Thomas, *Infrared Physics* **30**, 161 (1990).
137. A. Bauer and M. Godon, *J. Quant. Spectrosc. Radiat. Transfer* **46**, 211 (1991).
138. A. Bauer, M. Godon, J. Carlier, Q. Ma, and R. H. Tipping, *J. Quant. Spectrosc. Radiat. Transfer* **50**, 463 (1993).
139. A. Bauer, M. Godon, J. Carlier, and Q. Ma, *J. Quant. Spectrosc. Radiat. Transfer* **53**, 411 (1995).

140. A. Bauer, M. Godon, J. Carlier, and R. R. Gamache, *J. Mol. Spectrosc.* **176**, 45 (1996).
141. A. Bauer, M. Godon, J. Carlier, and R. R. Gamache, *J. Quant. Spectrosc. Radiat. Transfer*, 1997, in press.
142. D. C. Tobin, L. L. Strow, W. J. Lafferty, and W. B. Olson, *Applied Optics* **35**, 4724 (1996).
143. J.-M. Thériault, P. L. Roney, D. St.-Germain, H. E. Revercomb, R. O. Knuteson, and W. L. Smith, *Applied Optics* **33**, 323 (1994).
144. R. E. Roberts, J. E. A. Selby, and L. M. Biberman, *Applied Optics* **15**, 2085 (1976).
145. P. Varanasi, S. Chou, and S. S. Penner, *J. Quant. Spectrosc. Radiat. Transfer* **8**, 1537 (1968).
146. A. A. Victorova and S. A. Zhevakin, *Sov. Phys. Dokl.* **11**, 1059 (1966).
147. A. A. Victorova and S. A. Zhevakin, *Izv. VUZ, Radiophys.* **18**, 211 (1975).
148. C. Braun and H. Leidecker, *J. Chem. Phys.* **61**, 3104 (1974).
149. A. A. Vigasin and G. V. Chlenova, *Izvestiya, Atmos. and Ocean Phys.* **20**, 596 (1984).
150. A. A. Vigasin and G. V. Chlenova, *Izvestiya, Atmos. and Ocean Phys.* **22**, 21 (1986).
151. L. A. Curtiss, D. J. Frurip, and M. Blander, *J. Chem. Phys.* **71**, 2703 (1979).
152. Yu. E. Gorbaty and G. V. Bondarenko, *Molecular Physics* **72**, 639 (1991).
153. O. Matsuoka, E. Clementi, and M. Yoshimine, *J. Chem. Phys.* **64**, 1351 (1976).
154. G. H. F. Dierksen, W. P. Kraemer, and B. O. Ross, *Theor. Chimica Acta* **36**, 249 (1975).
155. L. A. Curtiss and M. Blander, *Chem. Rev.* **88**, 827 (1988).
156. A. Bleiber and J. Sauer, *Chem. Phys. Lett.* **238**, 243 (1995).
157. A. A. Vigasin, 1998, to be published.
158. R. C. Cohen, K. L. Busarow, K. B. Laughlin, G. A. Blake, M. Havenith, Y. T. Lee, and R. J. Saykally, *J. Chem. Phys.* **89**, 4494 (1988).
159. G. P. Montgomery, Jr., *Applied Optics* **17**, 2299 (1978).
160. M. E. Thomas and R. J. Nordstrom, *J. Quant. Spectrosc. Radiat. Transfer* **28**, 81 (1982).
161. M. E. Thomas and R. J. Nordstrom, *J. Quant. Spectrosc. Radiat. Transfer* **28**, 103 (1982).
162. J.-M. Hartmann, M. Y. Perrin, Q. Ma, and R. H. Tipping *J. Quant. Spectrosc. Radiat. Transfer* **49**, 675 (1993).

163. Q. Ma and R. H. Tipping, *J. Chem. Phys.* **93**, 6127 (1990).
164. Q. Ma and R. H. Tipping, *J. Chem. Phys.* **95**, 6290 (1991).
165. Q. Ma and R. H. Tipping, *J. Chem. Phys.* **96**, 818 (1992).
166. Q. Ma and R. H. Tipping, *J. Chem. Phys.* **96**, 8655 (1992).
167. R. H. Tipping and Q. Ma, *Atmosph. Res.* **36**, 69 (1995).
168. S. A. Clough, F. X. Kneizys, R. Davies, R. Gamache, and R. Tipping in *Atmospheric Water Vapor*, ed. A. Deepak, T. D. Wilkerson, and L. H. Ruhnke (Academic Press, N.-Y., 1980) p. 25.
169. S. A. Clough, F. X. Kneizys, and R. W. Davies *Atmos. Res.*, **23**, 229 (1989).
170. A. A. Vigasin, *Izvestiya, Atmos. and Ocean Phys.* **19**, 542 (1983) (in Russian).
171. Yu. S. Demchuk, S. O. Mirumyants, N. I. Moskalenko, and V. L. Filippov, *Optics and Spectroscopy* **69**, 39 (1990).
172. G. V. Yukhnevich and E. G. Tarakanova, *J. Molec. Struct.* **177**, 495 (1988).
173. S. Yu. Epifanov and A. A. Vigasin, *Molecular Physics* **90**, 101 (1997).

DIMERS IN EARTH'S AND PLANETARY ATMOSPHERES: THE $(H_2O)_2$, $(N_2)_2$, N_2–O_2, $(O_2)_2$, $(O_3)_2$, $(CO_2)_2$, $(H_2)_2$, AND Ar–N_2 CASES

Z. SLANINA

Department of Knowledge-Based Information Engineering, Toyohashi University of Technology, Toyohashi 441, Aichi, Japan

S. J. KIM

Department of Astronomy and Space Sciences, Kyunghee University, Yong In, Kyung Kee Do, Korea

K. FOX

Department of Physics and Astronomy, University of Tennessee, Knoxville, TN 37996, USA, and Division of Physical Sciences, Anne Arundel Community College, Arnold, MD 21012–1895, USA

F. UHLÍK

Faculty of Science, Charles University, Prague 2, Czech Republic

A. HINCHLIFFE

Department of Chemistry, UMIST, P.O. Box 88, Manchester M60 1QD, UK

During the last decade, there has been a considerable interest in molecular aggregates, complexes or clusters, and a considerable amount of data has been obtained in laboratory observation, for example in matrix isolations or nozzle beam expansion techniques, and also in computations. It has been soon recognized that such dimers, or even higher aggregates, can be formed in atmospheres of cosmic bodies. In addition to the homo-dimers, hetero-dimers were given some attention, too. It has been obvious that molecular complexes should equally well be significant not only to Earth but also to other sufficiently dense planetary atmospheres and that, in principle, they must be present in any atmosphere (though their amount can be rather negligible). There are several types of molecular complexes of obvious importance to Earth's atmosphere, for example: $(H_2O)_2$, $(N_2)_2$, N_2–O_2, $(O_2)_2$, or $(O_3)_2$. The water dimer is relevant to cometary atmospheres, too, especially to comet Halley. The paper surveys recent results on the subjects. The stress is on the computational approach which combines advanced quantum-chemical methods with statistical-mechanical treatments. However, the computational approach is placed in a wider context of contemporary studies. The structure, energetics, and harmonic vibrations of the complexes were evaluated at a correlated *ab initio* level. The computed dimerization equilibrium constants are of a focused interest as a key stability measure and an input information for the evaluation of the altitude population profiles in atmospheres. The mole fractions of each of the dimers, computed for their equilibrium mixture with N_2 and O_2 under the atmospheric conditions, exhibit a monotonous decrease with the altitude. The ozone dimer is treated in a similar way for Antarctic atmospheric conditions. The $(CO_2)_2$ case is especially important for the atmosphere of Venus, though it may also have some relevancy for Mars. A recent example of observed planetary dimer is supplied by $(H_2)_2$ in the atmosphere of Jupiter. The computations also predict that Ar–N_2 should be significant in the atmosphere of Titan. The survey pays a systematic attention to the present knowledge of the selected planetary complexes derived from computational chemistry treatments.

1 Introduction

The atmospheres of cosmic bodies represent vast laboratories for physicochemical exercises, especially for gas phase studies. Indeed, some species have been studied in atmosphere sooner than in laboratory (e.g., the water dimer). Others were first supposed in extraterrestrial space, and then observed in laboratory — larger carbon clusters like C_{60} are a well known example. The contemporary physical chemistry is both experimental and computational field. Consequently, species and properties which are difficult to measure are computed instead, and this holds true for the atmospheric species, too.

Quantum chemistry was born soon after quantum mechanics. Already in 1929 Dirac wrote:[1] "The underlying physical laws necessary for the mathematical theory of a large part of physics and the whole chemistry are thus completely known, and the difficulty is only that the exact application of these laws leads to equations much too complicated to be soluble." There has been a change since Dirac's time — application of high-performance computers to quantum-chemical problems has somewhat overcome the numerical difficulties, though the solutions offered by quantum chemistry are essentially approximative. Hence, they presuppose a sensible usage combined with a constant experimental verification. Quantum or computational chemistry is now playing the role of a real partner in molecular science where theory and experiment are mutually complementary tools.

Quantum-chemical computations have mostly been performed on stationary points of the related potential hypersurfaces. Standard techniques for stationary-point location, frequently with analytical first energy derivatives, have been applied. However, with respect to the computational demands, they are not always followed by a vibrational analysis, based on the second energy derivatives from an analytical or numerical differentiation. Nevertheless, the step is critical for simulation of vibrational properties. Semiempirical quantum-chemical methods, which are of course used in many active areas of contemporary research, are less important for description of atmospheric species, unless we are dealing with big clusters like $(H_2O)_{20}$. In fact, *ab initio* methods, even beyond the Hartree–Fock approach, represent a more common tool for the applications, and they will also prevail in our description of the selected topical atmospheric dimers. However, in order to respect temperature and pressure effects, quantum-chemical description has to be linked with statistical mechanics and thus, we shall be on the border line of the two fields.

2 The Water Dimer

Water clusters in general [2–5] and the water dimer in particular [6–10] have frequently been treated computationally. This interest comes not only from the role of the clusters in the atmosphere but also from their significancy in solutions. Recently, a new type of water–water interaction potentials has been introduced,[11–23] viz. the flexible or non-rigid potentials containing, in addition to the inter-molecular part, a sophisticated intra-molecular component derived from a quartic force field of the free monomeric units (thus surpassing the simpler (harmonic) free water molecule potentials [24–26] previously employed for the purpose, and representing an alternative to the local-mode surfaces.[27]) The new potentials are well suited for thermodynamic applications. In fact, altogether 12 modifications of the flexible potentials were originally studied on various aspects of the gas-phase water-dimer thermodynamics, and compared with observed data whenever available.[28–35] In this way various theoretical values were recommended [36] for a further use. In particular, populations of water dimer in saturated steam (over ice or liquid water) could be evaluated.

The so called central-force (CF) model (i.e., the potential energy as a function of only two-center distances) of water [11–13] can be considered as a precursor of the flexible potentials. In the approach, the inter- and intra-molecular potential parts were described in a uniform form, and three modifications (CF [11], CF1 [12], CF2 [13]) of the potential were suggested. In order to increase further the versatility of the potential Bopp, Jancsó and Heinzinger combined [14,15] the CF2 inter-molecular potential with a quartic spectroscopic force field of free water molecule.[17] In such a way, a potential was created which will be referred here as BJH/G. A modification of the monomeric force field, adjusted to liquid water conditions, was also suggested.[14] We shall consider the related water–water flexible potential of BJH-type, viz. under its acronym BJH/L. Finally, for a sake of completeness, the original CF1 and CF2 inter-molecular terms can be combined with the gas-phase intra-molecular part (CF1/G and CF2/G potentials). These four potential modifications represent the BJH family of flexible potentials.

There is, however, still another family of flexible water–water potentials, namely derived from the MCY inter-molecular potentials suggested by Matsuoka, Clementi and Yoshimine.[18] The MCY potentials are based on advanced *ab initio* calculations. Lie and Clementi [19] created so called MCYL potential by combining the MCYII inter-molecular potential with a quartic potential of a free water molecule from a quantum-chemical evaluation.[21] Because there are three other parametric modifications of the MCY inter-molecular part available (MCYI,[18] MCYB,[22] MCYC [23]), again altogether four members of the MCYL family of flexible potentials can be considered here (coded as MCY-X, where X = I, L, B or C). Finally, prior to introduction of the MCY-L flexible potential, a lower degree of flexibility was allowed to the MCY water–water potentials,[24–26] viz. by means of a simple harmonic intra-molecular force field. In this way, the third potential family (MCY semi-rigid) was created (its members labeled by MCYI, MCYII, MCYB and MCYC).

The 12 potentials were computationally treated in a common, unified way that was developed and described recently.[36] The local energy minimum possessing the usual C_s point group of symmetry was located on the potential hypersurfaces, using their analytically constructed energy derivatives. No a priori geometrical constraint was imposed and a redundancy-free coordinate set was applied. In spite of employing various starting structures no other local minimum was found in addition to the C_s one (see Fig. 1). In the local energy minimum harmonic vibrational analysis was carried out using the force constants constructed by a numerical differentiation of the analytical energy first derivatives. The calculated energetics, structure, and vibrational frequencies served for construction of the rigid-rotor and harmonic-oscillator (RRHO) partition functions and then, subsequently, for evaluation of various thermo-

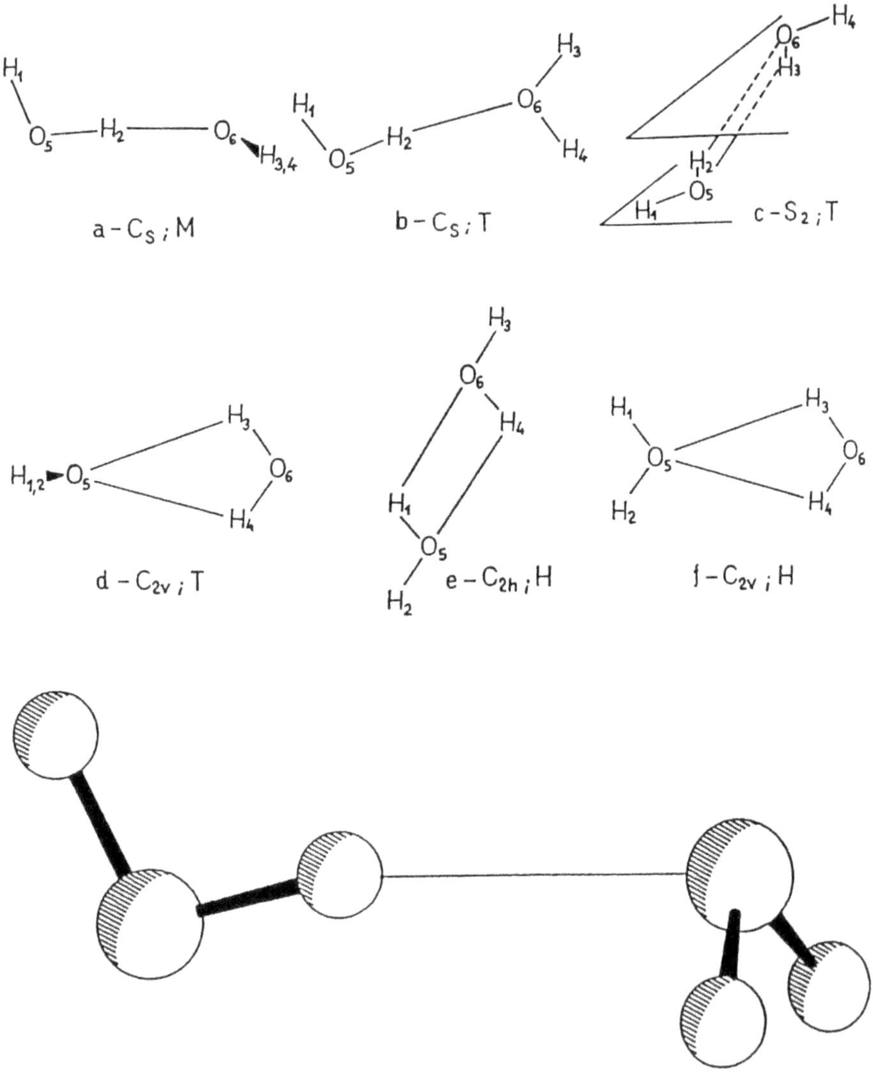

Figure 1: Schemes[24,25] of the stationary points found on the ab initio SCF CI hypersurface[18] of the water dimer; M denotes a minimum, T a transition state, H a higher type of stationary point. The minimum-energy structure is also enlarged in another perspective.

dynamic characteristics. Let us note that even though there is only a solitary relevant minimum on the hypersurface, there are still several interesting saddle points — see Fig. 1. However, they are not directly related to understanding the atmospheric conditions.

In terms of the non-rigid potential functions, structural, vibrational and energy data, and a complete thermodynamics of the gas-phase light water-dimer formation

$$2H_2O(g) = (H_2O)_2(g), \tag{1}$$

as well as of its deutero analogue

$$2D_2O(g) = (D_2O)_2(g) \tag{2}$$

were evaluated employing the partition functions of the RRHO quality. The comparison with observed data was carried out at the level of the standard enthalpy ΔH_T^o and entropy ΔS_T^o changes for processes (1) and (2). The observed ΔH_T^o and ΔS_T^o terms were obtained either from studies [33,34] of thermal conductivity of H_2O and D_2O vapor, or extracted [28,29,33] from different evaluations of the second virial coefficient of steam. There are also other, partial observed data available,[30–32] however, for the purpose they are not comprehensive enough.

The comparison was based on the sums of the squares of the differences between calculations and observations over five temperatures used in observation (i.e., the only one observed value for reaction (2) was treated as having the significancy of any one of the results for association (1)), namely for ΔH_T^o, ΔS_T^o, $T\Delta S_T^o$, and $\Delta G_T^o = \Delta H_T^o - T\Delta S_T^o$ terms. Within the BJH family of potentials the BJH/G results produced [36] the best agreement with observation for all the four levels of comparison. With the MCYL-type potentials, the situation was somewhat more complex: the smallest value of the $\Sigma\delta^2 H_i$ term was met for the MCY-B potential, while for the two entropy terms it was the MCY-I potential, and the lowest value of the $\Sigma\delta^2 G_i$ term was given by the MCY-L approach. Finally, in the MCY semi-rigid family of potentials the best reproduction of the observed terms was found [36] for the MCYB potential.

In order to allow for compensation of possible imperfections of the potentials, a refinement with respect to the observed K_p values was moreover considered. It was concluded that the MCYB, BJH/G, and MCY-B potential in the semi-flexible MCY, semi-flexible BJH, and flexible MCY family of potentials, respectively, should be preferred for a further use. These three potentials should thus be primarily applied to the water-dimer populations, too.

Let us consider a simple equilibrium monomer–dimer mixture in terms of mole fractions first. The mole fraction of the dimer is given $x_2 = n_2/(n_1 + n_2)$ where n_1 and n_2 denote the number of mols (i.e., mass divided by the molar mass) of the monomer and dimer, respectively, in their equilibrium mixture. The dimer mole fraction is different from the degree of association or extent of reaction (sometimes labeled by α). It holds[37] for the dimer mole fraction:

$$x_2 = \frac{2PK_p + 1 - (4PK_p + 1)^{1/2}}{2PK_p}, \tag{3}$$

where P stands for the total pressure of the mixture. In a limiting case $PK_p \ll 1$, Eq. 3 reduces to a simple form $x_2 \approx PK_p$.

In a real situation, however, higher water clusters are also present. In order to account for their effects an assumption has been introduced[38] that the trimerization, tetramerization, etc., equilibrium-constant values decrease along a geometrical series with a quotient z (which itself however depends on temperature). The assumption works well for the available data for higher water oligomers and, in principle, it allows for inclusion of a cluster of any dimension in our considerations. Within this unlimited cluster-dimension model, the dimeric mole fraction reads:

$$x_2 = K_p x_1^2 P, \tag{4}$$

where x_1, the monomeric mole fraction, is given by:

$$x_1 = \frac{-Q + [Q^2 + 4P(K_p/z - 1)/z]^{1/2}}{2P(K_p/z - 1)}, \tag{5}$$

while the term Q denotes:

$$Q = 1/z + P. \tag{6}$$

In the simple monomer–dimer model the two mole fractions follow the condition $x_1 + x_2 = 1$ while in the unlimited dimension approach all clusters are (formally) involved:

$$\sum_{i=1}^{\infty} x_i = 1. \tag{7}$$

Let us note, however, that actually two conventions are used in the literature, with the upper bound for the mole fractions either simply 1, or 100%. In our connections the total pressure P is exclusively the saturated pressure of steam over either liquid water[39,40] or ice.[41] It should be mentioned that the original

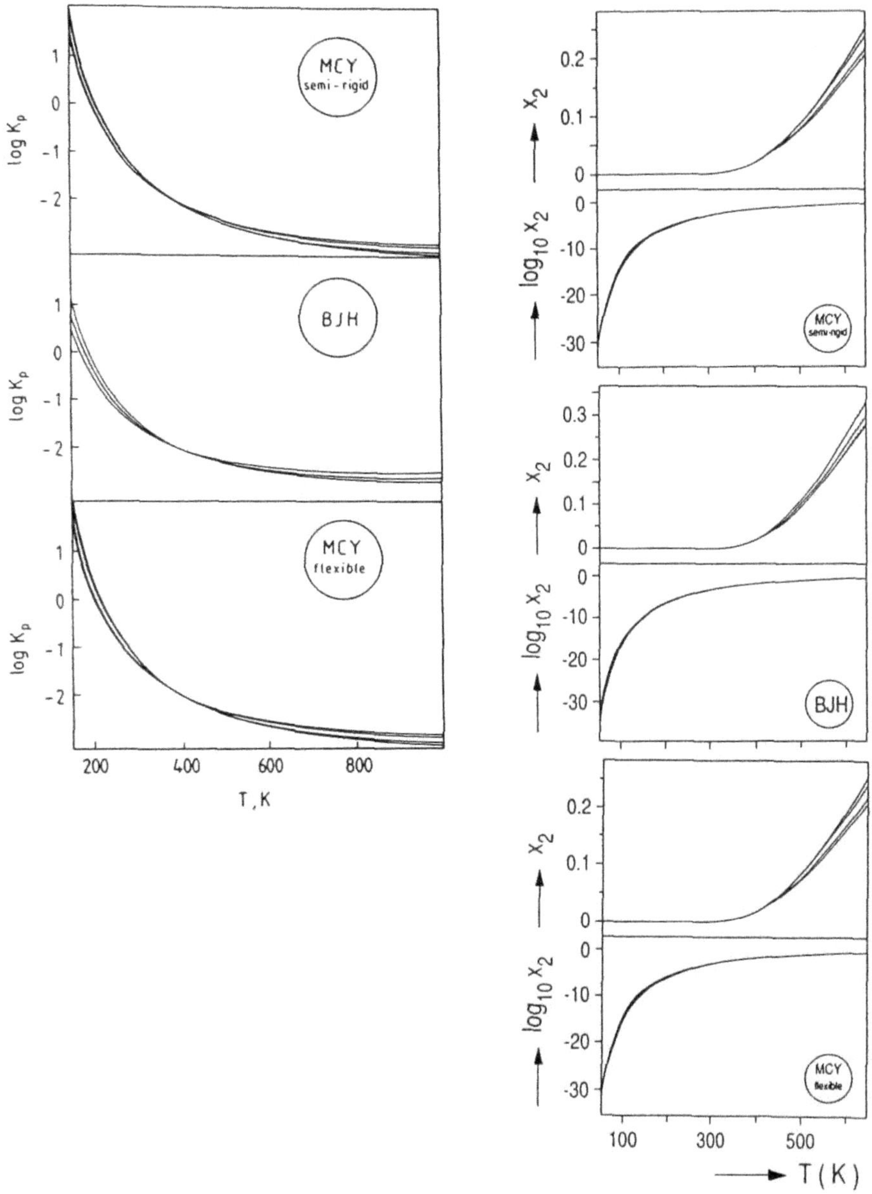

Figure 2: Temperature dependences of the water-dimerization equilibrium constant computed[42] in various potentials (left), and of the water-dimer mole fraction evaluated[43] for saturated steam (right).

formula[39] for the saturated pressure of steam is repeated in survey,[40] however, in a way which is not completely consistent. Hence, we follow the very original form[39] though there are still some minor uncertainties, for example, in relation to the practical temperature scale[39] versus the thermodynamic scale. The other formula, the ice formula was designed[41] for temperatures between the triple point and -100°C; its use also below the lower limit is an extrapolation only.

Fig. 2 presents temperature evolutions[42,43] of the dimeric mole fraction x_2 in the saturated steam over a wide temperature interval from 50 K till high temperatures close to the critical point as evaluated for the three families of flexible potentials (MCY semi-rigid, BJH, and MCY flexible). The qualitative picture does not differ in any of the potentials. The mole fraction x_2 increases with increasing temperature. In low temperature region there is a fast decrease of the x_2 term with decreasing temperature. It may be a surprising result but in fact it can easily be rationalized. The K_p values decrease with temperature for weak molecular complexes (see Fig. 2). However, the saturated pressure P increases with temperature. It is just the competition between these two terms which decides the final x_2 temperature behavior. The simple monomer–dimer treatment yields somewhat higher x_2 values compared to the unlimited dimension model (as the latter model has to accommodate all the higher clusters, too). However, the difference is felt at higher temperatures only — it can hardly be distinguished in Fig. 2, and thus, it is not particularly significant. The results produced by various potentials are not very different either. In order to have an idea on higher-cluster populations, one can consider the sum $\sum_{i=3}^{\infty} x_i$. For example, at 400 K the sum is still by the two orders of magnitude smaller than the x_2 term.

With respect to the applied combination of the sophisticated potentials with the observed data (i.e., the refinement), the derived x_2 values can be considered as the most reliable so far. There still may be a possibility of a further improvement, namely, a reevaluation of the observed-data analysis in the cases where an estimation of the B_f component of the second virial coefficient was applied (i.e., now using the most sophisticated B_f values). Alternatively, a newer, independent experimental information would be useful for a further refinement of the water potentials. Although associations are widely allowed in the model, each class of clusters is treated as an ideal gas characterized by its partial pressure. Consequently, the equilibrium constants involved are pressure independent which should not be the case at higher pressures (i.e at higher temperatures in the saturation regime). From this point of view the results are more reliable at lower temperatures.

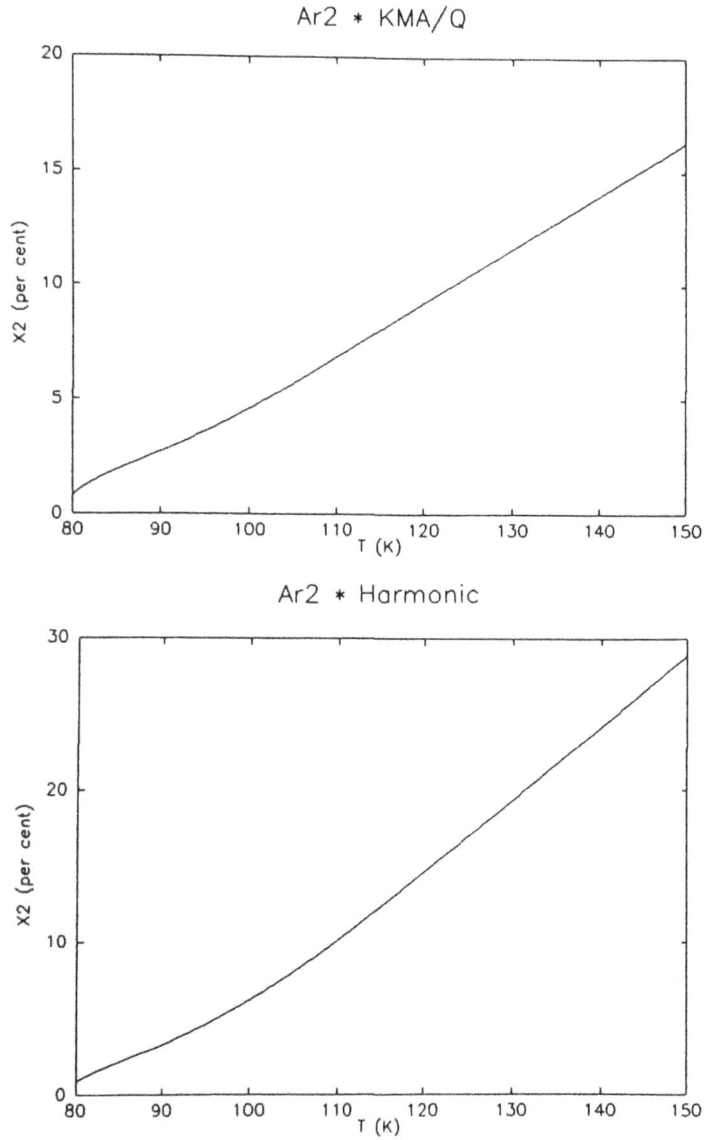

Figure 3: Temperature evolution[47] of the argon-dimer mole fraction under the saturation conditions computed in the best available approximation **KMA/Q** (top), and in the harmonic approximation (bottom).

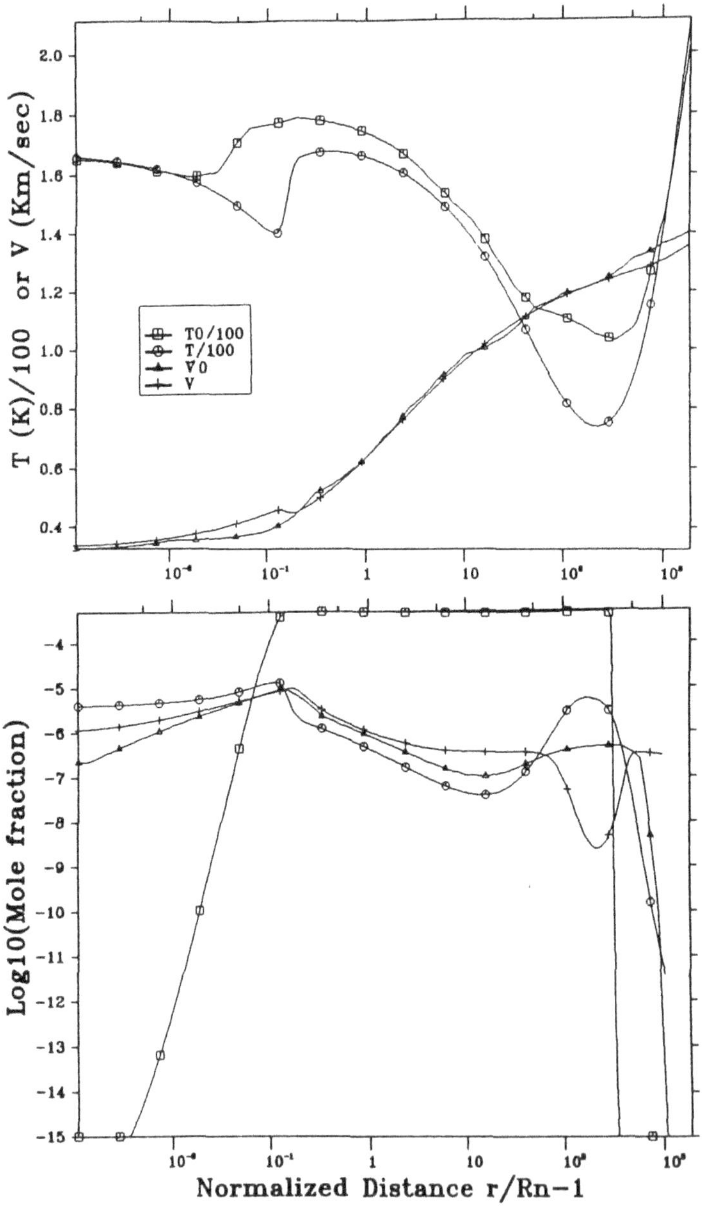

Figure 4: The altitude profiles [62] in comet P/Halley: temperature and velocity profiles (top), and the water-dimer mole fraction from various treatments (bottom).

The calculations employed the usual rigid rotor and harmonic oscillator partition functions. This should not be critical as the refinement applied should account for deviations from the model to some extent. One can expect that at higher temperatures, effects of the infinite harmonic-oscillator sum can play a role. However, information available from diatomic clusters [44,45] still does not allow a straightforward conclusion on the magnitude of these effects. Anyhow, this factor cannot change the conclusion on the proportions of the dimer populations in low-temperature saturated steam, i.e. in region which is of primary importance for the Earth's, planetary or cometary atmosphere applications. The water dimer has recently been subject of a constant interest and computations at various quantum-chemical levels have been performed,[46] nevertheless, this on-going computational activity does not change the already established picture.

We can further discuss the temperature–pressure interplay on the case of the argon saturated vapor. The argon vapor in equilibrium with the condensed Ar phase was also treated as the equilibrium mixture of the monomer and dimer according to Eq. 3. The dimerization equilibrium constant was recently calculated [45] in a number of approximations, employing several analytical potentials. As the problem has only one variable, rotational and vibrational partition functions can be constructed well beyond the RRHO approximation. Fig. 3 samples the results.[47] The label KMA/Q denotes the most sophisticated interaction potential (KMA) used in conjunction with numerically exact (Q – quantum) partition functions, i.e., a very accurate treatment. On the other hand, one can take a related simple harmonic potential together with the RRHO partition functions (Harmonic). As Fig. 3 shows, the dimeric mole fractions in the saturated argon vapor are not considerably different in both approaches. Moreover, the strong temperature increase of the dimeric fraction is clearly evident and therefore, it is certainly not an artifact of the RRHO approximation. Finally, it suggests that Ar_2 can be a relevant species for some planetary atmospheres, especially for Titan's atmosphere.

The saturated regime is a physically well defined situation and thus, it is useful as a reference standard. However, in real atmospheric conditions we can expect conditions below the saturation rather than the saturated or even super-saturated vapor. Hence, in order to evaluate realistic atmospheric dimeric populations [48–59] we have to take the actual temperature and concentration/pressure altitude profiles into consideration, if they are known. The step was also carried out for the Earth's atmosphere.[50,51] It was concluded that for a typical atmospheric temperature/humidity profile, the cluster populations in the Earth's atmosphere should decrease with increasing height, being of the order of magnitude of 0.1 mg/m^3 and 0.1 μg/m^3 for the water dimer and trimer, respectively, in the atmospheric region of 700–800 mbar. The corresponding dimeric mole fraction is of the order of 0.01%.

The abundance of the water dimer in a cometary atmosphere is a related topic, important for simulation of the gas-phase chemistry in comets and recondensation processes. The problem was treated [42,43,60–62] computationally on the case of comet P/Halley. Using our reliable water dimerization equilibrium constants from *ab initio* calculations we could simulate [62] the dimeric mole fractions for the cometary conditions (Fig. 4). It follows that for comet P/Halley at the time of the 1986 flyby missions, the computed dimer mole fraction is everywhere smaller than 10^{-5}. The computed dimer fractions are too small to have previously believed chemical implications. Nevertheless, the formation of the dimers even at the computed small abundance level should support homogeneous recondensation of the coma water. It is interesting to note that the future *in situ* mass spectrometric observations of the water dimer would require special precautions with respect to the presence of C_3, i.e., a species of about the same mass.

3 The $(N_2)_2$, N_2–O_2, and $(O_2)_2$ Species

The above three dimeric systems represent interesting species of the Earth's atmosphere. The complexes have some relevancy to atmospheric spectroscopy, and could play a role in atmospheric-chemistry processes and thermophysics, too. Although various their properties were investigated in the past [63–84] (for both neutral and charged species) their thermodynamic stabilities were not systematically studied. Hence, in a series of articles we evaluated structure, energetics, vibrations, and thermodynamics of formation of the gas phase $(N_2)_2$ (Refs. [85,86]), N_2–O_2 (Refs. [87,88]), and $(O_2)_2$ (Refs. [89,90]). The study produced *ab initio* computed dimerization equilibrium constants and was finalized by evaluation of the altitude concentration profiles for $(N_2)_2$, N_2–O_2, and $(O_2)_2$ in the Earth's atmosphere.

The computations used the standard 6-31+G* basis in the SCF approach (HF/6-31+G* treatment). Electron correlation was considered at the 6-31+G* level through the all-electron second and fourth order Møller-Plesset perturbation treatment (MP2/6-31+G*, MP4/6-31+G*). Within the HF/6-31+G* approach, and in some cases also at the MP2/6-31+G* level, geometry optimization and harmonic vibrational analysis were performed. The optimizations were carried out for selected (symmetry) structural types (see Fig. 5) in the smallest possible sets of internal coordinates. The energy gradient was constructed analytically. For the optimized structures, the harmonic vibrational analysis was carried out in the terms of the Cartesian force-constant matrix obtained from analytical differentiation. The vibrational analysis allowed to distinguish local energy minima and other types of stationary points by means of the nature of the force-matrix eigenvalues. The vibrational analysis indeed ruled out some structures as mere saddle points. Hence, the $(N_2)_2$, N_2–O_2, and $(O_2)_2$ systems are represented by four, two (the parallel structure is a saddle point), and three (the tetrahedral D_{2d} and the oblong D_{2h} structures are saddle points) isomers, respectively — see Fig. 5. The whole treatment was also applied to the related monomers. For the all-electron full fourth order Møller-Plesset correlation energy calculations the basis set superposition error (BSSE) was estimated. The BSSE term was evaluated using the approximative counterpoise method.[92]

We deal with three overall dimerization processes (though each of them is actually a superposition of at least two partial dimerizations):

$$2N_2(g) = (N_2)_2(g), \tag{8}$$

$$N_2(g) + O_2(g) = N_2 - O_2(g), \tag{9}$$

$$2O_2(g) = (O_2)_2(g), \tag{10}$$

described by the dimerization equilibrium constants:

$$K_{p,N_2} = \frac{x_{(N_2)_2}}{x_{N_2}^2 P}, \tag{11}$$

$$K_{p,N_2/O_2} = \frac{x_{N_2-O_2}}{x_{N_2} x_{O_2} P}, \tag{12}$$

$$K_{p,O_2} = \frac{x_{(O_2)_2}}{x_{O_2}^2 P}. \tag{13}$$

The equilibrium constants (11)–(13) are expressed in the terms of the equilibrium mole fractions of the components, x_i, and the total pressure of the mixture, P. No other atmospheric species are considered so that it holds for the mole fractions:

$$x_{N_2} + x_{O_2} + x_{(N_2)_2} + x_{N_2-O_2} + x_{(O_2)_2} = 1. \tag{14}$$

Finally, the known nitrogen to oxygen atmospheric ratio r should be fulfilled:

$$\frac{x_{N_2} + 2x_{(N_2)_2} + x_{N_2-O_2}}{x_{O_2} + 2x_{(O_2)_2} + x_{N_2-O_2}} = r. \tag{15}$$

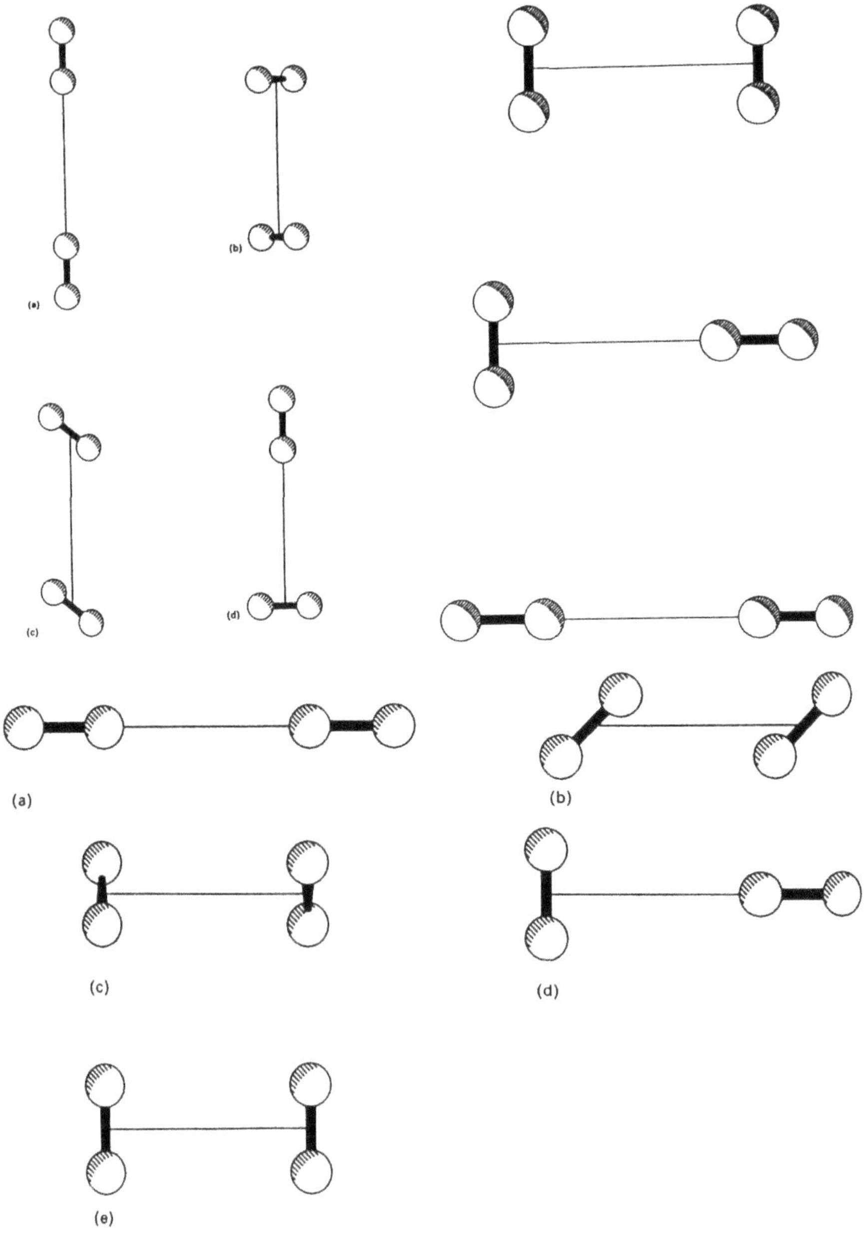

Figure 5: Schemes of the stationary points localized[85,88,89] on the interaction hypersurface of two nitrogen molecules (top left), of a nitrogen and an oxygen molecule (top right), and of two oxygen molecules (bottom).

A solution of Eqns. 11–15 allows for determination of the equilibrium mole fractions, supposing that the equilibrium constants and the P, r parameters are available.

The equilibrium constants were evaluated by means of the partition functions supplied with parameters from our *ab initio* computations.[85,87–89] With respect to the amount of the computed information available, the partition functions had to be again of the usual rigid-rotor and harmonic-oscillator quality. The dimerization energetics was treated at two different levels — with and without the BSSE correction — the term was either ignored or estimated by the counterpoise method.[92] Although the latter method is an approximation,[93,94] these two approaches should generally yield the upper and lower limits for the well depth.[95] In fact, the reported[91] mole fractions are entirely based on the energetics with the BSSE–correction inclusion. This approach reflects a prevailing belief in a good applicability of the technique. However, we should keep in mind that the dimerization energies are most probably underestimated and thus, the reported mole fractions represent a kind of the lower bounds. This expectation is also supported by an experimental estimate by Orlando *et al.*[77] who placed the dimerization enthalpy of the oxygen molecule close to -1.1 kcal/mol. The oxygen molecule is a particularly difficult case, and most probably a coupled-cluster approach would be needed to settle its dimerization energetics.

The computed equilibrium constants were combined with the observed profiles[96] of temperature and pressure in the interval of the altitude $z = 0$–90 km (in fact, the so called geopotential altitude was employed). The nitrogen/oxygen ratio r (Ref. [96]) was supposed to be constant over the region. The equilibrium constants in fact exhibit a convolutional nature as any of the dimers is actually formed by an equilibrium mixture of several isomeric structural forms, i.e., a mixture of at least two isomers. Clearly enough, the isomerism enhances the equilibrium-constant values.

The computed dimeric mole fractions exhibit[91] a smooth development along altitudes z — in all three cases the dimeric mole fractions decrease with the increasing altitude. It may be surprising as temperature decreases in lower parts of the atmosphere which should stabilize the dimers. However, also the pressure decreases which helps dissociation processes. Once again, it is the competition between the temperature and pressure effects which decides the final altitude decrease. There is also a uniform order in the relative populations of the dimers. The most populated N_2 dimer is followed by the heterodimer N_2–O_2 and finally, the O_2 dimer is the least populated species.

As already mentioned, the computations of the dimeric fractions were based, in agreement with a prevailing belief, on the interaction energies after inclusion of the BSSE correction, and therefore the corrected energies should rather represent a lower bound of the well depth.[95] Moreover, the vibrational part of the RRHO partition functions was actually neglected in the approach.[86,88,90] This step further stresses the expectation that the computed mole-fraction values really represent a lower bound of the dimeric populations. On the other hand, the interaction energies without the BSSE correction represent a kind of the upper bound. For example, at a temperature of 298 K and 1 atm pressure, the dimeric mole fractions for nitrogen and oxygen would reach quite impressive values of 8.6 and 1.6 %, respectively. On the other hand, the $(N_2)_2$ atmospheric mole fraction close to the surface is computed, in the energetics corrected for the BSSE term, as $8{\cdot}10^{-4}$ %, the N_2–O_2 fraction $8{\cdot}10^{-5}$ %, and the $(O_2)_2$ fraction $3{\cdot}10^{-5}$ %.

In the treatment other components of the Earth's atmosphere were ignored. It is reasonable as they form less than 1% of the total. Nevertheless, it should be realized that molecular complexes of N_2 and O_2 with those minor atmospheric components can be formed, too. Although the nitrogen/oxygen ratio r was considered as a constant in the study, it is not necessarily true at a high altitude. It is interesting to note that the computed dimeric mole fractions can be comparable with the mole fractions of some minor atmospheric components like He, CH_4 or SO_2.

This study was the first, purely computational evaluation of the altitude profiles of the three molecular complexes especially pertinent to the Earth's atmosphere. Although various improvements of the computational treatment could be suggested, the conclusion that the dimeric mole fractions decrease with increasing altitude should not be changed. The improvements of the treatment should concern the quality of both partition functions and interaction energies. A final solution of the problem would however be possible only within sophisticated potential hypersurfaces, which is a very remote target.

4 The Ozone Dimer

The ozone molecule, its properties,[97] and various potential agents for its decomposition have vigorously been studied in relationship with the phenomenon of ozone depletion.[98–100] Also quantum-chemical computations have been applied[101–113] in the broad research. In particular, many ozone-related species have vigorously been computed by Francisco, e.g. Refs.[108,110–113] From our point of view, it is pertinent to address the question of the ozone containing dimers,[114–117] and in particular of the ozone dimer, $(O_3)_2$. The ozone-dimer populations in the equilibrium with its monomer under the atmospheric conditions were evaluated[117] on the base of correlated *ab initio* computations[116] of

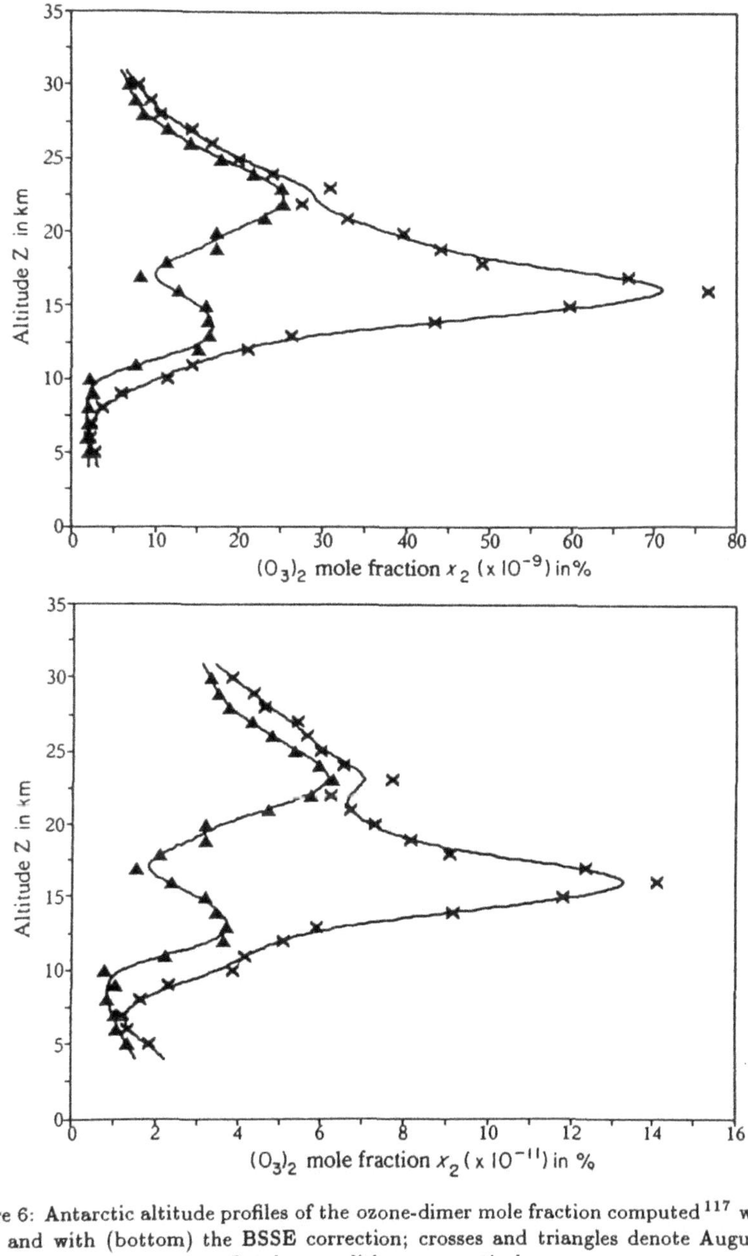

Figure 6: Antarctic altitude profiles of the ozone-dimer mole fraction computed[117] without (top) and with (bottom) the BSSE correction; crosses and triangles denote August and October conditions, respectively.

the dimer. The computations were performed in the diffuse-orbital augmented polarization 6-31+G* basis set with an inclusion of the electron correlation through the second order Møller-Plesset perturbation treatment (MP2). The full geometry optimization followed by harmonic vibrational analysis was carried out in the MP2/6-31+G* approach. The dimerization energy was again considered with and without the BSSE correction. Most probably, the true energy value is located between these two boundaries. Interestingly enough, the optimized structure of the ozone dimer is of a C_s symmetry and to some extent resembles the water-dimer structure (see Fig. 1).

The computed structural, vibrational and energy data serve for construction of partition functions in the RRHO approximation of the rigid rotator and harmonic oscillator and the dimerization equilibrium constant K_p. The dimeric mole fraction x_2 is again given by Eq. 3, however, the pressure P now refers to the partial pressure of ozone in the atmosphere. An Antarctic ozone pressure profile was taken from observations of Hoffmann *et al.*[118] In fact, two profiles were treated, namely from August and October, for a region of heights Z between 5 and 30 km. As related temperature profile is not given in Ref.,[118] a representative dependency was taken from Ref.[100] instead. The highest temperature used for the x_2 evaluations was 230 K (the lowest 183 K). Hence, the RRHO approximation should not be a critical factor.

Fig. 6 presents the altitude profiles of the ozone-dimer mole fraction in its equilibrium with the monomer calculated for both estimates of the dimerization energy. As the input pressures and temperatures were extrapolated from measured values, they contain an error drift and this factor is reflected in the computed x_2 dependences. Qualitatively, the x_2 temperature dependences are quite similar for both sources of the dimerization energy. However, there is a difference of two orders of magnitude. The true dimeric mole fractions x_2 are to be expected somewhere between them. The computed values of the x_2 term are quite small, the highest values found being of the order of 10^{-8} and 10^{-10} % — see Fig. 6. This a considerably smaller degree of dimerization than reported for saturated steam at similar temperatures.[43]

This first evaluation of the ozone dimerization degree could further be improved (when the needed computational power is available) by a larger basis set, higher order contributions to the correlation energy, and an inclusion of anharmonicity and non-rigidity effects into partition functions. Nevertheless, it should not dramatically changed the interval of the expected x_2 values for the ozone dimer.

5 $(CO_2)_2$ & Venus

Carbon dioxide has widely been recognized as an important atmospheric component in view of the greenhouse effects. It is thus not surprising that its dimer, $(CO_2)_2$, has also recently received ever increasing attention,[59,95,119–138] both experimental and theoretical. A comprehensive, combined research on the dimeric species has been performed in particular by Vigasin *et al.* — see for example Refs.[10,139–143] Among cosmic-body atmospheres, that of Venus is especially relevant regarding the carbon-dioxide dimer as the atmosphere is mostly[144] composed of CO_2 (96.5%), and is very dense.

Quantum-chemical studies of $(CO_2)_2$, performed at various methodological levels, supplied[95,125,136,138] reasonably consistent data on the dimer. Thus, it has became possible to re-evaluate[53,59,137,145] the previous results based on the simple Lennard-Jones potential.[146] The treatment is not particularly different from the water-dimer case as the temperature and pressure atmospheric profiles are available[144] for Venus. However, similarly to the nitrogen/oxygen molecular dimers, already Brigot *et al.*[125] predicted that there should actually be two different structures of the dimer, viz. parallel and T–form possessing C_{2h} and C_{2v} point-group of symmetry, respectively — see Fig. 7. In the newer treatment by Eggenberger *et al.*[136] the T–shaped conformation was identified as a transition state for interconversion between two parallel structures (i.e., the T species does not contribute into the dimer concentration). Still, the former result[125] requires a simultaneous consideration of both isomers. One can at first evaluate the partial equilibrium constants[147] $K_p(C_{2h})$ and $K_p(C_{2v})$, and then construct the overall, convolutional K_p term as sum of the partial ones:

$$K_p = K_p(C_{2h}) + K_p(C_{2v}). \tag{16}$$

The partial terms themselves are evaluated in exactly the same way as if no isomerism is present. Both isomers actually change their relative stabilities drastically with temperature, and close to the normal sublimation point of CO_2 an interchange actually happens[147] (see Fig. 7).

In the computations[59,137,145] only CO_2 and $(CO_2)_2$ were considered, all other components of the atmosphere of Venus were disregarded. In other, thermodynamic words, it is supposed that the equilibrium constant K_p has the same value in presence or absence of the minor components (primarily[144] N_2). The total atmospheric pressure P is however composed of contributions of all the components. Hence, the pressure was scaled down by the known[144] mixing ratio of carbon dioxide in the Venus atmosphere (96.5%).

As already mentioned, there is an older, independent treatment of the dimeric populations suggested by Stogryn and Hirschfelder[146] based on the second virial coefficients (derived from the simple Lennard-Jones potential in particular). They broke down the virial coefficients into three parts related to bound (b), metastable (m), and free molecules. The first two terms can be used for evaluation of the dimer concentrations. The mole fraction of dimers is given:

$$x_2 = -b_0(B_b^* + B_m^*)n/V, \tag{17}$$

where B_b^* and B_m^* are the reduced second virial coefficients, b_0 is a constant related to the Lennard-Jones potential, and n/V denotes the molar concentration per unit volume (for a detailed discussion of several computational aspects of this approach, see Ref.[59]) In any case, the Lennard-Jones potential is a one-dimensional potential function which is not particularly suitable for description of the interactions between polyatomic molecules (generally, a six-dimensional potential is needed).

The dimeric mole fractions were computed[137] for two quantum-chemical data sets[125,136] and also with the treatment by Stogryn and Hirschfelder.[146] For example, in going from the surface to 50 km in height, the dimeric mole fraction x_2 decreases by one order of magnitude. Both quantum-chemical approaches yield higher values than the Lennard-Jones potential. The quantum-chemical data set[125] with the dimer isomerism yields higher x_2 terms compared to the newer calculations.[136] There is, however, a common qualitative feature — a smooth decrease of the equilibrium dimeric fraction x_2 with increasing height. Temperatures in the atmosphere of Venus are relatively high, so that the applicability of the RRHO treatment can be questioned. It may seem probable that the RRHO approach can overestimate the K_p values though the overestimation degree is unclear. The overestimation is related to the infinite nature of the summation in the harmonic vibrational partition function.[45] The differences between both quantum-chemical treatments reflect uncertainties in the description of the CO_2-CO_2 interaction potential. In particular, the two-minimum nature of the potential remains a somewhat open question.

Our treatment nevertheless indicates that one should expect unusually high populations of the carbon dioxide dimers in the atmosphere of Venus (apparently the highest in our Solar system). The values are in particular enhanced by the high atmospheric pressures. On the surface of Mars, with much lower pressures, the dimeric fraction is computed[145] by three orders of magnitude lower. However, the carbon dioxide dimeric mole fraction should be at least 4% at the surface of Venus and, thus, it is important in the planning of future observations.

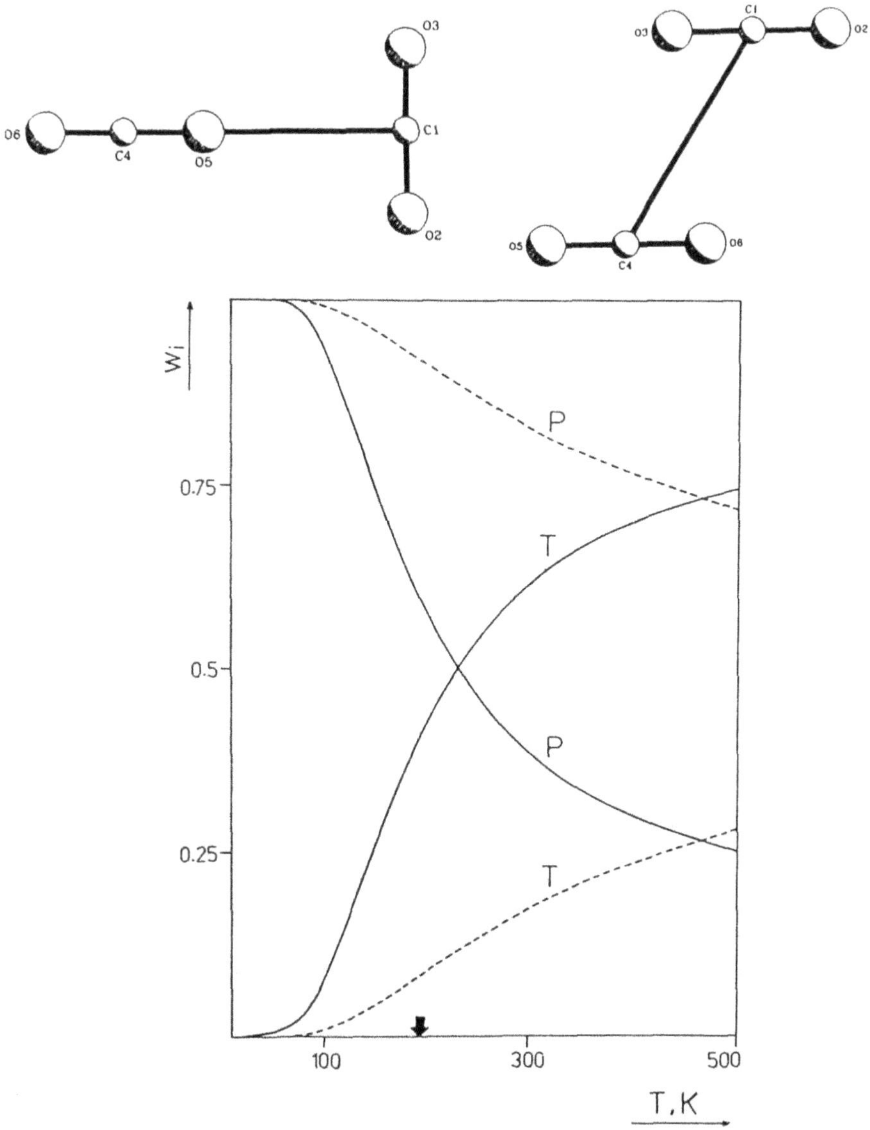

Figure 7: The computed T–shaped (T, top left) and parallel (P, top right) structures[138] of the carbon-dioxide dimer, and temperature evolution[147] of their relative concentrations (bottom, the dashed lines indicate the simple Boltzmann factors).

6 $(H_2)_2$ & Jupiter

The hydrogen-molecule dimer was historically the very first detected extraterrestrial dimer [52,148–150] (in the spectra of Jupiter), and a considerable amount of experimental and theoretical studies has been devoted to the floppy dimeric species.[55–57,151–154] Interestingly enough, $(H_2)_2$ is a relatively very weak complex, and apparently its pressure stabilization operates in the atmosphere of Jupiter. In fact, there are several stronger complexes in the literature, well characterized by computations, which could also have a role in atmospheres. A series of such advanced computational characterizations has been reported by Bone *et al.*[55–57,151–154]

The molecular-orbital calculations used in the evaluations of the $(H_2)_2$ mole-fraction employed [153,154] the standard 6-311G, 6-311G**, and 6-31++G** basis sets. Electron correlation was considered through the second order Møller-Plesset (MP2) perturbation technique. Within the correlated MP2/6-311G, MP2/6-311G**, and MP2/6-31++G** approaches, geometry optimization and harmonic vibrational analysis were carried out. The energetics of the system was further refined using the optimal MP2/6-31++G** geometry. The fourth order Møller-Plesset (MP4) correlation energy calculations were performed. Finally, the basis set superposition error (BSSE) was estimated for the MP4/6-31++G** treatment. The further steps were carried out as for the carbon-dioxide dimer.

In the computations the percentage of hydrogen in the atmosphere of Jupiter was taken [159,160] as 88.8%, and other components of the atmosphere were ignored. The pressure and temperature values for selected heights above the Jupiter 1 atm pressure level were taken from Ref. [160] (range from 0 to 124 km). Let us mention that two types of the energy-minimum structure of the dimer were found, either linear ($D_{\infty h}$) or so called T–shape (C_{2v}). However, the non-linear, T–shape structure comes in more sophisticated treatments. Moreover, the energy minima are so shallow that they can hardly accommodate even the computed harmonic vibrational ground state for inter-molecular motions. It does not prevent vibrational motion but this motion must be treated in a proper anharmonic potential. In order to reflect the feature at the partition-function level, we neglected the harmonic vibrational contribution, i.e. we worked with rotational contribution only. Finally, it is interesting that an experimental estimation [151] of the dimerization energy is in fact quite close to the computed MP4/6-31++G** value.

The dimeric populations computed for lower atmospheric levels (between 0.04 and 1.06 % for the lowest level considered) are certainly of interest of spectroscopic observations. The fact that the hydrogen-molecule dimer is in-

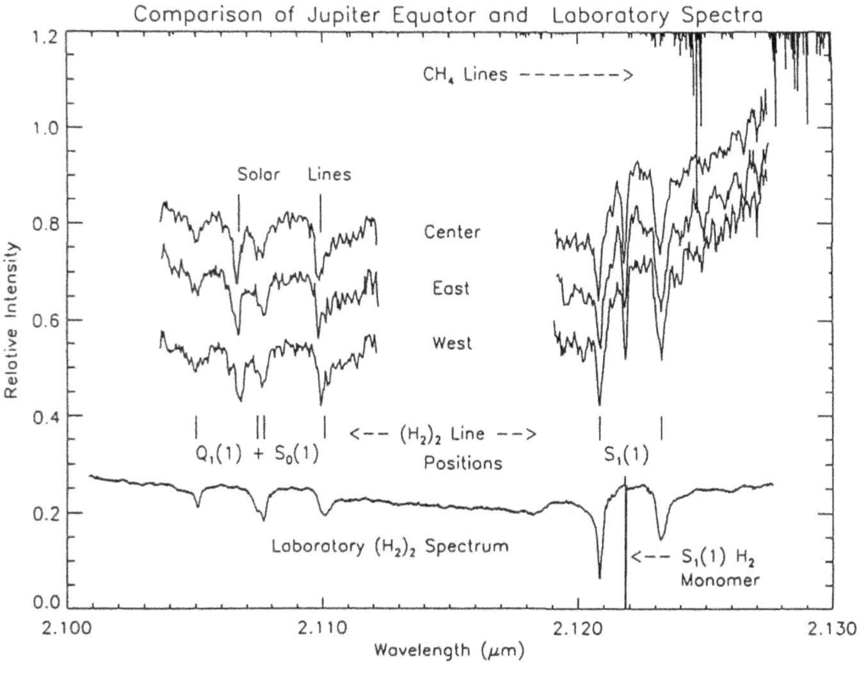

Figure 8: A spectral region of Jupiter compared[161] with a laboratory spectrum of the hydrogen-molecule dimer.

deed active in the spectra[161] of Jupiter (see Fig. 8) is consistent with a non-negligible dimeric fraction. Incidentally, all quantum-chemical estimates yield higher values than the treatment by Stogryn and Hirschfelder.[146] Still, all the dependences exhibit a uniform qualitative feature — a decrease of the equilibrium dimeric fraction x_2 with increasing height.

7 Ar–N_2 & Titan

Ar–N_2 is particularly relevant to the atmosphere of Titan with respect to its contents[144,162–164] of N_2 and Ar. The complex of Ar with N_2 has been studied by various experimental and computational techniques,[165–176] and recently also its altitude profile in Titan's atmosphere was computed.[177,178] The treatment was based on *ab initio* computations in the 6-31G* basis set with the second order perturbation treatment of electron correlation and found two different minimum energy structures, T–shape (C_{2v} symmetry) and $C_{\infty v}$ linear isomers. Although the linear species is lower in potential energy, the T–shape structure is in fact the ground state structure (i.e., the latter species is lower after addition of the zero point vibrational energy).

In fact, we deal with two dimerization processes:

$$Ar(g) + N_2(g) = Ar - N_2(C_{2v}, g), \tag{18}$$

$$Ar(g) + N_2(g) = Ar - N_2(C_{\infty v}, g). \tag{19}$$

Again, two partial process (18) and (19) are combined into the overall dimerization equilibrium. According to Samuelson *et al.*[163] Titan's atmosphere has two major components — N_2 and Ar with mole fractions of 82.2 and 11.6 %, respectively. All other components are neglected and the total observed pressure, P, is scaled by a reduction factor of 0.938. It holds for the mole fractions in the simplified model:

$$x_{N_2} + x_{Ar} + x_{Ar-N_2} = 1, \tag{20}$$

while the nitrogen/argon ratio still has to obey the observed term 82.2/11.6.

The equilibrium constants were evaluated either from the ground-state energy changes and the full RRHO partition functions, or the vibrational partition functions and the zero-point vibrational contributions were neglected. The observed profiles[164] of temperature and pressure in the region of the altitude 0–204 km were applied. The nitrogen/argon ratio[163] was supposed to be constant over the region.

At the lowest temperatures the isomeric dimer lower in energy must be dominant. However, in both treatments we can notice a relative stability interchange — at about 176 and 12 K. Hence, we in fact deal with a coexistence of the two structures. It again turns out that the dimeric mole fraction decreases with increasing altitude. Hence, the highest population of the heterodimer should be close to Titan's surface. The recommended value of $8 \cdot 10^{-2}$% could be of some spectroscopic significancy. One can imagine still a more general treatment within our simplified Titan's atmosphere — inclusion of two other dimers, Ar_2 and $(N_2)_2$. Such a treatment would at least be interesting from the point of view of thermodynamic methodology.

8 Concluding Remarks

Quantum chemistry was born almost instantly with quantum mechanics but reached its maturity only with the introduction of the high-performance computers. It is now much closer to the ideal formulated[179] by two of its pioneers, E. Clementi and C. A. Coulson, already a quarter of century ago: "We can calculate everything". Computational chemistry at present spreads this message to less and less traditional fields of its application. It could also be of a considerable help to atmospheric community for a more intimate insight into its remote objects of interest.

Acknowledgments

Z. S. highly appreciates a partial support from the Alexander von Humboldt–Stiftung and the Max–Planck–Institut für Chemie (Otto–Hahn–Institut) where a significant part of the reported research had been carried out, as well as valuable discussions with P. J. Crutzen and K. Heinzinger. F. U. appreciates a research fellowship from the European Environmental Research Organization. The authors also wish to thank the following organizations for kindly permitting the reprinting of copyrighted material: Academia, Academic Press, Elsevier Scientific Publishing Company, Kluwer Academic Publishers, Plenum Publishing Corporation, The American Astronomical Society.

References

1. P. A. M. Dirac, *Proc. Roy. Soc. A* **123**, 714 (1929).
2. E. Clementi, *Determination of Liquid Water Structure, Coordination Numbers for Ions and Solvation for Biological Molecules* (Springer-Verlag, Berlin, 1976).

3. J. R. Reimers, R. O. Watts, and M. L. Klein, *Chem. Phys.* **64**, 95 (1982).
4. J. L. Finney, J. E. Quinn, and J. O. Baum, *Water Sci. Rev.* **1**, 93 (1985).
5. F. Franks, *Anal. Sci.* **4**, 441 (1988).
6. N. D. Sokolov, *Dokl. Akad. Nauk SSSR* **58**, 611 (1947).
7. K. Morokuma and L. Pedersen, *J. Chem. Phys.* **48**, 3275 (1968).
8. R. W. Bolander, J. L. Kassner Jr., and J. T. Zung, *J. Chem. Phys.* **50**, 4402 (1969).
9. J. P. O'Connell and J. M. Prausnitz, *Ind. Eng. Chem. Fundam.* **8**, 453 (1969).
10. A. A. Vigasin, *Zh. Strukt. Khim.* **24**, 116 (1983).
11. H. L. Lemberg and F. H. Stillinger, *J. Chem. Phys.* **62**, 1677 (1975).
12. A. Rahman, F. H. Stillinger, and H. L. Lemberg, *J. Chem. Phys.* **63**, 5223 (1975).
13. F. H. Stillinger and A. Rahman, *J. Chem. Phys.* **68**, 666 (1978).
14. P. Bopp, G. Jancsó, and K. Heinzinger, *Chem. Phys. Lett.* **98**, 129 (1983).
15. G. Jancsó, P. Bopp, and K. Heinzinger, *Chem. Phys.* **85**, 377 (1984).
16. K. Heinzinger, P. Bopp, and G. Jancsó, *Acta Chim. Hung.* **121**, 27 (1986).
17. G. D. Carney, L. A. Curtiss, and S. R. Langhoff, *J. Mol. Spectr.* **61**, 371 (1976).
18. O. Matsuoka, E. Clementi, and M. Yoshimine, *J. Chem. Phys.* **64**, 1351 (1976).
19. G. C. Lie and E. Clementi, *Phys. Rev. A* **33**, 2679 (1986).
20. K. Refson, G. C. Lie, and E. Clementi, *J. Chem. Phys.* **87**, 3634 (1987).
21. R. J. Bartlett, I. Shavitt, and G. D. Purvis III, *J. Chem. Phys.* **71**, 281 (1979).
22. D. G. Bounds, *Chem. Phys. Lett.* **96**, 604 (1983).
23. V. Carravetta and E. Clementi, *J. Chem. Phys.* **81**, 2646 (1984).
24. Z. Slanina, *J. Chem. Phys.* **73**, 2519 (1980).
25. Z. Slanina, *Collect. Czech. Chem. Commun.* **45**, 3417 (1980).
26. Z. Slanina, *Chem. Phys. Lett.* **127**, 67 (1986).
27. J. R. Reimers and R. O. Watts, *Chem. Phys.* **85**, 83 (1984).
28. G. S. Kell, G. E. McLaurin, and E. Whalley, *J. Chem. Phys.* **48**, 3805 (1968).
29. G. S. Kell and G. E. McLaurin, *J. Chem. Phys.* **51**, 4345 (1969).
30. H. A. Gebbie, W. J. Burroughs, J. Chamberlain, J. E. Harries, and R. G. Jones, *Nature* **221**, 143 (1969).
31. F. T. Greene, T. A. Milne, A. E. Vandergrift, and J. Beachey, *Chem. Abstr.*.**72**, 93891r, 115230d (1970).

32. G. E. Ashwell, P. A. Eggett, R. Emery, and H. A. Gebbie, *Nature* **247**, 196 (1974).
33. L. A. Curtiss, D. J. Frurip, and M. Blander, *J. Chem. Phys.* **71**, 2703 (1979).
34. L. A. Curtiss, D. J. Frurip, and M. Blander in *Water and Steam, Their Properties and Current Industrial Applications*, eds. J. Straub and K. Scheffler (Pergamon Press, Oxford, 1980), p. 521.
35. G. S. Kell, G. E. McLaurin, and E. Whalley, *Proc. Roy. Soc. London, A* **425**, 49 (1989).
36. Z. Slanina, *Chem. Phys.* **150**, 321 (1991).
37. Z. Slanina, *Thermochim. Acta* **106**, 323 (1986).
38. Z. Slanina, *Z. Phys. D* **5**, 181 (1987).
39. O. C. Bridgeman and E. W. Aldrich, *J. Heat Transfer* **86C**, 279 (1964).
40. G. S. Kell in *Water, A Comprehensive Treatise, Vol. 1*, ed. F. Franks (Plenum Press, New York, 1972), p. 392.
41. G. Jancso, J. Pupezin, and W. A. Van Hook, *J. Phys. Chem.* **74**, 2984 (1970).
42. Z. Slanina and J.-F. Crifo, *Int. J. Thermophys.* **13**, 465 (1992).
43. Z. Slanina, F. Uhlík, and J.-F. Crifo, *J. Mol. Struct.* **270**, 1 (1992).
44. H. K. Shin, *Chem. Phys. Lett.* **47**, 225 (1977).
45. P. S. Dardi and J. S. Dahler, *J. Chem. Phys.* **93**, 3562 (1990).
46. S. J. Chakravorty and E. R. Davidson, *J. Phys. Chem.* **97**, 6373 (1993).
47. Z. Slanina, *Thermochim. Acta* **209**, 1 (1992).
48. J. M. Calo and J. H. Brown, *J. Chem. Phys.* **61**, 3931 (1974).
49. J. M. Calo and R. S. Narcisi, *Geophys. Res. Lett.* **7**, 289 (1980).
50. Z. Slanina, *Int. J. Thermophys.* **8**, 387 (1987).
51. Z. Slanina, *J. Atmos. Chem.* **6**, 185 (1988).
52. A. R. W. McKellar, *Astrophys. J.* **326**, L75 (1988).
53. K. Fox and S. J. Kim, *J. Quant. Spectrosc. Radiat. Transfer* **40**, 177 (1988).
54. W. Meyer, A. Borysow, and L. Frommhold, *Phys. Rev. A* **40**, 6931 (1989).
55. A. Borysow and L. Frommhold, *Astrophys. J.* **41**, L348 (1990).
56. A. R. W. McKellar and J. Schaefer, *J. Chem. Phys.* **95**, 3081 (1991).
57. A. Borysow, *Icarus* **92**, 273 (1991).
58. A. J. Colussi, S. P. Sander, and R. R. Friedl, *Chem. Phys. Lett.* **178**, 497 (1991).
59. Z. Slanina, K. Fox, and S. J. Kim, *J. Quant. Spectrosc. Radiat. Transfer* **47**, 91 (1992).
60. J. F. Crifo, *Icarus* **84**, 414 (1990).

61. Z. Slanina and J. F. Crifo, *Thermochim. Acta* **181**, 109 (1991).
62. J. F. Crifo and Z. Slanina, *Astrophys. J.* **383**, 351 (1991).
63. G. N. Lewis, *J. Am. Chem. Soc.* **46**, 2027 (1924).
64. A. R. W. McKellar, N. H. Rich, and H. L. Welsh *Can. J. Phys.* **50**, 1 (1972).
65. C. A. Long and G. Henderson, *Chem. Phys.* **2**, 485 (1973).
66. A. Horowitz, G. Von Helden, W. Schneider, F. G. Simon, P. J. Crutzen, and G. K. Moortgat, *J. Phys. Chem.* **92**, 4956 (1988).
67. A. Horowitz, W. Schneider, and G. K. Moortgat, *J. Phys. Chem.* **93**, 7859 (1989).
68. A. Horowitz, W. Schneider, and G. K. Moortgat, *J. Phys. Chem.* **94**, 2904 (1990).
69. R. M. Berns and A. van der Avoird, *J. Chem. Phys.* **72**, 6107 (1980).
70. B. Jönsson, G. Karlström, and S. Romano, *J. Chem. Phys.* **74**, 2896 (1981).
71. J. Tennyson and A. van der Avoird, *J. Chem. Phys.* **77**, 5664 (1982); **80**, 2986, (1984).
72. M. S. H. Ling and M. Rigby, *Mol. Phys.* **51**, 855 (1984).
73. P. J. Hay, R. T. Pack, and R. L. Martin, *J. Chem. Phys.* **81**, 1360 (1984).
74. G. Brocks and A. van der Avoird, *Mol. Phys.* **55**, 11 (1985).
75. H.-J. Böhm and R. Ahlrichs, *Mol. Phys.* **55**, 1159 (1985).
76. A. Borysow and L. Frommhold, *Astrophys. J.* **311**, 1043 (1986).
77. J. J. Orlando, G. S. Tyndall, K. E. Nickerson, and J. G. Calvert, *J. Geophys. Soc.* **96**, 755 (1991).
78. R. Hashemi, T. Jaffke, L. G. Christophorou, and E. Illenberger, *J. Phys. Chem.* **96**, 10605 (1992).
79. T. Jaffke, R. Hashemi, L. G. Christophorou, and E. Illenberger, *Z. Phys. D* **25**, 77 (1992).
80. E. J. Bieske, *J. Chem. Phys.* **99**, 8672 (1993).
81. J. B. Peel, *Chem. Phys. Lett.* **218**, 367 (1994).
82. L. A. Barnes and R. Lindh, *Chem. Phys. Lett.* **223**, 207 (1994).
83. D. Lessen, S. Nizkorodov, P. Knupfer, T. Ruchti, E. J. Bieske, and J. P. Maier, *Chem. Phys. Lett.* **226**, 187 (1994).
84. B. Bussery, S. Ya. Umanskii, M. Aubert-Frécon, and O. Bouty, *J. Chem. Phys.* **101**, 416 (1994).
85. F. Uhlík, Z. Slanina, and A. Hinchliffe, *J. Mol. Struct. Theochem* **282**, 271 (1993).
86. F. Uhlík, Z. Slanina, and A. Hinchliffe, *Thermochim. Acta* **223**, 1 (1993).

87. W. B. De Almeida and Z. Slanina, *J. Mol. Struct., Theochem* **285**, 77 (1993).
88. F. Uhlík, Z. Slanina, W. B. De Almeida, and A. Hinchliffe, *Thermochim. Acta* **225**, 1 (1993).
89. F. Uhlík, Z. Slanina, and A. Hinchliffe, *J. Mol. Struct., Theochem* **285**, 273 (1993).
90. F. Uhlík, Z. Slanina and A. Hinchliffe, *Thermochim. Acta* **228**, 9 (1993).
91. Z. Slanina, F. Uhlík, W. B. De Almeida, and A. Hinchliffe, *Thermochim. Acta* **231**, 55 (1994).
92. S. F. Boys and F. Bernardi, *Mol. Phys.* **19**, 553 (1970).
93. D. W. Schwenke and D. G. Truhlar, *J. Chem. Phys.* **82**, 2418 (1985).
94. I. Mayer and A. Vibok, *Chem. Phys. Lett.* **140**, 558 (1987).
95. R. G. A. Bone and N. C. Handy, *Theor. Chim. Acta* **78**, 133 (1990).
96. M. Dubin, N. Sissenwine, and H. Wexler, *U. S. Standard Atmosphere 1962* (NASA, USAF, USWB, 1962); N. Sissenwine, M. Dubin, and H. Wexler, *J. Geophys. Res.* **67**, 3627 (1962).
97. J. I. Steinfeld, S. M. Adler-Golden, and J. W. Gallagher, *J. Phys. Chem. Ref. Data* **16**, 911 (1987).
98. P. J. Crutzen, *Quart. J. Roy. Met. Soc.* **96**, 32 (1970).
99. F. S. Rowland, *Annu. Rev. Phys. Chem.* **42**, 731 (1991).
100. P. Hamill and O. B. Toon, *Phys. Today* **44**, 34 (1991).
101. S. D. Peyerimhoff and R. J. Buenker, *J. Chem. Phys.* **47**, 1953 (1967).
102. T. J. Lee, W. D. Allen, and H. F. Schaefer III, *J. Chem. Phys.* **87**, 7062 (1987).
103. T. J. Lee, *Chem. Phys. Lett.* **169**, 529 (1990).
104. T. J. Lee and G. E. Scuseria, *J. Chem. Phys.* **93**, 489 (1990).
105. A. Banichevich, S. D. Peyerimhoff, and F. Grein, *Chem. Phys. Lett.* **173**, 1 (1990).
106. Z. Slanina and F. Uhlík, *J. Phys. Chem.* **95**, 5432 (1991).
107. Z. Slanina, *Spectr. Lett.* **26**, 1965 (1993).
108. J. S. Francisco, *Mol. Phys.* **82**, 831 (1994).
109. Z. Slanina, F. Uhlík, and A. Hinchliffe, *Spectr. Lett.* **27**, 563 (1994).
110. J. S. Francisco, *J. Phys. Chem.* **99**, 13422 (1995).
111. J. S. Francisco, *Chem. Phys. Lett.* **260**, 485 (1996).
112. J. S. Francisco, *J. Chem. Phys.* **105**, 3338 (1996).
113. J. S. Francisco, *Chem. Phys.* **214**, 213 (1997).
114. C. V. Gillies, J. Z. Gillies, R. D. Suenram, F. J. Lovas, E. Kraka, and D. Cremer, *J. Am. Chem. Soc.* **113**, 2412 (1991).
115. J. Z. Gillies, C. W. Gillies, R. D. Suenram, F. J. Lovas, T. Schmidt, and D. Cremer, *J. Mol. Spectr.* **146**, 493 (1991).

116. Z. Slanina and L. Adamowicz, *J. Atmos. Chem.* **16**, 41 (1993).
117. F. Uhlík, Z. Slanina, and A. Hinchliffe, *Thermochim. Acta.* **232**, 1 (1994).
118. D. J. Hofmann, J. W. Harder, S. R. Rolf, and J. M. Rosen, *Nature* **326**, 59 (1987).
119. R. E. Leckenby and E. Robbins, *Proc. Roy. Soc. London A* **291**, 389 (1966).
120. L. Mannik, J. C. Stryland, and H. L. Welsh, *Can. J. Phys.* **49**, 3056 (1971).
121. S. E. Novick, P. B. Davies, T. R. Dyke, and W. Klemperer, *J. Am. Chem. Soc.* **95**, 8547 (1973).
122. L. Fredin, B. Nelander, and G. Ribbegård, *J. Mol. Spectrosc.* **53**, 410 (1974).
123. M. Hashimoto and T. Isobe, *Bull. Chem. Soc. Japan.* **47**, 40 (1974).
124. A. Koide and T. Kihara, *Chem. Phys.* **5**, 34 (1974).
125. N. Brigot, S. Odiot, S. H. Walmsley, and J. L. Whitten, *Chem. Phys. Lett.* **49**, 157 (1977).
126. R. Guasti, V. Schettino, and N. Brigot, *Chem. Phys.* **34**, 319 (1978).
127. A. E. Barton, A. Chablo, and B. J. Howard, *Chem. Phys. Lett.* **60**, 414 (1979).
128. N. Brigot, S. Odiot, and S. H. Walmsley, *Chem. Phys. Lett.* **88**, 543 (1982).
129. Z. Slanina, *J. Mol. Struct.* **94**, 401 (1983).
130. Z. Slanina, *Surf. Sci.* **157**, 371 (1985).
131. K. Hiraoka, T. Shoda, K. Morise, S. Yamabe, E. Kawai, and K. Hirao, *J. Chem. Phys.* **84**, 2091 (1986).
132. A. J. Illies, M. L. McKee, and H. B. Schlegel, *J. Phys. Chem.* **91**, 3489 (1987).
133. K. W. Jucks, Z. S. Huang, D. Dayton, R. E. Miller, and W. J. Lafferty, *J. Chem. Phys.* **86**, 4341 (1987).
134. M. A. Walsh, T. H. England, T. R. Dyke, and B. J. Howard, *Chem. Phys. Lett.* **142**, 265 (1987).
135. K. W. Jucks, Z. S. Huang, R. E. Miller, G. T. Fraser, A. S. Pine, and W. J. Lafferty, *J. Chem. Phys.* **88**, 2185 (1988).
136. R. Eggenberger, S. Gerber, and H. Huber, *Mol. Phys.* **72**, 433 (1991).
137. Z. Slanina, K. Fox, and S. J. Kim, *Thermochim. Acta* **200**, 33 (1992).
138. Z. Slanina, S. J. Kim, and K. Fox, *Vibr. Spectr.* **4**, 251 (1993).
139. T. G. Adiks, G. V. Tchlenova, and A. A. Vigasin, *Infrared Phys.* **29**, 575 (1989).
140. A. A. Vigasin and G. V. Tchlenova, *Sol. Syst. Res.* **24**, 129 (1990).

141. A. A. Vigasin, N. N. Filippov, and G. V. Tchlenova, *Opt. Spectr.* **72**, 56 (1992).
142. A. A. Vigasin, E. G. Tarakanova, and G. V. Tchlenova, *J. Quant. Spectrosc. Radiat. Transfer* **50**, 695 (1993).
143. A. A. Vigasin, T. G. Adiks, and G. V. Yukhnevich, *J. Quant. Spectrosc. Radiat. Transfer* **52**, 295 (1994).
144. A. Seiff in *Venus*, eds. D. M. Hunten, L. Colin, T. M. Donahue, and V. I. Moroz (The University of Arizona Press, Tucson, 1983).
145. Z. Slanina, *Thermochim. Acta* **195**, 45 (1992).
146. D. E. Stogryn and J. O. Hirschfelder, *J. Chem. Phys.* **31**, 1531 (1959); **33**, 942 (1960).
147. Z. Slanina, *Contemporary Theory of Chemical Isomerism* (D. Reidel, Dordrecht, 1986).
148. R. Hanel, B. Conrath, M. Flasar, V. Kunde, P. Lowman, W. Maguire, J. Pearl, J. Pirraglia, R. Samuelson, D. Gautier, P. Gierasch, and S. Kumar, *Science* **204**, 972 (1979).
149. A. R. W. McKellar, *Can. J. Phys.* **62**, 760 (1984).
150. L. Frommhold, R. Samuelson, and G. Birnbaum, *Astrophys. J.* **283**, L79 (1984).
151. U. Buck, F. Huisken, A. Kohlhase, D. Otten, and J. Schaefer, *J. Chem. Phys.* **78**, 4439 (1983).
152. L. M. Trafton and J. K. G. Watson, *Astrophys. J.* **385**, 320 (1992).
153. Z. Slanina, S. J. Kim, and K. Fox, *J. Geophys. Res.* **99**, 3753 (1994).
154. Z. Slanina, S. J. Kim, and K. Fox, *Thermochim. Acta* **220**, 1 (1993).
155. R. G. A. Bone, C. W. Murray, and R. D. Amos, *Chem. Phys. Lett.* **161**, 166 (1989).
156. R. G. A. Bone, R. D. Amos, and N. C. Handy, *J. Chem. Soc. Faraday Trans.* **86**, 1931 (1990).
157. R. G. A. Bone, *Chem. Phys.* **178**, 255 (1993).
158. J. S. Craw, R. G. A. Bone, and G. B. Bacskay, *J. Chem. Soc. Faraday Trans.* **89**, 2363 (1993).
159. J. W. Chamberlain and D. M. Hunten, *Theory of Planetary Atmospheres* (Academic Press, New York, 1987).
160. S. J. Kim, P. Drossart, J. Caldwell, and J.-P. Maillard, *Icarus* **84**, 54 (1990).
161. S. J. Kim, L. M. Trafton, T. R. Geballe, and Z. Slanina, *Icarus* **113**, 217 (1995).
162. V. G. Kunde, A. C. Aikin, R. A. Hanel, D. E. Jennings, W. C. Maguire, and R. E. Samuelson, *Nature* **292**, 686 (1981).

163. R. E. Samuelson, R. A. Hanel, V. G. Kunde, and W. C. Maguire, *Nature* **292**, 688 (1981).
164. S. J. Kim, *Unpublished results*, 1982.
165. M. D. Pattengill, R. A. La Budde, R. B. Bernstein, and C. F. Curtiss, *J. Chem. Phys.* **55**, 5517 (1971).
166. C. A. Long, G. Henderson, and G. E. Ewing, *Chem. Phys.* **2**, 485 (1973).
167. G. Henderson and G. E. Ewing, *Mol. Phys.* **27**, 903 (1974).
168. P. G. Kistenmaker and A. E. de Vries, *Chem. Phys.* **7**, 371 (1975).
169. G. Rotzoll, *Chem. Phys. Lett.* **88**, 179 (1982).
170. R. Candori, F. Pirani, and F. Vecchiocattivi, *Chem. Phys. Lett.* **102**, 412 (1983).
171. M. S. Bowers, K. T. Tang, and J. P. Toennies, *J. Chem. Phys.* **88**, 5465 (1988).
172. F. R. W. McCourt, F. B. van Duijneveldt, T. van Dam, and R. R. Fuchs, *Mol. Phys.* **61**, 109 (1987).
173. A. R. W. McKellar, *J. Chem. Phys.* **88**, 4190 (1988).
174. W. Rijks, M. Gerritsen, and P. E. S. Wormer, *Mol. Phys.* **66**, 929 (1989).
175. A. G. Ayllón, J. Santamaria, S. Miller, and J. Tennyson, *Mol. Phys.* **71**, 1043 (1990).
176. W. Jäger and M. C. L. Gerry, *Chem. Phys. Lett.* **196**, 274 (1992).
177. Z. Slanina, S. J. Kim, and K. Fox, *J. Mol. Struct., Theochem* **288**, 17 (1993).
178. Z. Slanina, S. J. Kim, and K. Fox, *Thermochim. Acta* **232**, 111 (1994).
179. A. Hinchliffe, *Computational Quantum Chemistry* (John Willey & Sons, Chichester, 1988), p. 98.

FULLERENES & OTHER CARBON AGGREGATES AND THE DIFFUSE INTERSTELLAR BANDS

Z. SLANINA

Department of Knowledge-Based Information Engineering,
Toyohashi University of Technology, Toyohashi 441, Aichi, Japan

L. ADAMOWICZ

Department of Chemistry, The University of Arizona,
Tucson, AZ 85721, USA

J.-P. FRANÇOIS

Departement SBG, Limburgs Universitair Centrum, Universitaire Campus,
B-3590 Diepenbeek, Belgium

E. ŌSAWA

Department of Knowledge-Based Information Engineering,
Toyohashi University of Technology, Toyohashi 441, Aichi, Japan

The first spectral diffuse interstellar bands were observed as early as 1921, however, their origin is not entirely clarified even at present. Well over a hundred of these broad interstellar features have been listed in the visible and near-infrared spectral regions. It is probable that they come from more than one specific carrier, and various carriers have been considered like dust grains, carbon chains, polycyclic aromatic species, fullerenes, endohedral fullerenes. Recently, highly unsaturated hydrocarbon species have attracted attention as particularly promising candidates for the carriers.
Historically, astrophysical spectral problems had accelerated studies of purely carbonaceous molecules. The laboratory experiments designed to simulate the formation of carbon molecules and graphitic grains in space lead to the discovery and synthesis of fullerenes, respectively, one of the most celebrated events of modern science. Still, there is no final proof that fullerenes are present in interstellar space. Nevertheless, fullerene science has developed dramatically with far reaching consequences for chemistry, molecular physics, and materials science.
The article surveys relevant theoretical and experimental aspects of small carbon aggregates as well as fullerenes. The small clusters had been observed and computed well before the fullerene era. In spite of the long research interest, they are still not well understood. They are apparently present in early stages of the fullerene synthesis and, with a few hydrogen atoms (possibly only one), they can be responsible for the diffuse interstellar bands. Considerably more results have been accumulated on fullerenes, not only on C_{60} but also on higher fullerenes, giant fullerenes, and then on nanotubes. The tremendous developments, actually rooted in the diffuse interstellar bands and other astrophysical spectral problems, represent an instructive lesson how it is difficult to evaluate interlinks and make predictions in fundamental research.
Small carbon clusters C_n were observed already more than fifty years ago. Since the fullerene discovery they have always been considered as a logical first step in fullerene synthesis. Existence of both linear and cyclic species has been established, the linear even-numbered structures being in triplet electronic state. The linear structures prevail for low n values, then the cyclic species become more important, and finally polyhedral structures. Although C_{60} and C_{70} have primarily attracted research interest, higher fullerenes are of growing importance, too. They are experimentally and computationally characterized well over n = 90. Computations play a particularly important role in the field owing to the structural complexity.

1 Introduction

Hahn *et al.* in 1942 studied [1,2] isotopic composition of some minerals using mass spectrometry as this type of analysis was still possible [3] in the difficult war times. They evaporated geological materials in an electric arc and also reported that the electrodes produced species from C_1 till C_{15}. In fact, C_3 was even before that observed in emissions from comets [4] though its clear identification came later on. Experimental studies of carbon vapor had continued [5-7] in the fifties and sixties. For example, von Hintenberger *et al.* could observe [7] charged aggregates till C_{33} (see Fig. 1). The small carbon aggregates were also computed that time, first by Pitzer and Clementi, [8] and later on by Hoff-

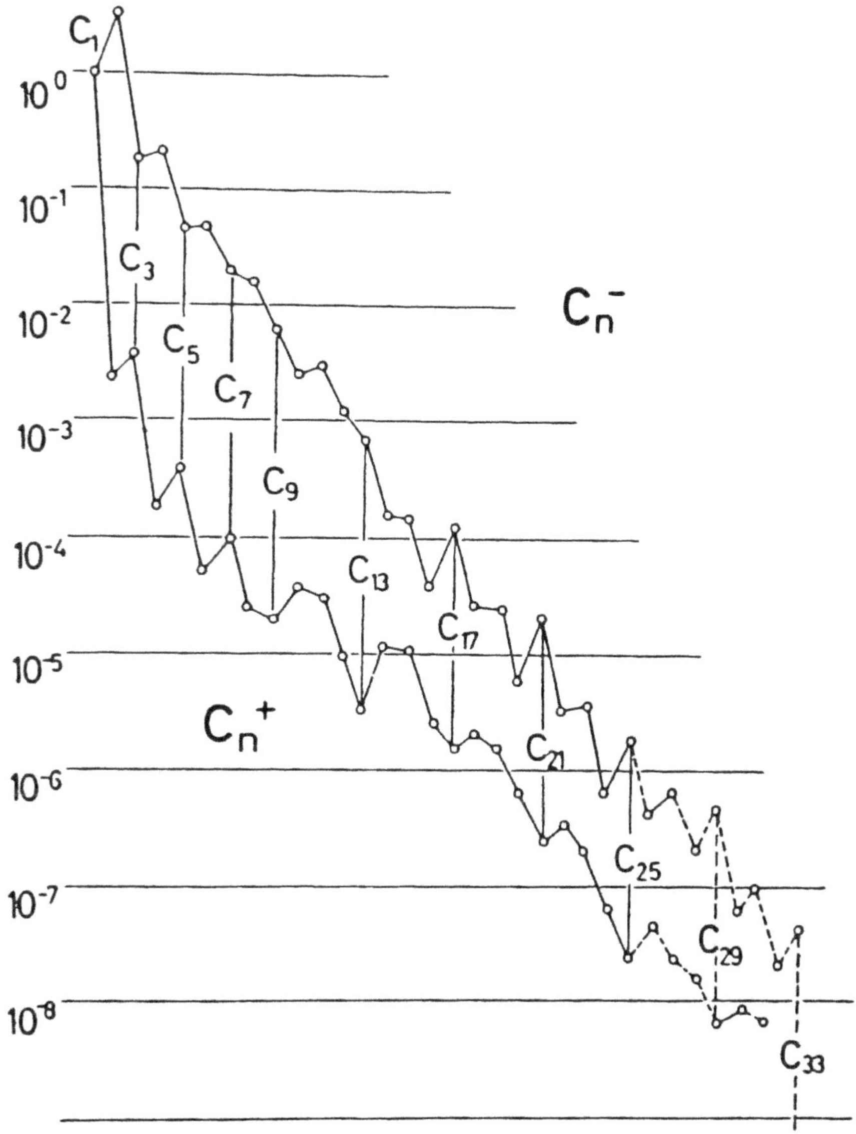

Figure 1: Charged carbon aggregates till C_{33} reported by von Hintenberger *et al.*[7] in 1963.

mann[9] though only as linear species. Pitzer and Clementi applied[8] a simple Hückel-type approach and recognized that the linear molecules with n even would be in a triplet electronic state. Later on, cyclic and polyhedral species were added, treated[10] at the MINDO/2 level, and a possibility of isomerism demonstrated[11,12] for $n = 4$–7 (Fig. 2).

There was another and more important line of the carbon-cluster research. Much bigger structures were considered,[13] and polyhedral C_{60} suggested[14,15] as a stable species with unique electronic properties. Even Hückel calculations on C_{60} were performed.[16–18] They proved that C_{60} had a closed-shell electronic configuration and a relatively large HOMO-LUMO gap. Unfortunately, the theoretical conclusions were not immediately employed in observations.

In fact, they were even unknown to the C_{60} discoverers in September 1985. One of them, H. W. Kroto, had studied[19,20] rotational spectra of species of astrophysical importance. The studies resulted in observations of polyynylcyanides like[21] HC_5N in interstellar space (for related instructive spectra[19] of HC_3N — see Fig. 3). Such long-chain molecules, even composed of carbon atoms only, might provide an explanation[22] for the diffuse interstellar bands[23] and further laboratory data were needed. This research need lead to the famous experiment by Kroto, Heath, O'Brien, Curl and Smalley[24] in 1985 in which still considerably larger species were observed, namely C_{60} and C_{70}. They were detected by mass spectroscopy, however only as gas-phase species. Their polyhedral, highly symmetric structures were hypothesized and their name fullerenes suggested. The discovery represents the first milestone of fullerene science.

Some five years more were needed to prepare fullerenes as a solid-state material. Huffman and Krätschmer had produced and studied[25,26] small carbon particles as potential interstellar dust material and even a possible source of the diffuse interstellar bands. They produced the particles in a carbon evaporator in which graphitic rods were evaporated by resistive heating under an inert gas atmosphere (He or Ar). During their research they had optimized the production conditions, and in particular they recognized a desirable helium pressure. The research activity lead to the famous fullerene synthesis reported by Krätschmer, Lamb, Fostiropoulos and Huffman[27] in 1990. The synthesis represents the second milestone of fullerene science.[28–65]

2 The Diffuse Interstellar Bands

There has been several decades old astrophysical spectral phenomenon which is still not sufficiently understood — the diffuse interstellar bands.[20,23,66–79] The first diffuse interstellar bands were observed[71] already in 1921, however,

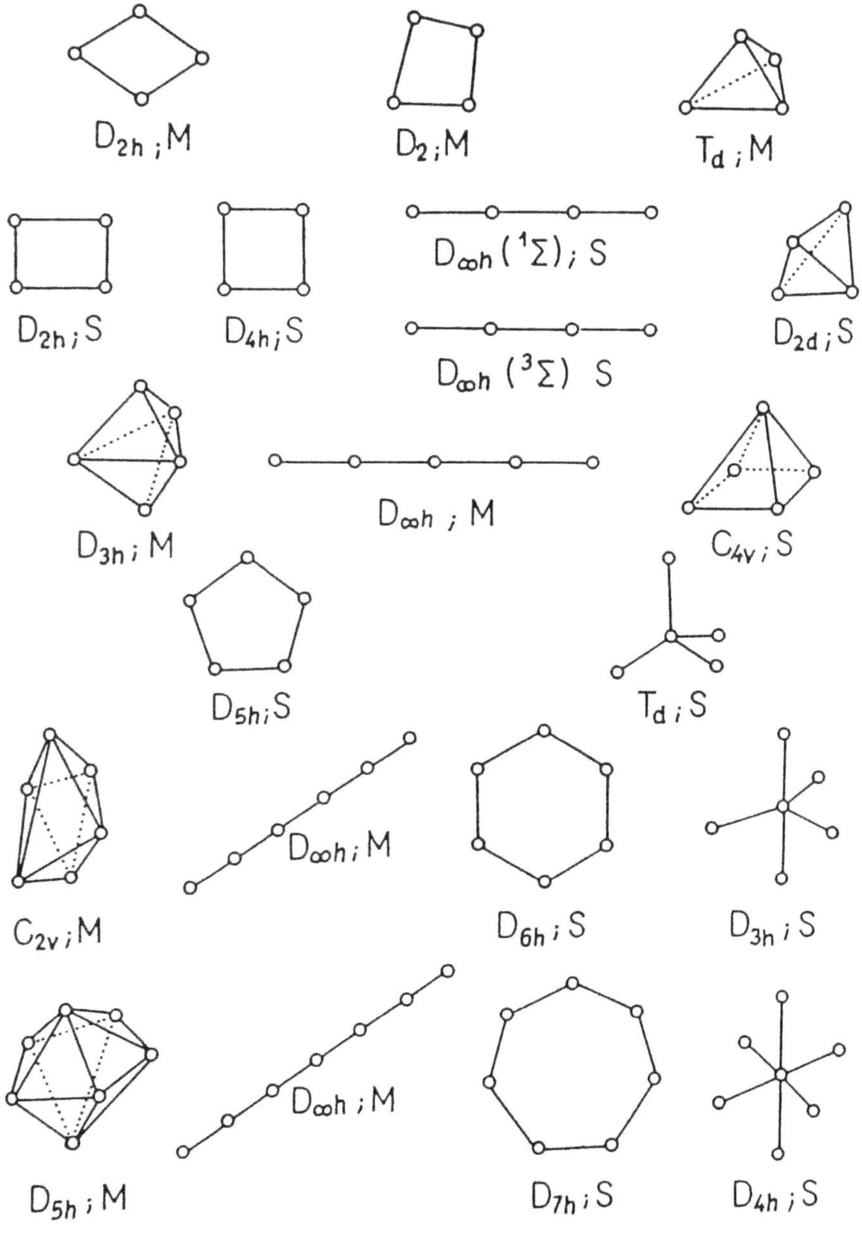

Figure 2: MINDO/2 structures [10–12] of small linear, cyclic and polyhedral carbon aggregates (M — minimum, S — saddle point).

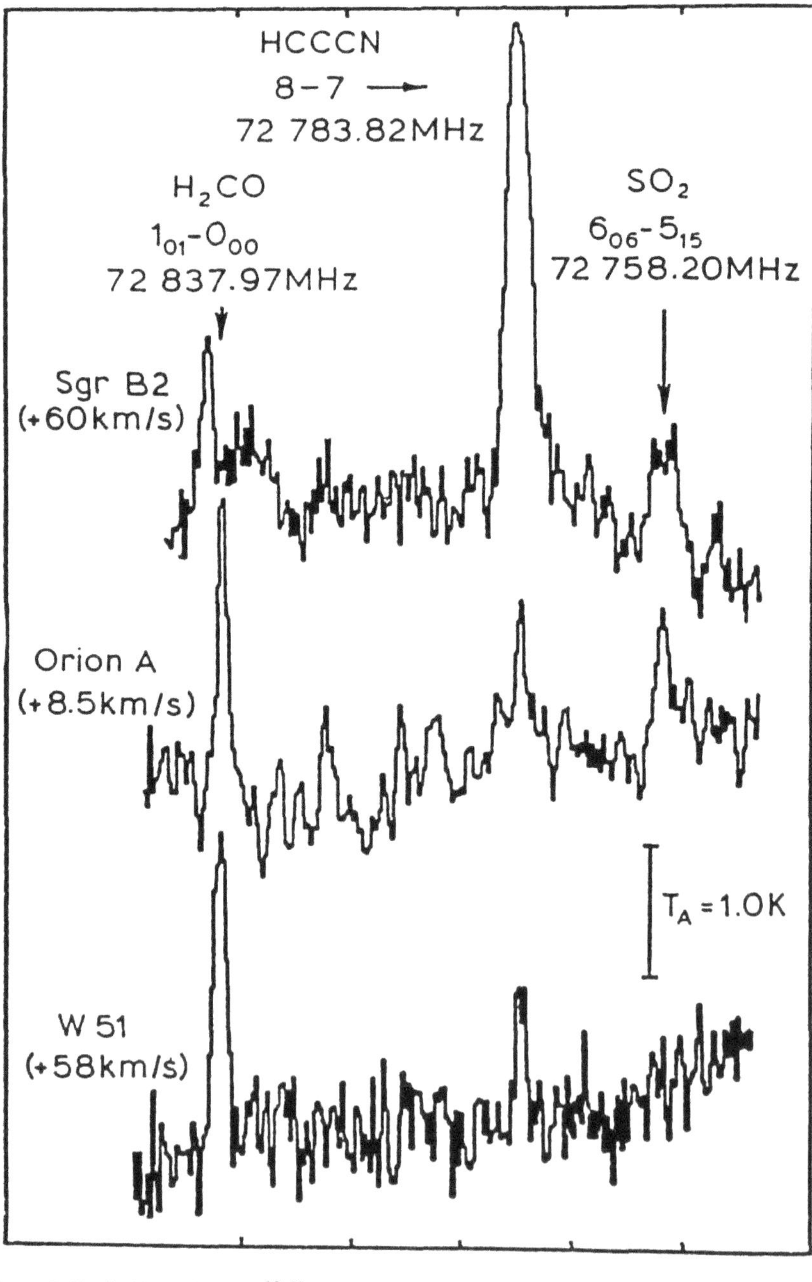

Figure 3: Radio lines observed[19,41] by Snyder *et al.* indicate the polyynylcyanide abundance in Orion and Sagittarius.

their origin is not entirely clarified even at present. They do not exhibit clearly defined vibrational or rotational fine structure; this factor could speak against small molecules. Hypotheses on their origin have oscillated between molecules and dust particles (as after all is illustrated by the history of fullerene discovery and synthesis). Potential candidates should however have no ultraviolet absorption lines in the regions where no diffuse interstellar bands have been observed yet. There are at present over one hundred of these broad interstellar bands in the visible and near-infrared region. Features of the bands suggest that the bands probably do not come from a single carrier.

Polycyclic aromatic hydrocarbons (PAHs) have represented one plausible candidate[23,73] (PAHs are also present in interplanetary dust particles[80]). Purely carbonaceous chains are another frequently considered carrier.[20,22] Neutral C_{60} had been considered as a potential candidate, however, its visible absorption spectrum does not show any match with the known diffuse interstellar bands. C_{60}^+ has also been thought to be a promising carrier. Foing and Ehrenfreund indeed demonstrated[81] that two diffuse interstellar bands were coincident with the spectral features of C_{60}^+ in a Ne matrix (see Fig. 4). However, as pointed out by Maier,[82] there could be some identification problems. The results are nevertheless encouraging as higher fullerenes, endohedral fullerenes,[68] chemically derivatized fullerenes, and other fullerene-related structures[78,79] could serve as the carriers, too. Still, there is another class of compounds, highly unsaturated hydrocarbons, with considerably higher degree of coincidence with the known diffuse interstellar bands.

2.1 Highly Unsaturated Hydrocarbons

Fulara *et al.* in a series of papers[83–91] reported observations of C_nH_m ($2 \le n \le 12$; $m = 1,2$) and related highly unsaturated hydrocarbons. Their laboratory spectra contain many features known from the diffuse interstellar bands. They measured the electronic absorption spectra of mass-selected species deposited in a neon matrix at a temperature of 5 K. Anions of C_nH_m, where $2 \le n \le 12$ and $m = 1,2$, were generated in a discharge from a mixture of diacetylene with argon. In fact, species containing odd numbers of carbon atoms were not produced in a sufficient abundance. It was also observed that the monohydride ions C_nH^- are much more abundant than C_n^- or $C_nH_2^-$. Mass-selected species $C_nH_m^-$ were co-deposited with neon on a metal mirror held at 5 K. It is supposed that predominantly neutral species result as the anions lose an electron during the deposition. The absorption spectrum was measured in the 230–1150 nm region. Fig. 5 presents an example.[83] Four spectral curves A–D in Fig. 5 correspond to C_nH_m with n = 6, 8, 10,

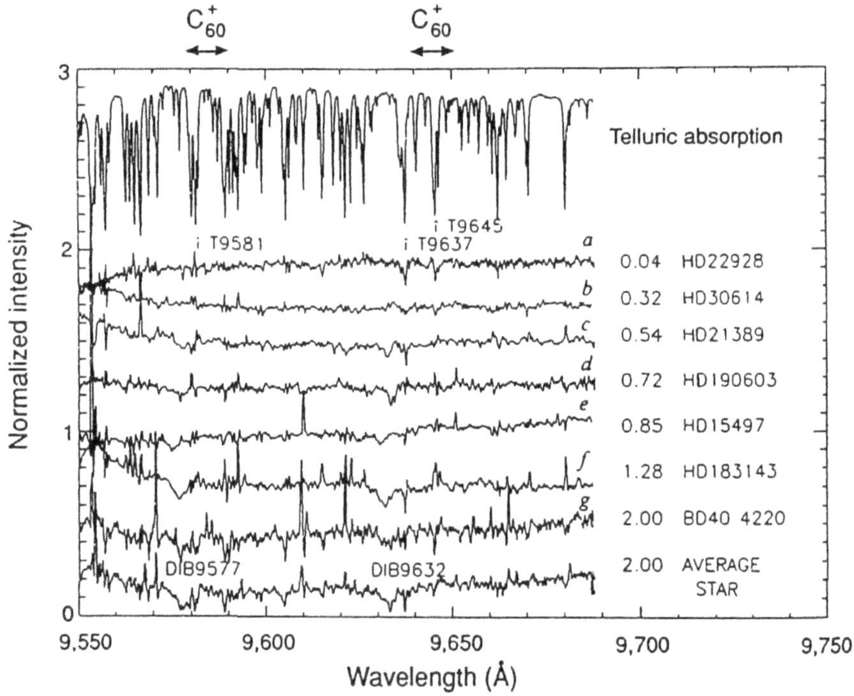

Figure 4: Comparison of two spectral features of C_{60}^{+} in a Ne matrix with the diffuse interstellar features performed by Foing and Ehrenfreund.[81]

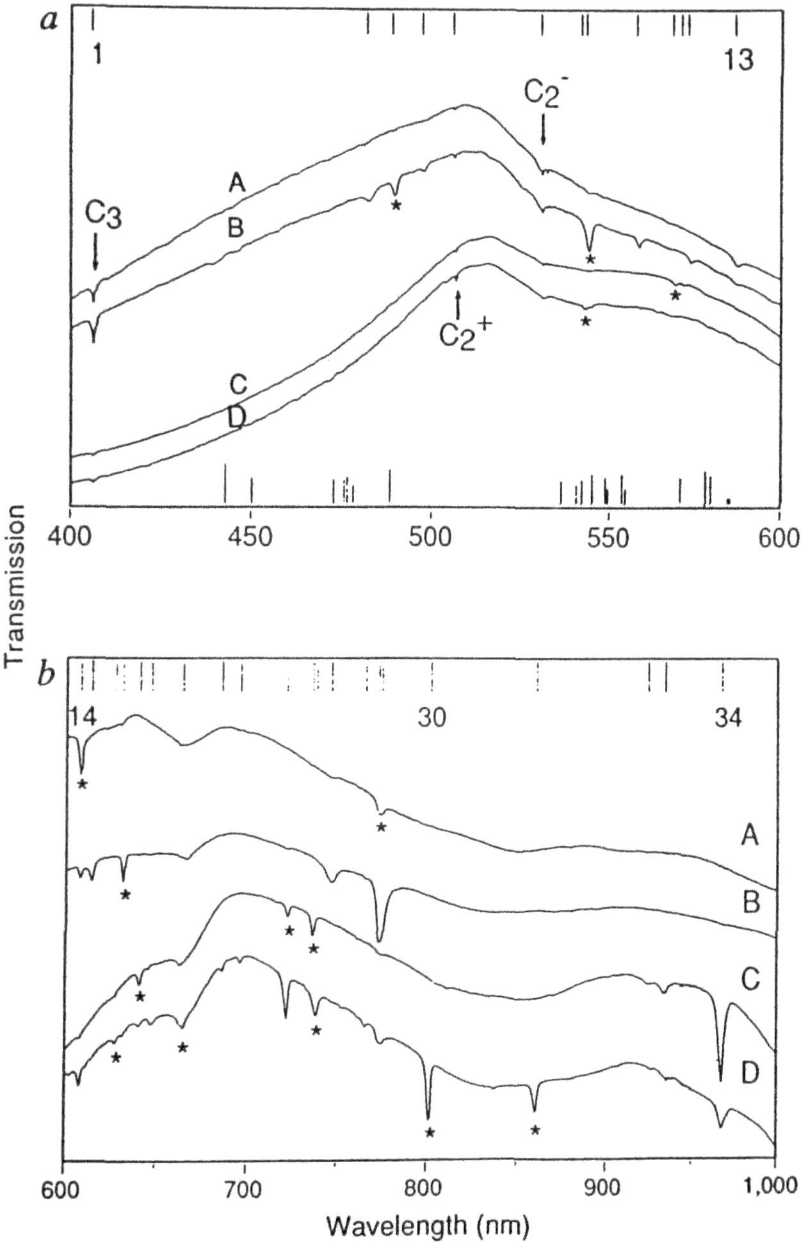

Figure 5: Spectra of C_nH_m reported by Fulara *et al.*[83] — curves A–D correspond to $n = 6$, 8, 10, 12, respectively; coincidence with the diffuse interstellar bands (vertical bars) marked by a star.

12, respectively. Coincidence with the observed diffuse interstellar bands is marked by a star. Altogether sixteen interstellar bands fit [83] to the laboratory spectra in the region 400–1000 nm. It can be concluded that the highly unsaturated C_nH_m species should represent a subset of the species responsible for the diffuse interstellar bands.

The recorded electronic spectra could not provide geometrical structure of the C_nH_m aggregates though both chain and ring structures seem plausible. [92] Quantum-chemical computations have therefore been performed. [93,94] In particular, two members of the new series, C_6H (Fig. 6) and C_8H (Fig. 7) were computed, [93,94] not only the neutral radical but also the related anion and cation species. The computations were performed at an *ab initio* correlated level, namely with the MP2/6-31G* approach. The standard 6-31G* basis set is applied and the electron correlation is treated by means of all-electron second order Møller-Plesset (MP2) perturbation treatment. Open shell species are computed using the related unrestricted approach.

For the C_6H radical (Fig. 6), two local energy minima, (quasi)-linear and cyclic, were located and the linear structure exhibited lower energy. While the ground state for the neutral species is strictly linear, the ground state for the anion exhibits a quasi-linearity. The strictly linear structure for the anion possesses in fact two imaginary vibrational frequencies. With the cation, however, a linear minimum-energy structure could not be found. Hence, only the cyclic form for the cation could be characterized.

Fig. 7 surveys the MP2/6-31G* computed C_8H structures. For both the anion and cation there is a quasi-linear local energy minimum. For the neutral species the related structure is however strictly linear, a quasi-linear species for the radical is higher in energy by about 22 kJ/mol. The cyclic structures for the radical and anion should be out of question as they are too high in energy and they are not energy minima. Hence, there are some similarities between the C_6H and C_8H neutral and anionic systems. However, for the latter stoichiometry a quasi-linear cation could be found. The geometrical structures represent a prerequisite for future computations of the electronic spectra.

3 Small Carbon Clusters

Although small carbon clusters had been studied [28] well before the fullerene discovery they have certainly received substantial impetus from the discovery. They have been treated by both computational [95–113] and experimental [114–137] means and have been considered as a part of the fullerene formation mechanism. Already Raghavachari and Binkley a decade ago formulated [96] the basic

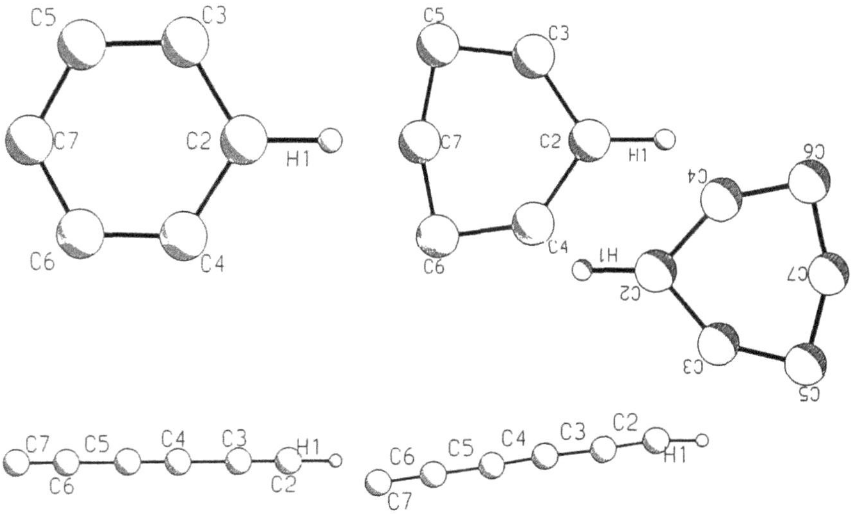

Figure 6: MP2/6-31G* computed [94] structures for the C_6H radical (top), C_6H^-, C_6H^+ (bottom).

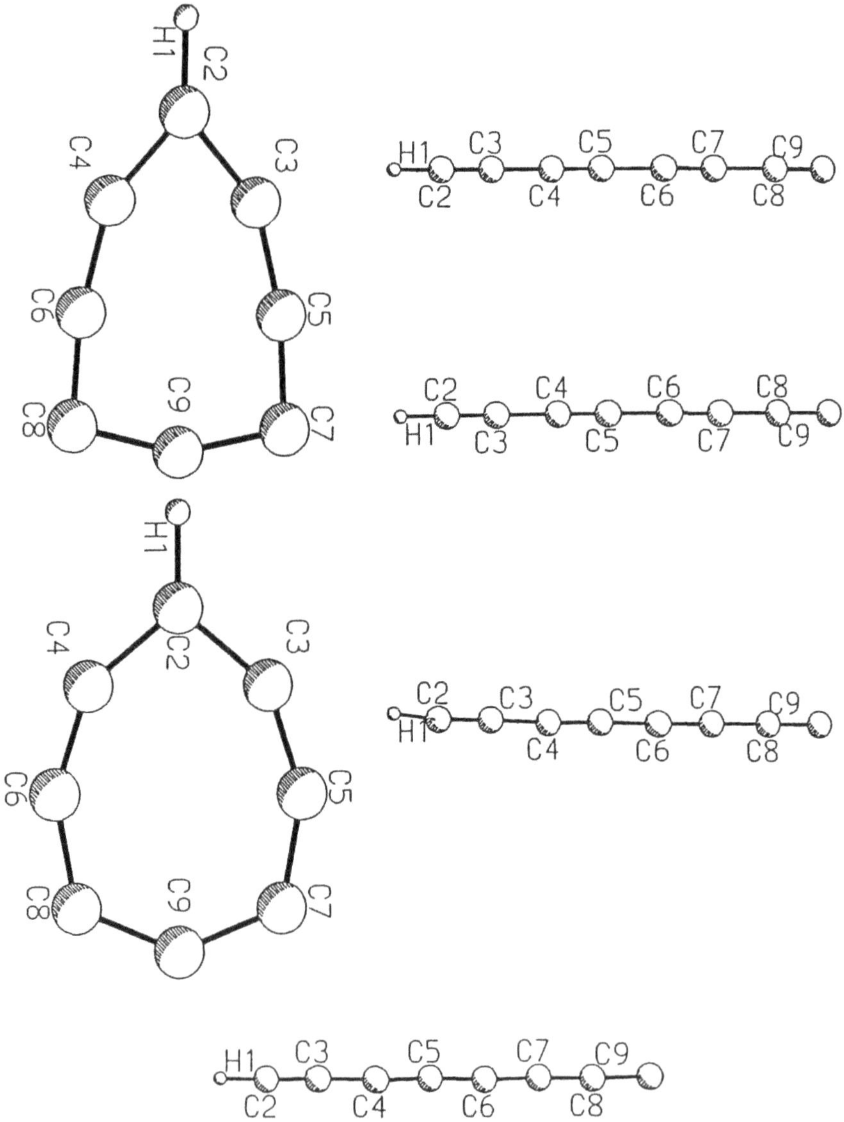

Figure 7: MP2/6-31G* computed[94] structures for the C_8H radical (top), C_8H^-, C_8H^+ (bottom).

structural features of the systems: (i) linear isomers exist for both even- and odd-numbered species with the even species being in the triplet electronic state, (ii) planar cyclic isomers exist for the even species (see [112] Fig. 8). In fact, the *ab initio* conclusions are not dramatically different from the old MINDO/2 results in Fig. 2 though polyhedral species are not relevant for too small n. However, their study was performed at the HF/6-31G* level only so that a further check with a correlated treatment had been necessary. Moreover, ion-chromatography observations by von Helden *et al.*[134,136,137] showed that most carbon clusters had more than one stable form. Fig. 9 documents [137] that starting from about $n = 7$ cyclic rings appear in addition to the linear forms (though there is a difference between the anion and cation series).

For the first observed cyclic case, C_7, the planar ring form was confirmed[109] at the MP2/6-31G* level. The ring isomer was located less than 2 kcal/mol above the linear form (17 kcal/mole at the MP4/6-31G* level). Apparently, the HF/6-31G* treatment does not allow for such low energy species (therefore, the structural survey in Fig. 8 is completed with small odd-membered rings [138]). The cyclic, albeit somewhat non-planar, structure for C_9 was also computed[139] in the MP2/6-31G* approximation. It is predicted 6 kcal/mol above the linear form.

Raghavachari and Binkley's conclusion [96] (ii) has also been discussed for even-numbered clusters because the C_8 cycle is computed [110] to be nonplanar at the MP2/6-31G* level. In fact, the non-planar D_{2h} cycle is the ground state of the system while the linear form is located 34 kcal/mol higher in energy. The nonplanarity of C_8 allows for an inversion motion with an inversion barrier [110,140] of 13 kcal/mol which would clearly be of spectroscopic significancy. The activated complex for the ring inversion motion has an interesting, low symmetry C_{2h} form. However, the non-planar C_8 cycle was not confirmed in the recent coupled-cluster study [113] by Martin and Taylor and thus, the MP2/6-31G* applicability needs further critical tests. Hence, experimental structural and thermochemical data would be of a great use.

The MP2/6-31G* calculations are also available[93,94,141] for C_{10}, C_{11}, C_{12}, and C_{13} — the cyclic form is located 131, 48, 102, and 23 kcal/mol, respectively, below the parent linear form. The separation energy for C_{10} is however considerably higher than the term reported[108] by Parasuk and Almlöf. On the other hand, the best C_{11} separation energy computed[106] in the HF geometries is 34 kcal/mol. Interestingly enough, for C_{13} there could still be a local increase of the linear-form relative stability. This behavior would also agree with the recent observations of Giesen *et al.*[142] Although the presently applicable level of computational approximations is probably not sufficient [107,143] for a really precise evaluation of the separation energetics, it should not automatically be

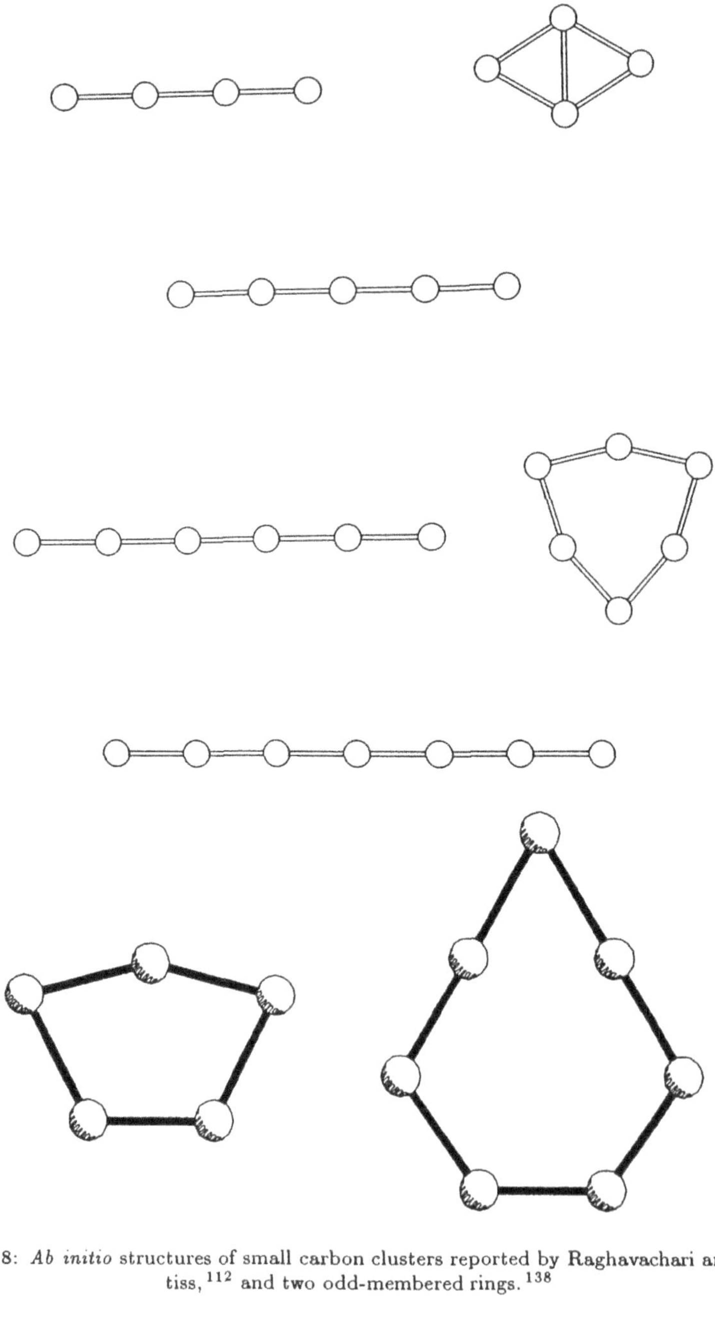

Figure 8: *Ab initio* structures of small carbon clusters reported by Raghavachari and Curtiss,[112] and two odd-membered rings.[138]

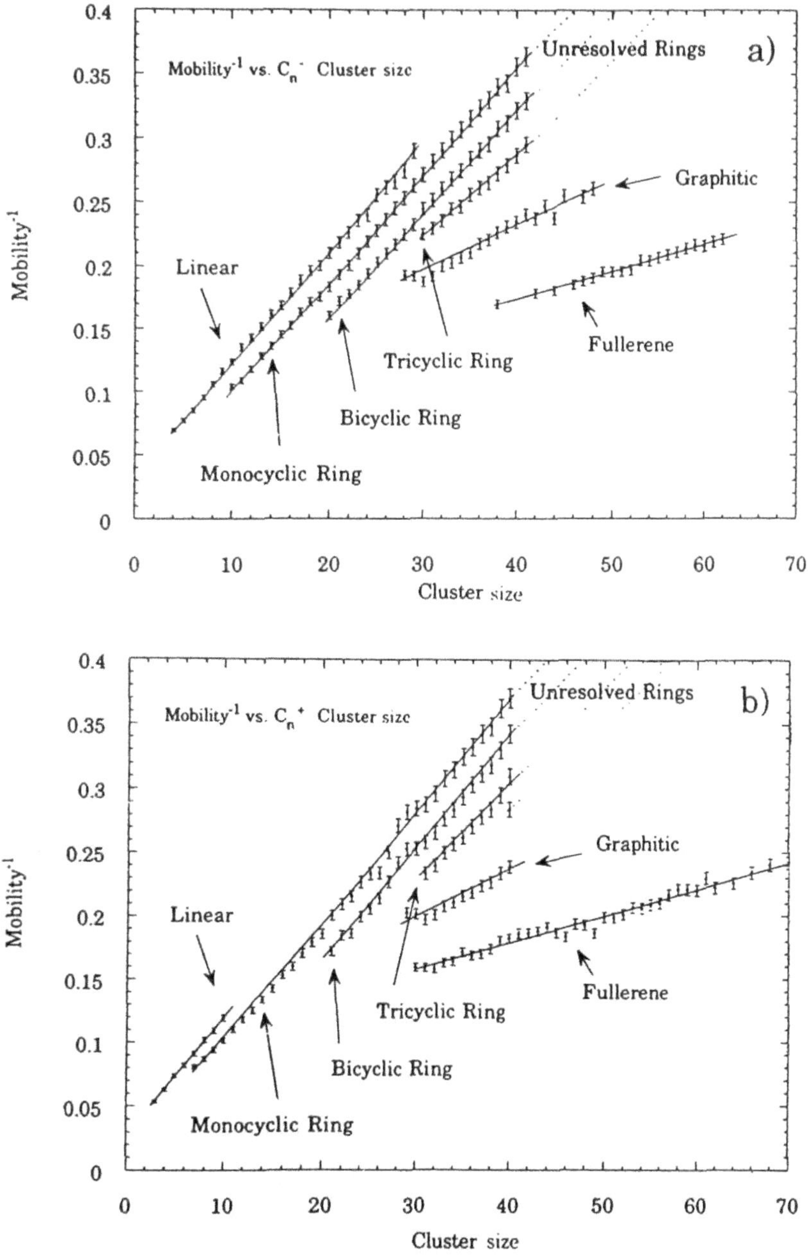

Figure 9: Ion chromatography reported by Gotts *et al.*[137] with indication of various structural families.

true for the geometrical and vibrational findings, too. Clearly enough, a further computational effort is needed in order to get a complete picture of small carbon aggregates.

In overall, we can say that the ongoing computations qualitatively agree with the findings from ion chromatography [134,136,137] though its results somewhat vary with charge. These computations agree with the observations that C_{11} is the first (or, at least first odd-numbered) species for which the cyclic structure becomes dominant. This is the first of two important turning points on the route towards fullerenes. The other, the switch from cyclic to polyhedral species, is still not sufficiently well studied but is thought to be somewhere [144] around $n = 45$.

An instructive example of the present limitations of computational chemistry is supplied by the energy differences between the isomers of C_4. The linear and cyclic (or rhombic — it can also be considered bicyclic) isomers were computed several times. Scattered values of the energy differences between isomers were reported (for a review, see Ref. 28). The cyclic vs. linear difference varied from -16 to 1 kcal/mol. The calculations usually placed the singlet cyclic form at the lower energy. It was suggested [28] to estimate the rhombus energy preference as about 5 kcal/mol though further investigations would be needed to resolve the question.

The unclear computational picture for C_4 indeed prompted another approach to the problem, [145] based on available experimental equilibrium constants for C_4 formation. [6,146] This system is in fact exceptionally well described because two different observations provided the equilibrium constant at two different temperatures (in contrast to higher clusters where the constant is available for one temperature, at most). Using the computed entropy, the energetics of the isomers could be extracted. It was thus concluded [145] that cyclic vs. linear difference was 2.4 kcal/mol (or 2.9 kcal/mol), *i.e.*, the linear form was slightly lower in energy. In fact, the energy term represents the difference in ground-state energies. The quantum-chemical calculations usually supply only the potential energy difference (*i.e.*, without the zero-point vibrational energy). The 2.4 kcal/mol value corresponds [105] to a change of 3.1 kcal/mol in the potential energy. This thermochemical estimate however falls outside the predictions of most quantum-chemical computations and might be considered as just another computational artifact. However, an extended computation by Nygren and Pettersson, [147] with a basis sets including f functions, placed the linear structure 3.6 kcal/mol below the cycle, *i.e.*, the same result as from the thermochemical fit. Density functional theory probably gives a numerical artifact in this case — Johnson [148] reports that the rhombic structure is higher by 20.6 kcal/mol in the B-LYP approximation with the 6-311+G(2df) basis set.

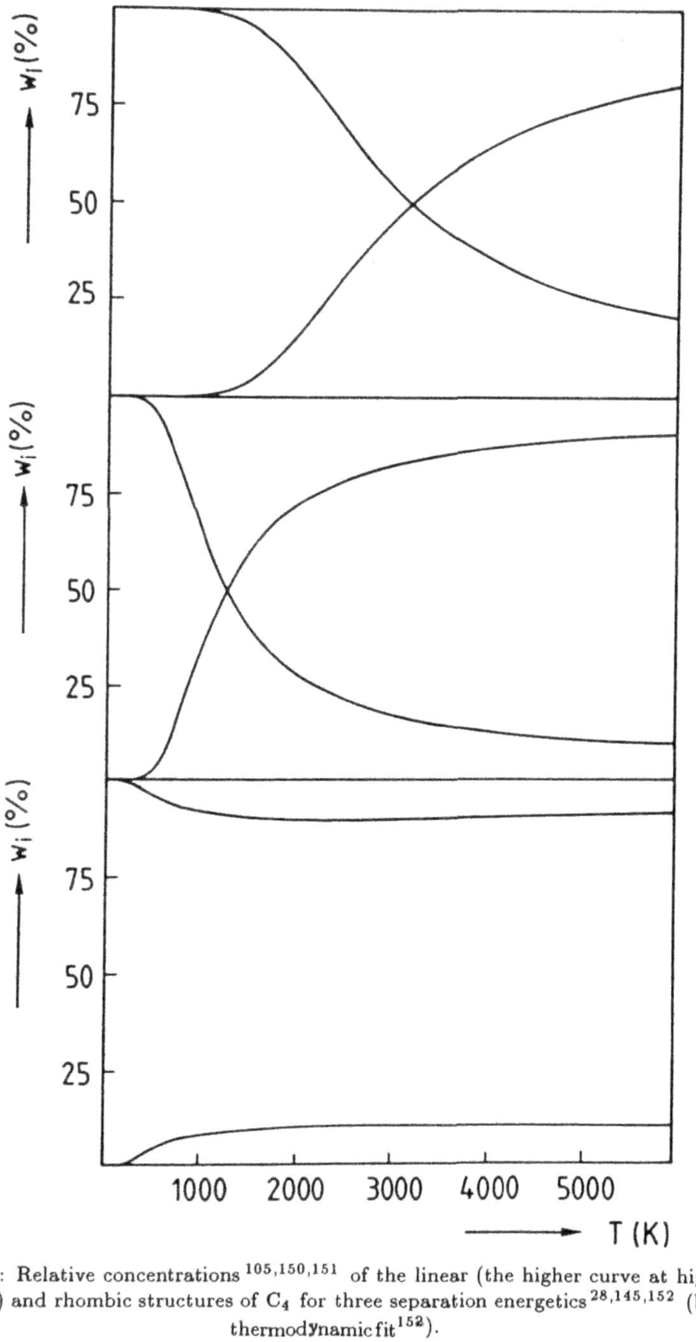

Figure 10: Relative concentrations [105,150,151] of the linear (the higher curve at high temperatures) and rhombic structures of C_4 for three separation energetics [28,145,152] (bottom: thermodynamic fit [152]).

At the G1 and G2 levels of theory, the rhombus form is found to be more stable by 3.4 and 5.1 kcal/mol, respectively, by Raghavachari and Curtiss.[112] They guess that the G2 value is too large since the method tends to underestimate the stability of triplet states. Another sophisticated calculation by Watts *et al.*[149] concluded the cyclic isomer to be more stable by about 1 kcal/mol.

In overall, the energy difference between the C_4 isomers seems to be small, and this observation might lead to a straightforward but wrong conclusion that in a real experimental situation both species would be present in comparable concentrations. However, while energy only sometimes works against the cycle, the role of entropy is always the same.[118,150,151] Entropy calculations can indeed influence the stability interrelationships considerably, especially at higher temperatures (linear structures are typically favored by this contribution). This is documented[105] in Fig. 10 on temperature evolution of the relative concentrations of the C_4 isomers evaluated for three selected values of the separation energy.[28,145,152] If the thermodynamic fit[145] is used, the highest population of the cycle obtainable in the system is reached at around 2700 K, and then it is only 5–10%. However, even if the cycle is the ground state, the linear isomer still dominates at high temperatures. This conclusion agrees with the experimental finding of only one (and not two) peaks in ion chromatography.[134]

We can provide still another example of a small carbon cluster where energy itself, without the contribution of entropy, is not conclusive and can be misleading when predicting relative stabilities of isomeric carbon clusters at very high temperatures. Raghavachari *et al.*[95] found, in their refined evaluation of the energy difference between the cyclic and linear forms of C_6, an energy separation of -11 kcal/mol with the cycle representing the ground state for the system. At non-zero temperatures, the picture is however considerably different as demonstrated[105] in Fig. 11. At a temperature of observation[6] (2500 K) the linear species is, in fact, computed[153,154] to form more than 95% of the equilibrium mixture. The result is not very different for the other separation energetics considered in Fig. 11, and also Martin and Taylor confirmed[113] the finding recently. This thermodynamic reasoning again explains why only one peak was recorded[134] in the ion chromatography for C_6 (though it is pertinently mentioned[134] that isomers of C_5^+ and C_6^+ could probably be resolved).

4 C_{60} and C_{70} Fullerenes

In their historical experiment Kroto *et al.*[24] evaporated a graphite disk into a high-density helium flow by a focused pulsed laser. The resulting carbon

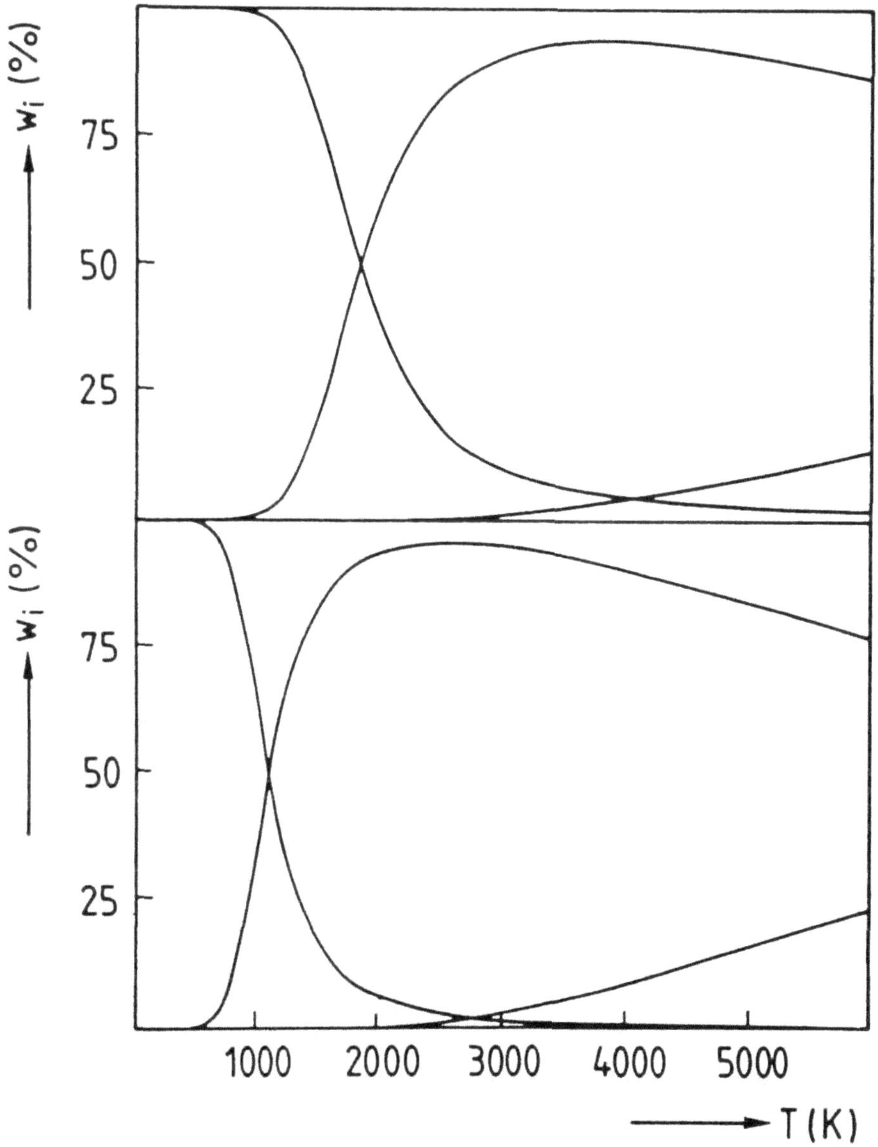

Figure 11: Relative concentrations [105,153,154] of the linear (the highest curve at high temperatures), cyclic (the lowest at high temperatures), and another linear structures of C_6 for two separation energetics. [95]

clusters were expanded in a supersonic molecular beam, photoionized by a laser, and detected by mass spectrometry. Fig. 12 shows their three cluster distributions resulting from variations in the evaporization conditions and thus, temperature-pressure regime. Only a spheroidal structure appeared to satisfy all sp^2 valences of C_{60} and C_{70}. Their choice of the truncated icosahedron for C_{60} and a similar cage for C_{70} was soon supported by computations [155,156] and confirmed by ^{13}C NMR spectra [157] after five years (Fig. 12).

Fullerenes are defined as purely carbonaceous aggregates C_n (n even) built from five- and six-membered rings and three-coordinated carbon atoms. Topological reasoning, based on Euler's formula (in fact, for convex bodies, otherwise the rules can be different), can prove that any such polyhedron always contains exactly 12 five-membered rings while the number of six-membered rings is simply given as $n/2 - 10$ ($n \geq 20$; $\neq 22$). There is an exceptional case of $n = 22$ (no fullerene cage) — it points out that the 12 pentagon rule is a necessary, not a sufficient condition. However, the limitation to n being even is not relevant now as large carbon cages with n odd are known from both experiment [158–160] and theory. [161] Also, the limitation to the five- and six-membered rings can be relaxed in order to allow for other types of rings. Such more general cages have been called [162] *quasi*-fullerenes.

A high symmetry has frequently been associated with fullerenes, and especially the icosahedral I_h symmetry with C_{60}, although species with no symmetry are also possible. Babić *et al.* [163] discussed fullerenes in the original pentagon/hexagon sense till C_{70} and constructed polyhedrons with the following symmetries: C_1, C_i, C_s, C_2, C_3, C_{2v}, C_{3v}, C_{2h}, C_{3h}, D_2, D_3, D_5, D_{2h}, D_{3h}, D_{5h}, D_{6h}, D_{2d}, D_{3d}, D_{5d}, D_{6d}, S_4, S_6, T, T_d, and I_h. Of course, not all fullerenes with these symmetries will necessarily have a corresponding local energy minimum on the C_n hypersurfaces because they are derived purely from geometrical reasoning.

The Jahn-Teller distortion is an important factor which causes symmetry reduction. C_{20} serves as a good example [164] of a species that undergoes such distortion. It is, in fact, the smallest possible fullerene and it contains 12 pentagons and no hexagons. The high symmetry polyhedron built from 12 regular pentagons is the dodecahedron. In the molecule, however, the frontier orbitals are degenerate and partially filled, requiring a symmetry lowering in accord with the Jahn-Teller theorem. At the HF/6-31G* level, for example, the symmetry is lowered to C_2 according to computations of Raghavachari *et al.* [165] C_{60} is the smallest possible fullerene with all pentagons isolated, *i.e.*, each pentagon is surrounded by five hexagons. The structural pattern has been well recognized since the very first days and later formalized in the so called isolated-pentagon rule [166] (IPR). The limited computational experience

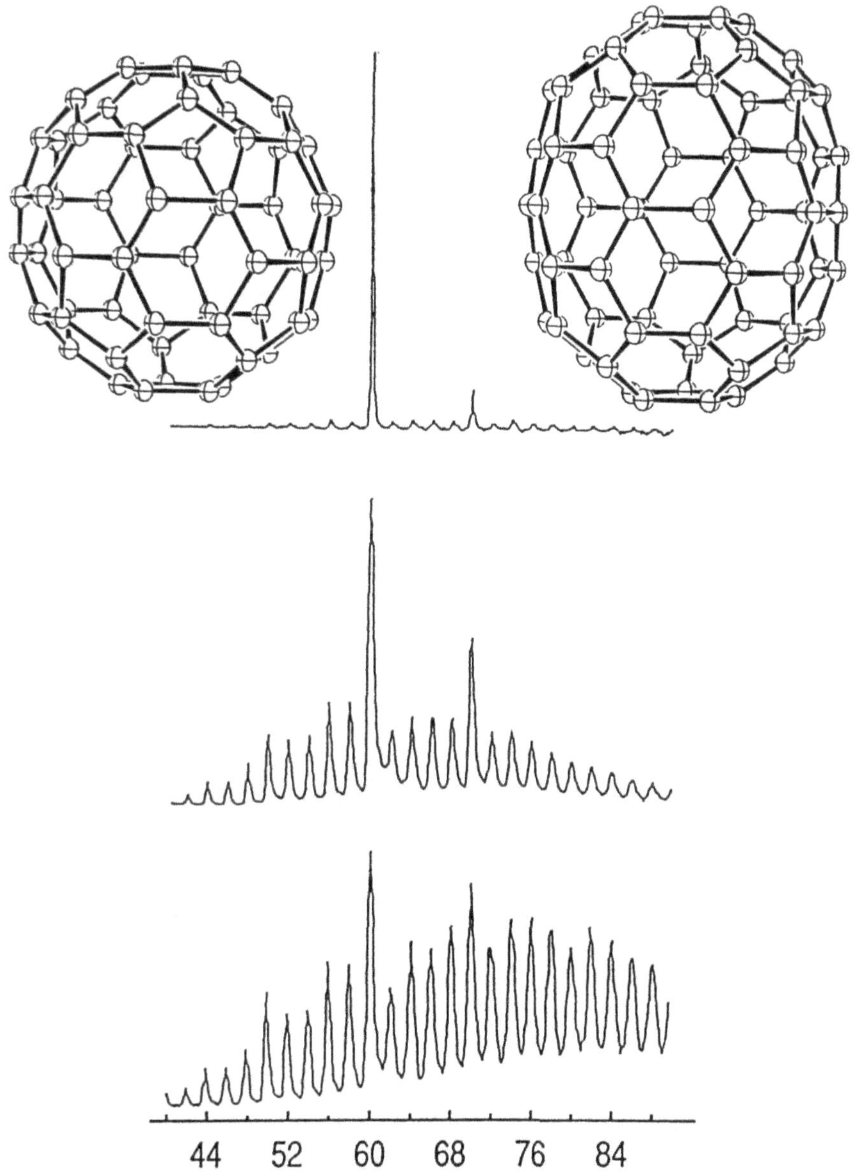

Figure 12: Cluster distributions resulting from variations of the graphite evaporization conditions in the historical experiment by Kroto *et al.*,[24] and the AM1 computed structures[155,156] of C_{60} (left, otherwise the dominant signal in each spectrum) and C_{70} (the second highest peak).

available [167–171] supports the belief that the IPR structures are more stable than structures with connected pentagons.

Stability of fullerenes represents an interesting physico-chemical problem as it does not come from their thermodynamic properties. C_{60} is particularly stable because it is formed with higher speed and/or is better protected kinetically against decomposition than any other fullerene under the special synthetic conditions. However, even C_{60} is much higher in energy or Gibbs energy than graphite or diamond. Once fullerenes are created, they condense on the cooler walls owing to their high sublimation temperatures. Soot covers them and effectively removes from the reaction zone.

4.1 Fullerenes in Nature

The coincidence between two diffuse interstellar bands with the spectral features of C_{60}^+ in a Ne matrix concluded [81] by Foing and Ehrenfreund (Fig. 4) is a rather indirect evidence of the fullerene presence in nature. They still estimate [172] that up to 0.9% of interstellar carbon could be in the form of C_{60}^+. This estimate indicates that C_{60}^+ may play an important role in interstellar chemistry. There are however geological and meteoritic materials containing small amounts of fullerenes. In fact, already back in 1986 Heymann suggested [173,174] that so called trapped inert-gas atoms in carbonaceous chondrite meteorites were actually locked up in fullerene cages.

Buseck *et al.* [175] studied shungite, a usual carbonaceous rock found near the town of Shunga in Karelia, Russia, classified as a meta-anthracite coal. Shungite is thought to have formed from carbonaceous material under extreme pressures. They found fullerenes while examining high-resolution transmission electron microscopy images and subsequently confirmed the presence of C_{60} and C_{70} by mass spectrometry. The mass spectra were measured under conditions sufficient to resolve the $^{13}C/^{12}C$ isotopic ratios and they corresponded to the normal range of isotopic values. The occurrence of fullerenes in shungite was somewhat surprising. Either the original carbonaceous material already contained fullerenes, or they must have formed during its metamorphism under high pressure by an unknown mechanism.

Lightning produces a variety of unusual conditions and the ground area where the lightning strikes tends to melt. The geological products of lightning are called fulgurites, they typically consist of glass. Daly *et al.* [176] analyzed a sample extracted from a fulgurite by mass spectrometry and found peaks from C_{60}, C_{70}, and other fullerenes. Interestingly enough, also peaks corresponding to C_{46}^+ to C_{58}^+ were observed. Although these peaks are possible fragmentation products during the laser desorption process, the fragmentation could also have occurred from secondary lightning return strokes. Incidentally, Daly *et al.*[176] argue rather against an lightning-related origin of fullerenes in shungite.

Becker *et al.*[177] studied deposits from an impact crater in Canada, the Sudbury crater, the second largest proven crater on Earth. The Sudbury crater is an elliptical basin 60 km long and 27 km wide. It is supposed that the Sudbury crater was produced by a large meteorite or comet impact about 1850 million years ago. The impact deposits with a relatively large amount of carbon (0.5 to 1.0%) were analyzed by mass spectrometry and C_{60} and C_{70} were identified. The C_{60} content is estimated at a few parts per million. The oxidation of the fullerenes during the 1850 million years period could be prevented by the presence of sulfur in the surrounding sulfide-silicate matrix. Becker *et al.*[178] searched for noble gases trapped inside the fullerene cage. They concluded that fullerenes from the Sudbury impact structure contained trapped helium with a $^3He/^4He$ ratio of 5.5 to $5.9 \cdot 10^{-4}$. The $^3He/^4He$ ratios in the Sudbury fullerenes are similar to those found in meteorites and in some interplanetary dust particles. This observation suggests that the helium within the C_{60} molecules at Sudbury is of extraterrestrial origin.

Carbon rich samples from two sites in New Zealand were analyzed by Heymann *et al.*[179] They used samples from the soot rich seams at the Cretaceous-Tertiary (K-T) boundary. The abundant soot is thought to have been derived from worldwide wildfires after a large meteor impact in the Yucatán Peninsula of Mexico about 65 million years ago. High-pressure liquid chromatography with UV spectral detection of toluene extracts revealed the presence of C_{60} at concentrations of 0.1 to 0.2 parts per million. C_{70} was found only at one site while it was virtually absent in the samples from the other site. In overall, the results [175–182] indicate that fullerenes have long been part of our natural environment. On the other hand, Heymann[183] could not detect fullerenes in available lunar samples.

Di Brozolo *et al.*[184] reported the observation of fullerenes in a chondritic carbonaceous residue of a tiny (0.14 mm diameter) impact crater on the Long Duration Exposure Facility (LDEF) spacecraft. The tiny impact crater was found in the aluminium panels of the collector tray. Mass spectrometry and Raman spectroscopy indicate the presence of fullerenes in the crater and in adjacent ejecta. Fig. 13 shows the spectra obtained[184] by the laser ionization time-of-flight mass spectrometry from the rim of the LDEF crater, from synthetic fullerene, and from a point remote from the rim of the LDEF crater. The origin of the fullerenes in the crater is unclear. Experiments with high-speed impact of carbonaceous materials on aluminium did not show a fullerene formation. However, tests with a still higher speed are needed. On the other hand, fullerenes can survive such high-speed impacts into aluminium targets,

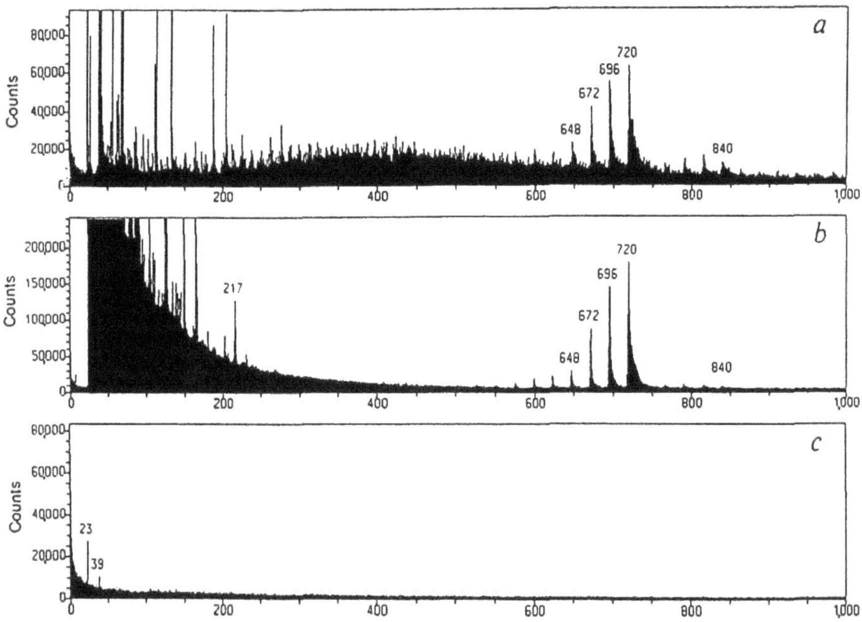

Figure 13: Mass spectra obtained by Di Brozolo *et al.*[184] in their study of the impact crater on the LDEF spacecraft: *a* — the rim of the crater, *b* — synthetic fullerene, *c* — a point remote from the rim.

suggesting that space fullerenes contained in a carbonaceous micrometeorite could survive the LDEF impact.

Fullerenes have for long time been searched [174] in meteorites. Finally, Becker *et al.* [185] reported the fullerene detection in Allende meteorite though in trace amounts. The Allende meteorite has been known to contain several well-ordered graphite particles. Mass spectrometry estimated the C_{60} content for the Allende meteorite as 0.1 parts per million. Failure to detect fullerenes in previous studies of meteorites could be because of their inhomogeneous distribution. Recently, Heymann [174] analyzed samples from the Allende meteorite for the presence of extractable fullerenes by liquid chromatography. The analysis could detect C_{60} only in traces less than 0.001 parts per million. This finding could however be linked to heterogeneity of the Allende meteorite with respect to fullerenes. Becker *et al.* also discussed [186] a possible presence of nested shells (carbon onions) in meteorites.

4.2 Seminatural Occurrence of Fullerenes

Soot and other heat-treated carbon materials represent potential candidates for a fullerene presence even if they are produced by industrial activities rather than by nature. In a large-scale screening study [187–189] fullerenes have been detected in Chinese ink sticks. Chinese ink sticks have been manufactured from soots prepared by slow burning of pine wood or various seed oils. The studied samples were pulverized, extracted with boiling toluene and analyzed by liquid chromatography.

It was discovered that the observed fullerene contents decreased with the estimated age of the analyzed Chinese ink sticks. In fact, the decrease follows the first order kinetics (Fig. 14). Several factors can be considered like growth of soot particles, migration of fullerenes from the surface, or their oxidative decomposition. The migration of fullerenes towards the surface allows for their sublimation or oxidation. Fullerenes can also be degraded by light. [190] Interestingly enough, C_{70} is decomposed [191] more efficiently than C_{60} at some conditions. Heymann and Chibante [192–195] pointed out ozone as a potent agent for fullerene destruction. The destruction of C_{60} in toluene proceeds at least five orders of magnitude faster with O_3 than it does with O_2. The crystalline samples exposed to O_3 at room temperature lost fullerene relatively slowly — 80% after seven days. Moreover, activated charcoal is oxidized by O_3 and also acts as a catalyst for the decomposition of ozone. Therefore Chibante and Heymann [194] suggest that the fullerenes in shungite may be protected against ozone by the carbonaceous material in which they are contained. Incidentally, ozone levels can be increased considerably even in the troposphere during a

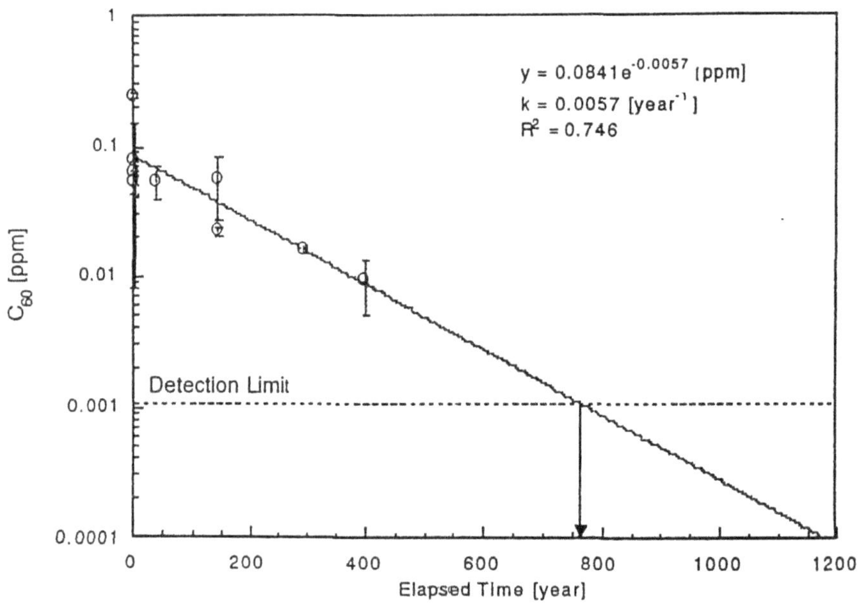

Figure 14: Kinetic analysis [189] of the C_{60} concentration in Chinese ink sticks.

smog-formation situation.

5 Higher Fullerenes

The fullerene family[196] is not limited to its two oldest members, C_{60} and C_{70}. At present over twenty stable fullerenes C_n have been identified [197–199] with n varying from 60 to 94. In principle, any of the higher fullerenes can contribute to the diffuse interstellar bands if present in interstellar space. Let us therefore review briefly the higher fullerenes. Their studies are considerably simplified by the IPR concept [200] and topological procedures. [201–204] The lowest IPR stoichiometry which allows for a structural multiplicity[205] is $n = 76$. Beyond $n = 76$ isomerism of the IPR structures should generally play a significant role in observations. In fact, several such mixtures of fullerene isomers have been computed and an agreement with experiments found: C_{78} (*e.g.* [206–209]), C_{80} (*e.g.*, [210–213] though experiment [213] has discovered only one C_{80} species so far), C_{82} (*e.g.*, [214–219] at present at least two C_2 isomers were confirmed experimentally[197]), C_{84} (*e.g.* [220–224]), C_{86} (*e.g.* [225–228]), C_{88} (*e.g.* [226–228]), and C_{90} (*e.g.*[229,230]).

Combined quantum-chemical and statistical-mechanical computations have clearly shown that temperature effects are critically important in understanding higher fullerenes. Although the inter-isomeric separation energies (or enthalpies) are important as a starting point, they alone cannot predict the relative stabilities of the IPR isomers at elevated temperatures. As the temperatures reached in fullerene syntheses are high, entropy contributions can even over-compensate the enthalpy terms. This energy-entropy interplay and other important aspects of computational chemistry of higher fullerenes are reviewed in this chapter.

Let us illustrate the enthalpy-entropy interplay on a relatively simple system, C_{50}. Three minimum-energy structures were found [231,232] in MNDO study, each of them built from pentagonal and hexagonal rings only (Fig. 15). The polyhedra resulting from the MNDO geometry optimization exhibit the D_{5h}, D_3, and C_{2v} point group of symmetry. The D_{5h} structure is lowest in energy, being followed after some 30 and 1600 kJ/mol by the D_3 and C_{2v} isomers, respectively. Fig. 16 presents [233] temperature evolution of the relative equilibrium concentrations in a wide temperature interval. Effectively, the system behaves like two-component mixture — populations of the C_{2v} species are negligible throughout. The simple Boltzmann factors are also shown in Fig. 16 — they do not represent a good approximation of the exact concentration terms. The most interesting feature of the relative-concentration temperature evolution is the relative-stability crossing between the D_{5h} and D_3 structures

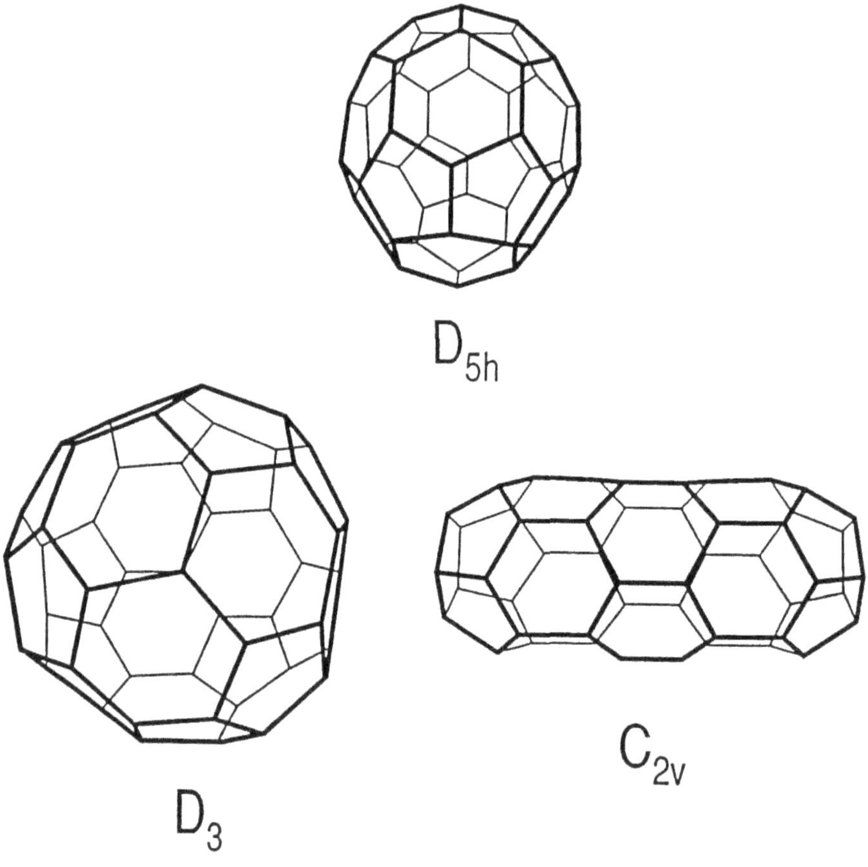

Figure 15: Schemes of three C_{50} isomers with D_{5h}, D_3, and C_{2v} symmetries MNDO computed by Bakowies et al.[231,232]

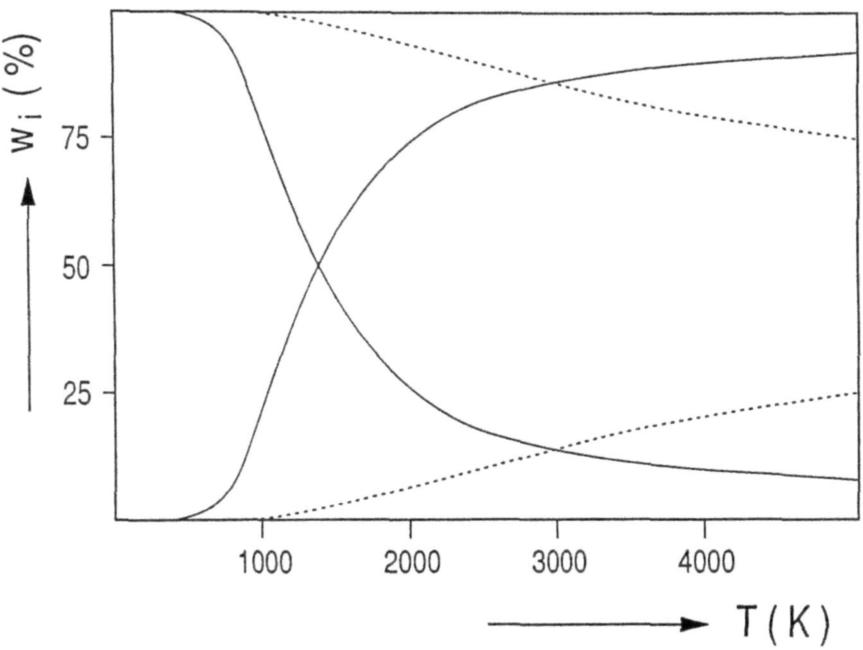

Figure 16: Temperature dependences[233] of the relative concentrations w_i and simple Boltzmann factors (dashed curves) for the D_{5h} (the decreasing curves) and D_3 C_{50} isomers from Fig. 15 (C_{2v} cannot be seen).

at about 1390 K. Hence, from the thermodynamic-equilibrium point of view the D_3 species should be preferentially formed in high temperature region, *i.e.*, under the fullerene-synthesis conditions.

Although there are no observed data on C_{50} available yet, similar computations on the isomeric fullerene systems C_{76}, C_{78}, C_{80}, C_{82}, and C_{84} have demonstrated that the inter-isomeric thermodynamic equilibrium presumption is actually well working. Hence, still higher systems could be submitted for computations. C_{86} is the next system in the row. [225–228] There are nineteen topologically different C_{86} structures (Fig. 17) which obey the IPR requirement. They were computed primarily with the newest tool of semiempirical quantum chemistry, the SAM1 method. [234] Nevertheless, also the AM1, HF/STO-3G, and STO/3-21G treatments consistently produce a C_2 species (numbered **17** in Fig. 17) as the ground state of the IPR C_{86} set.

Fig. 18 surveys the temperature development of the relative concentrations or mole fractions in the nineteen-membered isomeric set. At very low temperatures the ground-state structure is of course prevailing, however, its mole fraction decreases relatively fast. At first, a C_s species reaches close to 20% (it has its maximum at a temperature of 1503 K). There is a third significant species (C_3 symmetry). Although it is higher in energy, its entropy contribution allows for equimolarity (about 29% each) with the ground state at a temperature of 1739 K. Let us mention however that at this temperature two additional species exhibit relative concentrations of about 10%. Achiba *et al.* [197] concluded from ^{13}C NMR spectra presence of two C_2 isomers in their C_{86} sample (the abundance of the isomers was about 4:1). Our computations for the C_2 and C_s structures around 1500 K would agree with the observation; at higher temperatures we would predict coexistence of more than two isomers. It is however true that the computed symmetry is C_s, not C_2. This minor difference between theory and experiment has not been clarified yet.

6 Modified Fullerenes

In addition to pristine fullerenes, their various chemical modifications should be considered as possible carriers of the diffuse interstellar bands, too. Kroto and Jura [68] suggested complexed exohedral and endohedral analogues of fullerenes as another class of species relevant to astrophysical spectra. If graphite rods are packed with a dopant and then used in the fullerene synthesis, various endohedral fullerenes can be produced: $Li@C_{60}$, $Rb@C_{60}$, $K@C_{60}$, $Na@C_{60}$, $N@C_{60}$, $La@C_{60}$, $La@C_{70}$, $La@C_{74}$, $La_2@C_{80}$, $Sc_3@C_{82}$, etc. The encapsulated atoms are not necessarily located in the center and therefore the spectral symmetry selection rules can be broken. Moreover, intrinsic fullerene transi-

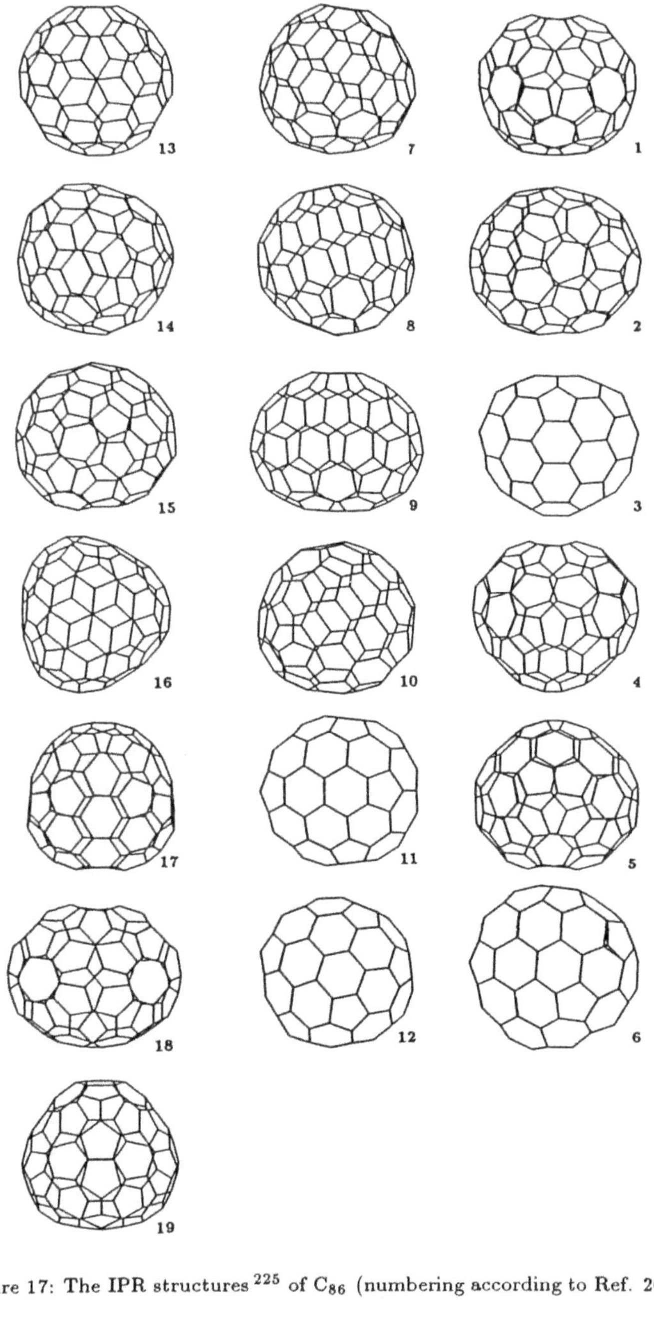

Figure 17: The IPR structures[225] of C_{86} (numbering according to Ref. 204).

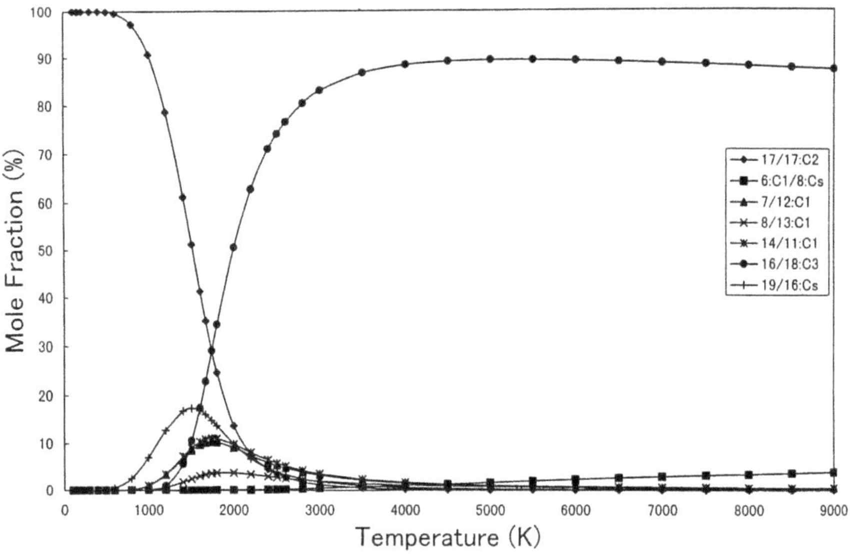

Figure 18: The SAM1 computed[225] relative concentrations of C_{86} isomers from Fig. 17 (numbering according to Refs. 202 and 204).

tions are perturbed by the complexed atom and intrinsic atomic transitions are perturbed by the fullerene cage. Various such fullerene complexes may exist in significant amounts in interstellar space and may be responsible for some unidentified interstellar features.

Webster[235] calculated vibrational spectrum of the fully saturated hydrocarbon $C_{60}H_{60}$. The computed infrared active frequencies agreed well with some unidentified astronomical infrared emission features. Recently, the computational treatment was extended[236] to the family $C_{60}H_m$. The lowest strongly infrared-active wavelength is found to be a strong function of m, increasing from about 19 μm for $m = 0$ to 23 μm for $m = 60$. It follows that the radiation from such mixed hydrogenated fullerenes is likely to consist of a broad band. This finding agrees with an emission band centered around 21 μm and known only in carbon-rich objects.

Multishell fullerenes or carbon onions have been studied by Henrard *et al.*[78,79] as possible carriers of some interstellar ultraviolet features. The theoretical treatment used allows for simulation of various modifications produced by a material filling the inner cavity or by a coating layer. It follows that coated carbon-onion particles have optical properties consistent with the UV interstellar-dust absorption data.

7 Concluding Remarks

As this survey shows there have been considerably rich and productive implications of the diffuse interstellar bands problem. During the years between the first and second fullerene milestone, when C_{60} could not be produced in larger amounts, its supposed astrophysical role had been a major driving force for the research activities towards its synthesis. Finally, the neutral C_{60} did not fit and C_{60}^+ could explain[81] only two diffuse interstellar bands. Anyhow, the research avalanche of fullerene science has already produced more than 10000 papers. One of the fathers of the fullerene synthesis, D. R. Huffman, wrote[35] in the year 1991: "*In presenting our story to audiences ranging from public school classes to scientists at symposiums on C_{60}, I have often claimed one of its morals to be, 'Everything has not yet been discovered.' The mystery of the diffuse interstellar bands continues to proclaim this fact.*"

Acknowledgments

The reported research has in part been supported by the Ministry of Education, Science and Culture in Japan. The authors also wish to thank the following organizations for kindly permitting the reprinting of copyrighted material: Academia; Akadémiai Kiadó; Butterworths; Elsevier Scientific Publishing Company; Kluwer Academic Publishers; Marcel Dekker, Inc.; Macmillan Magazines Ltd.; Springer-Verlag; The Electrochemical Society, Inc.; Verlag der Zeitschrift für Naturforschung.

References

1. O. Hahn, F. Strassman, J. Mattauch, and H. Ewald, *Naturwiss.* **30**, 541 (1942).
2. J. Mattauch, H. Ewald, O. Hahn, and F. Strassmann, *Z. Phys.* **20**, 598 (1943).
3. K. Heinzinger and Z. Slanina, *MPG-Spiegel* No. 2, 11 (1993).
4. P. Swings, C. T. Elvey, and H. W. Babcock, *J. Astrophys.* **94**, 320 (1941).
5. R. E. Honig, *J. Chem. Phys.* **22**, 126 (1954).
6. J. Drowart, R. P. Burns, G. DeMaria, and M. G. Inghram, *J. Chem. Phys.* **31**, 1131 (1959).
7. H. von Hintenberger, J. Franzen, and K. D. Schuy, *Z. Naturforsch.* A **18**, 1236 (1963).
8. K. S. Pitzer and E. Clementi, *J. Am. Chem. Soc.*, **81**, 4477 (1959). Large Molecules in Carbon Vapor.
9. R. Hoffmann, *Tetrahedron*, **22**, 521 (1966). Extended Hückel Theory V. Cumulunes, Polyenes, Polyacetylenes, and C_n.
10. Z. Slanina, *Radiochem. Radioanal. Lett.* **22**, 291 (1975).
11. Z. Slanina, *Ph.D. Thesis* (Czech. Acad. Sci., Prague, 1974, p. 105).
12. Z. Slanina, *Contemporary Theory of Chemical Isomerism* (Kluwer, Dordrecht, 1986).
13. D. E. H. Jones, *New Scientist* **32**, 245 (1966).
14. E. Ōsawa, *Kagaku*, **25**, 854 (1970); *Chem. Abstr.*, **74**, 75698v (1971).
15. Z. Yoshida and E. Ōsawa, *Aromaticity* (Kagakudojin, Kyoto, 1971), p. 174.
16. D. A. Bochvar and E. G. Gal'pern, *Dokl. Akad. Nauk SSSR* **209**, 610 (1973).
17. I. V. Stankevich, M. V. Nikerov, and D. A. Bochvar, *Russ. Chem. Rev.* **53**, 640 (1984).
18. R. A. Davidson, *Theor. Chim. Acta* **58**, 193 (1981).
19. H. W. Kroto, *Int. Rev. Phys. Chem.* **1**, 309 (1981).
20. H. W. Kroto, *Angew. Chem., Int. Ed. Engl.* **31**, 111 (1992).
21. L. W. Avery, L. W. Broten, J. M. MacLeod, T. Oka, and H. W. Kroto, *Astrophys. J.* **205**, L173 (1976).

22. A. E. Douglas, *Nature* **269**, 130 (1977).
23. E. Herbig, *Astrophys. J.* **196**, 129 (1975).
24. H. W. Kroto, J. R. Heath, S. C. O'Brien, R. F. Curl, and R. E. Smalley, *Nature* **318**, 162 (1985).
25. D. R. Huffman and W. Krätschmer, *Materials Res. Soc. Proc.* **206**, 601 (1991).
26. W. Krätschmer, *Z. Phys.* D **19**, 405 (1991).
27. W. Krätschmer, L. D. Lamb, K. Fostiropoulos, and D. R. Huffman, *Nature*, **347**, 354 (1990).
28. W. Weltner, Jr. and R. J. van Zee, *Chem. Rev.* **89**, 1713 (1989).
29. R. F. Curl and R. E. Smalley, *Science* **242**, 1017 (1988).
30. H. W. Kroto, *Science* **242**, 1139 (1988).
31. R. C. Haddon, *Accounts Chem. Res.* **21**, 243 (1988).
32. D. A. Jelski and T. F. George, *J. Chem. Educ.* **65**, 879 (1988).
33. D. J. Klein and T. G. Schmalz in *Quasicrystals, Networks, and Molecules of Fivefold Symmetry*, ed. I. Hargittai (VCH Publishers, New York, 1990).
34. E. Ōsawa, *Kagaku* **45**, 552 (1990).
35. D. R. Huffman, *Phys. Today* **44**, 22 (1991).
36. H. W. Kroto, A. W. Allaf, and S. P. Balm, *Chem. Rev.* **91**, 1213 (1991).
37. J. R. Heath in *ACS Symposium Series No 481 - Fullerenes. Synthesis, Properties, and Chemistry of Large Carbon Clusters*, eds. G. S. Hammond and V. J. Kuck (ACS, Washington, 1992).
38. J. E. Fischer, P. A. Heiney, D. E. Luzzi, and D. E. Cox in *ACS Symposium Series No 481 Fullerenes. Synthesis, Properties, and Chemistry of Large Carbon Clusters*, eds. G. S. Hammond and V. J. Kuck (ACS, Washington, 1992).
39. Z. Slanina, *Chem. Listy* **86**, 327 (1992).
40. R. E. Smalley, *Accounts Chem. Res.* **25**, 98 (1992).
41. J. P. Hare and H. W. Kroto, *Accounts Chem. Res.* **25**, 106 (1992).
42. J. E. Fischer, P. A. Heiney, and A. B. Smith III, *Accounts Chem. Res.* **25**, 112 (1992).
43. F. Diederich and R. L. Whetten, *Accounts Chem. Res.* **25**, 119 (1992).
44. R. C. Haddon, *Accounts Chem. Res.* **25**, 127 (1992).
45. P. J. Fagan, J. C. Calabrese, and B. Malone, *Accounts Chem. Res.* **25**, 134 (1992).
46. J. H. Weaver, *Accounts Chem. Res.* **25**, 143 (1992).
47. J. M. Hawkins, *Accounts Chem. Res.* **25**, 150 (1992).
48. F. Wudl, *Accounts Chem. Res.* **25**, 157 (1992).

49. S. W. McElvany, M. M. Ross, and J. H. Callahan, *Accounts Chem. Res.* **25**, 162 (1992).
50. R. D. Johnson, D. S. Bethune, and C. S. Yannoni, *Accounts Chem. Res.* **25**, 169 (1992).
51. T. Braun, *Angew. Chem., Int. Ed. Eng.*, **31**, 588 (1992).
52. D. Koruga, S. Hameroff, J. Withers, R. Loutfy, and M. Sundareshan, *Fullerene C_{60} — History, Physics, Nanobiology, Nanotechnology* (Elsevier, Amsterdam, 1993).
53. T. G. Schmalz and D. J. Klein in *Buckminsterfullerenes*, eds. W. E. Billups and M. A. Ciufolini (VCH Publishers, New York, 1993).
54. G. E. Scuseria in *Buckminsterfullerenes*, eds. W. E. Billups and M. A. Ciufolini (VCH Publishers, New York, 1993).
55. C. T. White, J. W. Mintmire, R. C. Mowrey, D. W. Brenner, D. H. Robertson, J. A. Harrison, and B. I. Dunlap in *Buckminsterfullerenes*, eds. W. E. Billups and M. A. Ciufolini (VCH Publishers, New York, 1993).
56. R. C. Haddon and K. Raghavachari in *Buckminsterfullerenes*, eds. W. E. Billups and M. A. Ciufolini (VCH Publishers, New York, 1993).
57. M. L. Cohen and V. H. Crespi in *Buckminsterfullerenes*, eds. W. E. Billups and M. A. Ciufolini (VCH Publishers, New York, 1993).
58. S. C. Erwin in *Buckminsterfullerenes*, eds. W. E. Billups and M. A. Ciufolini (VCH Publishers, New York, 1993).
59. M. S. Dresselhaus, G. Dresselhaus, and P. C. Eklund, *J. Mater. Res.* **8**, 2054 (1993).
60. J. Cioslowski, *Rev. Comput. Chem.* **4**, 1 (1993).
61. C. Z. Wang, B. L. Zhang, K. M. Ho, and X. Q. Wang, *Int. J. Mod. Phys.* B **7**, 4305 (1993).
62. W. Andreoni in *Electronic Properties of Fullerenes*, eds. H. Kuzmany, J. Fink, M. Mehring, and S. Roth (Springer-Verlag, Berlin, 1993).
63. J. González, F. F. Guinea, and M. A. H. Vozmediano, *Int. J. Mod. Phys.* B **7**, 4331 (1993).
64. J. Cioslowski, *Electronic Structure Calculations on Fullerenes and Their Derivatives* (Oxford University Press, Oxford, 1995).
65. Z. Slanina, S.-L. Lee, and C.-H. Yu, *Rev. Comput. Chem.* **8**, 1 (1996).
66. T. P. Snow, C. G. Seab, and C. L. Joseph, *Astrophys. J.* **335**, 185 (1988).
67. T. P. Snow and C. G. Seab, *Astron. Astrophys.* **213**, 291 (1989).
68. H. W. Kroto and M. Jura, *Astron. Astrophys.* **263**, 275 (1992).
69. M. Jura, *Phil. Trans. R. Soc. Lond.* A **343**, 63 (1993).
70. C. T. Pillinger, *Phil. Trans. R. Soc. Lond.* A **343**, 73 (1993).
71. T. P. Snow, *Nature* **366**, 407 (1993).

72. M. Kolbuszewski, *Astrophys. J.* **432**, L63 (1994).
73. T. P. Snow, E. L. O. Bakes, R. H. Buss, Jr., and C. G. Seab, *Astron. Astrophys.* **296**, L37 (1995).
74. R. P. A. Bettens and E. Herbst, *Int. J. Mass Spectr. Ion Process.* **149/150**, 321 (1995).
75. T. P. Snow and A. N. Witt, *Science* **270**, 1455 (1995).
76. E. Herbst, *Annu. Rev. Phys. Chem.* **46**, 27 (1995).
77. R. P. A. Bettens and E. Herbst, *Astrophys. J.* **468**, 686 (1996).
78. L. Henrard, P. Senet, Ph. Lambin, and A. A. Lucas, *Fullerene Sci. Technol.* **4**, 131 (1996).
79. L. Henrard, Ph. Lambin, and A. A. Lucas, *to be published.*
80. S. J. Clementt, C. R. Maechling, R. N. Zare, P. D. Swan, and R. M. Walker, *Science* **262**, 721 (1993).
81. B. H. Foing and P. Ehrenfreund, *Nature* **369**, 296 (1994).
82. J. P. Maier, *Nature* **370**, 423 (1994).
83. J. Fulara, D. Lessen, P. Freivogel, and J. P. Maier, *Nature* **366**, 439 (1993).
84. P. Freivogel, J. Fulara, and J. P. Maier, *Astrophys. J.* **431**, L151 (1994).
85. P. Freivogel, J. Fulara, D. Lessen, D. Forney, and J. P. Maier, *Chem. Phys.* **189**, 335 (1994).
86. J. P. Maier, P. Freivogel, J. Fulara, and D. Lessen in *The Diffuse Interstellar Bands*, eds. A. G. G. M. Tielens and T. P. Snow (Kluwer Academic Publishers, Dordrecht, 1995).
87. D. Forney, P. Freivogel, J. Fulara, and J. P. Maier, *J. Chem. Phys.* **102**, 1510 (1995).
88. D. Forney, J. Fulara, P. Freivogel, M. Jakobi, D. Lessen, and J. P. Maier, *J. Chem. Phys.* **103**, 48 (1995).
89. P. Freivogel, J. Fulara, M. Jakobi, D. Forney, and J. P. Maier, *J. Chem. Phys.* **103**, 54 (1995).
90. J. Fulara, P. Freivogel, D. Forney, and J. P. Maier, *J. Chem. Phys.* **103**, 8805 (1995).
91. M. Schäfer, M. Grutter, J. Fulara, D. Forney, P. Freivogel, and J. P. Maier, *Chem. Phys. Lett.* **206**, 406 (1996).
92. M. Doverstål, B. Lindgren, U. Sassenberg, and H. Yu, *Z. Phys.* D **19**, 447 (1991).
93. Z. Slanina, S.-L. Lee, M. Smigel, J. Kurtz, and L. Adamowicz in *Recent Advances in the Chemistry and Physics of Fullerenes and Related Materials, Vol. 1*, eds. K. M. Kadish and R. S. Ruoff (The Electrochemical Society, Pennington, 1994).

94. Z. Slanina, S.-L. Lee, J. Kurtz, and L. Adamowicz in *Recent Advances in the Chemistry and Physics of Fullerenes and Related Materials, Vol. 2*, eds. K. M. Kadish and R. S. Ruoff (The Electrochemical Society, Pennington, 1995).
95. K. Raghavachari, R. A. Whiteside, and J. A. Pople, *J. Chem. Phys.* **85**, 6623 (1986).
96. K. Raghavachari and J. S. Binkley, *J. Chem. Phys.* **87**, 2191 (1987).
97. J. M. L. Martin, J.-P. François, and R. Gijbels, *J. Chem. Phys.* **90**, 3403 (1989).
98. V. Parasuk and J. Almlöf, *J. Chem. Phys.* **91**, 1137 (1989).
99. C. Liang and H. F. Schaefer III, *J. Chem. Phys.* **93**, 8844 (1990).
100. J. M. L. Martin, J.-P. François, and R. Gijbels, *J. Chem. Phys.* **93**, 8850 (1990).
101. J. Kurtz and L. Adamowicz, *Astrophys. J.* **370**, 784 (1991).
102. J. M. L. Martin, J.-P. François, and R. Gijbels, *J. Chem. Phys.* **94**, 3753 (1991).
103. J. M. L. Martin, J.-P. François, and R. Gijbels, *J. Comput. Chem.* **12**, 52 (1991).
104. D. W. Ewing, *Z. Phys.* D **19**, 419 (1991).
105. Z. Slanina, J. M. Rudziński, and E. Ōsawa, *Z. Phys.* D **19**, 431 (1991).
106. J. M. L. Martin, J.-P. François, R. Gijbels, and J. Almlöf, *Chem. Phys. Lett.* **187**, 367 (1991).
107. J. D. Watts and R. J. Bartlett, *Chem. Phys. Lett.* **190**, 19 (1992).
108. V. Parasuk and J. Almlöf, *Theor. Chem. Acta* **83**, 227 (1992).
109. Z. Slanina, J. Kurtz, and L. Adamowicz, *Chem. Phys. Lett.* **196**, 208 (1992).
110. Z. Slanina, J. Kurtz, and L. Adamowicz, *Mol. Phys.* **76**, 387 (1992).
111. L. A. Curtiss and K. Raghavachari in *Quantum Mechanical Electronic Structure Calculations with Chemical Accuracy*, ed. S. R. Langhoff (Kluwer, Dordrecht, 1995).
112. K. Raghavachari and L. A. Curtiss in *Quantum Mechanical Electronic Structure Calculations with Chemical Accuracy*, ed. S. R. Langhoff (Kluwer, Dordrecht, 1995).
113. J. M. L. Martin and P. R. Taylor, *J. Phys. Chem.* **100**, 6047 (1996).
114. W. Krätschmer, N. Sorg, and D. R. Huffman, *Surf. Sci.* **156**, 814 (1985).
115. S. W. McElvany, W. R. Creasy, and A. O'Keefe, *J. Chem. Phys.* **85**, 632 (1986).
116. S. W. McElvany, B. I. Dunlap, and A. O'Keefe, *J. Chem. Phys.* **86**, 715 (1987).

117. W. Krätschmer and K. Nachtigall in *Polycyclic Aromatic Hydrocarbons and Astrophysics*, eds. A. Léger, L. d'Hendecourt and N. Boccara (Reidel, Dordrecht, 1987).
118. R. J. van Zee, R. F. Ferrante, K. J. Zerinque, W. Weltner, Jr., and D. W. Ewing, *J. Chem. Phys.* **88**, 3465 (1988).
119. S. W. McElvany, *J. Chem. Phys.* **89**, 2063 (1988).
120. J. R. Heath, A. L. Cooksy, H. W. Gruebele, C. A. Schmuttenmaer, and R. J. Saykally, *Science* **244**, 564 (1989).
121. M. Vala, T. M. Chandrasekhar, J. Szczepanski, R. van Zee, and W. Weltner, Jr., *J. Chem. Phys.* **90**, 595 (1989).
122. M. Vala, T. M. Chandrasekhar, J. Szczepanski, and R. Pellow, *High Temp. Sci.* **27**, 19 (1989).
123. L. N. Shen and W. R. M. Graham, *J. Chem. Phys.* **91**, 5115 (1989).
124. H. M. Cheung and W. R. M. Graham, *J. Chem. Phys.* **91**, 6664 (1989).
125. J. Kurtz and D. R. Huffman, *J. Chem. Phys.* **92**, 30 (1990).
126. J. R. Heath, R. A. Sheeks, A. L. Cooksy, and R. J. Saykally, *Science* **249**, 895 (1990).
127. J. Szczepanski and M. Vala, *J. Phys. Chem.* **95**, 2792 (1991).
128. L. N. Shen, P. A. Withey, and W. R. M. Graham, *J. Chem. Phys.* **94**, 2395 (1991).
129. J. R. Heath, A. van Orden, E. Kuo, and R. J. Saykally, *Chem. Phys. Lett.* **182**, 17 (1991).
130. J. R. Heath and R. J. Saykally, *J. Chem. Phys.* **94**, 1724 (1991).
131. J. R. Heath and R. J. Saykally, *J. Chem. Phys.* **94**, 3271 (1991).
132. P. A. Withey, L. N. Shen, and W. R. M. Graham, *J. Chem. Phys.* **95**, 820 (1991).
133. Q. Jiang and W. R. M. Graham, *J. Chem. Phys.* **95**, 3129 (1991).
134. G. von Helden, M.-T. Hsu, P. R. Kemper, and B. T. Bowers, *J. Chem. Phys.* **95**, 3835 (1991).
135. D. Zajfman, H. Feldman, O. Heber, D. Kella, D. Majer, Z. Vager, and R. Naaman, *Science*, **258**, 1129 (1992).
136. G. von Helden, P. R. Kemper, N. G. Gotts, and M. T. Bowers, *Science* **259**, 1300, (1993).
137. N. G. Gotts, G. von Helden, and M. T. Bowers, *Int. J. Mass Spectr. Ion Process.* **149/150**, 217 (1995).
138. Z. Slanina, F. Uhlík, and L. Adamowicz, *J. Radioanal. Nucl. Chem.* **219**, 69 (1997).
139. Z. Slanina, S.-L. Lee, J.-P. François, J. Kurtz, L. Adamowicz, and M. Smigel, *Mol. Phys.* **81**, 1489 (1994).

140. Z. Slanina, S.-L. Lee, J.-P. François, J. Kurtz, and L. Adamowicz, *Chem. Phys. Lett.* **223**, 397 (1994).
141. Z. Slanina, S.-L. Lee, M. Smigel, J. Kurtz, and L. Adamowicz in *Science and Technology of Fullerene Materials*, eds. P. Bernier, D. S. Bethune, L. Y. Chiang, T. W. Ebbesen, R. M. Metzger, and J. W. Mintmire (Materials Research Society, Pittsburgh, 1995).
142. T. F. Giesen, A. van Orden, H. J. Hwang, R. S. Fellers, R. A. Provençal, and R. J. Saykally, *Science* **265**, 756 (1994).
143. D. A. Plattner and K. N. Houk, *J. Am. Chem. Soc.* **117**, 4405 (1995).
144. G. von Helden, N. G. Gotts, and M. T. Bowers, *Nature* **363**, 60 (1993).
145. Z. Slanina, *Chem. Phys. Lett.* **173**, 164 (1990).
146. M. W. Chase, Jr., C. A. Davies, J. R. Downery Jr., D. J. Frurip, R. A. McDonald, and A. N. Syverud, *J. Phys. Chem. Ref. Data* **14**, (1985), Supplement 1. *JANAF Thermochemical Tables, Third Edition, Vols. 1,2.*
147. M. A. Nygren and L. G. M. Pettersson, *Chem. Phys. Lett.* **191**, 473 (1992).
148. B. G. Johnson in *Modern Density Functional Theory: A Tool for Chemistry*, eds. J. M. Seminario and P. Politzer (Elsevier, Amsterdam, 1995).
149. J. D. Watts, J. Gauss, J. F. Stanton, and R. J. Bartlett, *J. Chem. Phys.* **97**, 8372 (1992).
150. Z. Slanina, *Chem. Phys. Lett.* **161**, 265 (1989).
151. Z. Slanina, *Chem. Phys. Lett.* **166**, 317 (1990).
152. D. Michalska, H. Chojnacki, B. A. Hess, Jr., and L. J. Schaad *Chem. Phys. Lett.* **141**, 376 (1987).
153. Z. Slanina, *Chem. Phys. Lett.* **142**, 512 (1987).
154. Z. Slanina, *Thermochim. Acta* **127**, 237 (1988).
155. J. M. Rudziński, Z. Slanina, E. Ōsawa, and T. Iizuka, *8th IUPAC Conf. Phys. Org. Chem., Tokyo* 262 (1986).
156. Z. Slanina, J. M. Rudziński, M. Togasi, and E. Ōsawa, *J. Mol. Struct. (THEOCHEM)* **61**, 169 (1989).
157. R. Taylor, J. P. Hare, A. K. Abdul-Sada, and H. W. Kroto, *J. Chem. Soc., Chem. Commun.* 1423 (1990).
158. J. F. Christian, Z. Wan, and S. L. Anderson, *J. Phys. Chem.* **96**, 3574 (1992).
159. S. W. McElvany, J. H. Callahan, M. M. Ross, L. D. Lamb, and D. R. Huffman, *Science* **260**, 1632 (1993).
160. J.-P. Deng, D.-D. Ju, G.-R. Her, C.-Y. Mou, C.-J. Chen, Y.-Y. Lin, and C.-C. Han, *J. Phys. Chem.* **97**, 11575 (1993).

161. Z. Slanina and S.-L. Lee, *J. Mol. Struct. (THEOCHEM)* **304**, 173 (1994).
162. *The Chemistry of Fullerenes*, ed. R. Taylor (World Scientific, Singapore, 1995) p. 1.
163. D. Babić, D. J. Klein, and C. H. Sah, *Chem. Phys. Lett.* **211**, 235 (1993).
164. Z. Slanina and L. Adamowicz, *Fullerene Sci. Technol.* **1**, 1 (1993).
165. K. Raghavachari, D. L. Strout, G. K. Odom, G. E. Scuseria, J. A. Pople, B. Johnson, and P. W. Gill, *Chem. Phys. Lett.* **214**, 357 (1993).
166. T. G. Schmalz, W. A. Seitz, D. J. Klein, and G. E. Hite, *J. Am. Chem. Soc.* **110**, 1113 (1988).
167. M. L. McKee and W. C. Herndon, *J. Mol. Struct. (THEOCHEM)* **153**, 75 (1987).
168. L. Goodwin, *Phys. Rev.* B **44**, 11432, 1991.
169. R. E. Stanton, *J. Phys. Chem.* **96**, 111 (1992).
170. K. Raghavachari and C. M. Rohlfing, *J. Phys. Chem.* **96**, 2463 (1992).
171. R. L. Murry, J. R. Colt, and G. E. Scuseria, *J. Phys. Chem.* **97**, 4954 (1993).
172. P. Ehrenfreund and B. H. Foing, *Planet. Space Sci.* **43**, 1183 (1995).
173. D. Heymann, *Proc. Lunar Planet. Sci. Conf.* **17**, E135 (1986).
174. D. Heymann, *Meteoritics* **30**, 436 (1995).
175. P. R. Buseck, S. J. Tsipursky, and R. Hettich, *Science* **257**, 215 (1992).
176. T. K. Daly, P. R. Buseck, P. Williams, and C. F. Lewis, *Science* **259**, 1599 (1993).
177. L. Becker, J. L. Bada, R. E. Winans, J. E. Hunt, T. E. Bunch, and B. M. French, *Science* **265**, 642 (1994).
178. L. Becker, R. J. Poreda, and J. L. Bada, *Science* **272**, 249 (1996).
179. D. Heymann, L. P. F. Chibante, R. R. Brooks, W. S. Wolbach, and R. E. Smalley, *Science* **265**, 645 (1994).
180. D. Heymann, W. S. Wolbach, L. P. F. Chibante, R. R. Brooks, and R. E. Smalley, *Geochim. Cosmochim. Acta* **58**, 3531 (1994).
181. D. Heymann, L. P. F. Chibante, and R. E. Smalley, *J. Chromatography* A **689**, 157 (1995).
182. D. Heymann, A. Korochantsev, M. A. Nazarov, and J. Smit, *Cretaceous Res.* **17**, 367 (1996).
183. D. Heymann, *Meteoritics Planet. Sci.* **31**, 362 (1996).
184. F. R. di Brozolo, T. E. Bunch, R. H. Fleming, and J. Macklin, *Nature* **369**, 37 (1994).
185. L. Becker, J. L. Bada, R. E. Winans, and T. E. Bunch, *Nature* **372**, 507 (1994).

186. L. Becker, G. D. McDonald, and J. L. Bada, *Nature* **361**, 595 (1993).
187. A. Yamasaki, T. Iizuka, and E. Ōsawa, *Fullerene Sci. Technol.* **3**, 529 (1995).
188. E. Ōsawa, Y. Hirose, A. Kimura, and M. Shibuya, *Fullerene Sci. Technol.* **5**, 177 (1997).
189. E. Ōsawa, Y. Hirose, A. Kimura, M. Shibuya, and H. Takezawa, *Fullerene Sci. Technol.* **5**, 1045 (1997).
190. R. Taylor, J. P. Parsons, A. G. Avent, S. P. Rannard, T. J. Dennis, J. P. Hare, H. W. Kroto, and D. R. M. Walton, *Nature* **351**, 277 (1991).
191. L. Juha, V. Hamplová, P. Kubát, E. Koudoumas, and S. Couris, *Chem. Phys. Lett.* **231**, 314 (1994).
192. D. Heymann and L. P. F. Chibante, *Rec. Trav. Chim. Pays-Bas* **112**, 531 (1993).
193. D. Heymann and L. P. F. Chibante, *Rec. Trav. Chim. Pays-Bas* **112**, 639 (1993).
194. L. P. F. Chibante and D. Heymann, *Geochim. Cosmochim. Acta* **57**, 1879 (1993).
195. D. Heymann and L. P. F. Chibante, *Chem. Phys. Lett.* **207**, 339 (1993).
196. M. S. Dresselhaus, G. Dresselhaus, and P. C. Eklund, *Science of Fullerenes and Carbon Nanotubes* (Academic Press, San Diego, 1996).
197. Y. Achiba, K. Kikuchi, Y. Aihara, T. Wakabayashi, Y. Miyake, and M. Kainosho in *Science and Technology of Fullerene Materials*, eds. P. Bernier, D. S. Bethune, L. Y. Chiang, T. W. Ebbesen, R. M. Metzger, and J. W. Mintmire (Materials Research Society, Pittsburgh, 1995).
198. Y. Achiba, K. Kikuchi, Y. Aihara, T. Wakabayashi, Y. Miyake, and M. Kainosho in *The Chemical Physics of Fullerenes 10 (and 5) Years Later*, ed. W. Andreoni (Kluwer Academic Publishers, Dordrecht, 1996).
199. Y. Achiba, *Kagaku* **52** (5), 15 (1997).
200. H. W. Kroto, *Nature* **329**, 529 (1987).
201. D. E. Manolopoulos and P. W. Fowler, *J. Chem. Phys.* **96**, 7603 (1992).
202. P. W. Fowler and D. E. Manolopoulos, *An Atlas of Fullerenes* (Clarendon Press, Oxford, 1995).
203. M. Yoshida and E. Ōsawa, *Bull. Chem. Soc. Jpn.* **68**, 2073 (1995).
204. M. Yoshida, *Ph.D. Thesis* (Toyohashi University of Technology, Toyohashi, 1996).
205. Z. Slanina, M.-L. Sun, S.-L. Lee, and L. Adamowicz in *Recent Advances in the Chemistry and Physics of Fullerenes and Related Materials, Vol. 2*, eds. K. M. Kadish and R. S. Ruoff (The Electrochemical Society, Pennington, 1995).

206. F. Diederich, R. L. Whetten, C. Thilgen, R. Ettl, I. Chao, and M. M. Alvarez, *Science* **254**, 1768 (1991).
207. K. Kikuchi, N. Nakahara, T. Wakabayashi, S. Suzuki, H. Shiromaru, Y. Miyake, K. Saito, I. Ikemoto, M. Kainosho, and Y. Achiba, *Nature* **357**, 142 (1992).
208. D. Bakowies, A. Geleßus, and W. Thiel, *Chem. Phys. Lett.* **197**, 324 (1992).
209. Z. Slanina, J.-P. François, D. Bakowies, and W. Thiel, *J. Mol. Struct. (THEOCHEM)* **279**, 213 (1993).
210. S. J. Woo, E. Kim, and Y. H. Lee, *Phys. Rev.* B **47**, 6721 (1993).
211. K. Nakao, N. Kurita, and M. Fujita, *Phys. Rev.* B **49**, 11415 (1994).
212. M.-L. Sun, Z. Slanina, S.-L. Lee, F. Uhlík, and L. Adamowicz, *Chem. Phys. Lett.* **246**, 66 (1995).
213. F. H. Hennrich, R. H. Michel, A. Fischer, S. Richard-Schneider, S. Gilb, M. M. Kappes, D. Fuchs, M. Bürk, K. Kobayashi, and S. Nagase, *Angew. Chem., Int. Ed. Engl.* **35**, 1732 (1996).
214. K. Kikuchi, N. Nakahara, T. Wakabayashi, M. Honda, H. Matsumiya, T. Moriwaki, S. Suzuki, H. Shiromaru, K. Saito, K. Yamauchi, I. Ikemoto, and Y. Achiba, *Chem. Phys. Lett.* **188**, 177 (1992).
215. S. Nagase, K. Kobayashi, T. Kato, and Y. Achiba, *Chem. Phys. Lett.* **201**, 475 (1993).
216. G. Orlandi, F. Zerbetto, and P. W. Fowler, *J. Phys. Chem.* **97**, 13575 (1993).
217. S. Nagase and K. Kobayashi, *Chem. Phys. Lett.* **214**, 57 (1993).
218. Z. Slanina, S.-L. Lee, K. Kobayashi, and S. Nagase, *J. Mol. Struct. (THEOCHEM)* **339**, 89 (1995).
219. M. Saunders, H. A. Jiménez-Vázquez, R. J. Cross, W. E. Billups, C. Gesenberg, A. Gonzalez, W. Luo, R. C. Haddon, F. Diederich, and A. Herrmann, *J. Am. Chem. Soc.* **117**, 9305 (1995).
220. F. Diederich, R. Ettl, Y. Rubin, R. L. Whetten, R. Beck, M. Alvarez, S. Anz, D. Sensharma, F. Wudl, K. C. Khemani, and A. Koch, *Science* **252**, 548 (1991).
221. R. D. Beck, P. S. John, M. M. Alvarez, F. Diederich, and R. L. Whetten, *J. Phys. Chem.* **95**, 8402 (1991).
222. K. Kikuchi, N. Nakahara, M. Honda, S. Suzuki, K. Saito, H. Shiromaru, K. Yamauchi, I. Ikemoto, T. Kuramochi, S. Hino, and Y. Achiba, *Chem. Lett.* 1607 (1991).
223. K. Raghavachari and C. M. Rohlfing, *J. Phys. Chem.* **95**, 5768 (1991).
224. Z. Slanina, J.-P. François, M. Kolb, D. Bakowies, and W. Thiel, *Fullerene Sci. Technol.* **1**, 221 (1993).

225. Z. Slanina, S.-L. Lee, M. Yoshida, and E. Ōsawa, *Chem. Phys.* **209**, 13 (1996).
226. Z. Slanina, S.-L. Lee, M. Yoshida, and E. Ōsawa in *Physics and Chemistry of Fullerenes and Their Derivatives*, eds. H. Kuzmany, J. Fink, M. Mehring, and S. Roth (World Scientific, Singapore, 1996).
227. Z. Slanina, S.-L. Lee, M. Yoshida, and E. Ōsawa in *Recent Advances in the Chemistry and Physics of Fullerenes and Related Materials, Vol. 3*, eds. K. M. Kadish and R. S. Ruoff (The Electrochemical Society, Pennington, 1996).
228. Z. Slanina, S.-L. Lee, and L. Adamowicz, *Int. J. Quantum Chem.* **63**, 529 (1997).
229. Z. Slanina, X. Zhao, S.-L. Lee, and E. Ōsawa in *Recent Advances in the Chemistry and Physics of Fullerenes and Related Materials, Vol. 4*, eds. K. M. Kadish and R. S. Ruoff (The Electrochemical Society, Pennington, 1997).
230. Z. Slanina, X. Zhao, S.-L. Lee, and E. Ōsawa, *Chem. Phys.* **219**, 193 (1997).
231. D. Bakowies and W. Thiel, *J. Am. Chem. Soc.* **113**, 3704 (1991).
232. D. Bakowies and W. Thiel, *Chem. Phys.* **151**, 309 (1991).
233. Z. Slanina, L. Adamowicz, D. Bakowies, and W. Thiel, *Thermochim. Acta* **202**, 249 (1992).
234. M. J. S. Dewar, C. Jie, and J. Yu, *Tetrahedron* **49**, 5003 (1993).
235. A. Webster, *Nature* **352**, 412 (1991).
236. A. Webster, *Mon. Not. R. Astron. Soc.* **277**, 1555 (1995).

PLANETARY ATMOSPHERES: THE ROLE OF COLLISION-INDUCED ABSORPTION

L.M. TRAFTON
McDonald Observatory, University of Texas at Austin, Austin, TX 78712, USA

Collision-induced absorption plays several roles in planetary atmospheres. It provides the thermal opacity of the giant planets whereby H_2 and He radiate most of the absorbed sunlight and internal heat to space. This process governs the vertical temperature structure in their upper atmospheres. CIA also is expected to have played a role in the formation and early evolution of the giant planets by governing the rate of cooling of the protoplanetary material and the resultant contraction to their present sizes. It may also have played a role in an early massive atmosphere of Triton if this satellite was captured. CIA from H_2, N_2, and CH_4 in Titan's atmosphere is primarily responsible for the warm surface and the increase of atmospheric temperature with depth. On Venus, CIA of CO_2 is thought to contribute to the continuum spectrum of the nightside thermal emission in its atmospheric spectral windows. On Triton and Pluto, the spectral profile of the 2.15 μm band of N_2 ice provides the temperature of the ice which constrains the ice vapor pressure and atmospheric density. Since these atmospheres are supported entirely by the equilibrium vapor pressure, their mass varies sensitively with seasonal changes in the insolation. Finally, CIA may play a significant role in the thermal opacity of Jupiter-sized extrasolar planets. However, the thermal opacity of brown dwarfs and hot super-Jupiter sized extrasolar planets is likely to be dominated by H_2O and CH_4 absorption.

1 Introduction

On a global scale, planetary atmospheres are largely in hydrostatic equilibrium. This means that pressure supports the weight of the overlying atmosphere. This pressure is the source of the collision-induced absorption (CIA). This absorption is significant in planetary atmospheres in spite of its low absorption coefficient because of the characteristically long path lengths. The surface atmospheric density varies considerably among solar system planets. The terrestrial planets have relatively thin atmospheres above solid surfaces. These atmospheres are all secondary; they evolved long ago out of the rocks and oceans, long after the primary atmospheres escaped during their early impact heating in proximity to the Sun. The giant planets, on the other hand are largely fluid bodies with small rocky cores; they retained their primordial H_2 and He during their formation and early evolution due to their distance from the Sun. Deep in these bodies, the density and so the CIA rises to high values. The temperature also rises to high values since, with the possible exception of

Uranus, these planets are thought to generate internal heat through gravitationally driven separation of phases in the metallic hydrogen–helium system. This increases the heat flux that must be radiated to space by the atmospheric CIA thermal opacity.

CIA dominates the radiative transfer of heat in the atmospheres of all the major planets, and Titan. The temperatures of the observable parts of these atmospheres range from over 250 K for Jupiter to about 45 K for Neptune. Black body radiation at these temperatures lies mostly in the wavelength range from 10 to 100 μm, matching the CIA opacity range of H_2 and its enhancement by He. These gases therefore govern the radiative transfer of heat in these atmospheres. CIA also plays a role in the radiative transfer of heat on Venus, particularly in the radiation of heat from deep levels below the clouds to space. By contrast, on Mars, the Earth, and satellites except Titan, the atmosphere is too thin for CIA to play a significant role in the radiation balance.

There have now been detected around other stars a brown dwarf [a] and several planet-like objects. Some of these exceed Jupiter's mass and lie very close to the primary star. The role of CIA in the hotter of such objects may be considerably subdued owing to H_2O and CH_4 opacity and the reduced overlap between a hot planetary thermal spectrum and the H_2–He CIA opacity. However, CIA may play an important role in the cooler, Jupiter-like atmospheres.

Following the next section on relevant historical studies of CIA, we discuss the role of CIA in warming various planetary atmospheres, then summarize results from the Voyager mission. Finally, we summarize recent results from the post-Voyager era, and conclude with a discussion of the prospective role of CIA in newly-discovered, large extrasolar planets.

2 Historical Background

Because only the overtone bands of H_2 had wavelengths short enough to be accessible to photographic spectroscopy, the earliest astronomical investigations of CIA were carried out using these bands. The first observation of CIA in a planetary atmosphere, or even for an astronomical object, was made by Kuiper, who obtained a spectrum of Uranus containing a diffuse feature at 8260 Å. This spectrum was analyzed by Herzberg [29] who compared it with a laboratory spectrum of H_2 and identified the feature as the $S_3(0)$ pressure-induced overtone of H_2. This was the first evidence that H_2 is a major constituent of the atmospheres of the outer planets. Later, Spinrad [49] noticed a diffuse feature at 6420 Å in Uranus' spectrum and suggested that it was the $S_4(0)$ pressure-induced overtone of H_2. However, higher resolution spectra showed

[a] An object formed like a star but with too little mass to trigger nuclear heating.

that this feature was an artifact of the CH_4 absorption; the $S_4(1)$ feature was found to be much weaker than the diffuse feature at 6420 Å despite weaker background CH_4 absorption at its wavelength, which should have improved its visibility.[4,55]

The first analysis of Uranus' diffuse feature at 8260 Å was done by Poll[47] who also calculated the matrix elements of the polarizability of H_2. With the matrix elements of the H_2 quadrupole moment calculated by Birnbaum and Poll,[6] he computed the absorption coefficients for both the 8260 Å and 6420 Å bands and estimated the H_2 abundance over an assumed cloud deck. His estimate was inaccurate because his analysis weighted fits to the 6420 Å feature, which was later found to be largely CH_4 absorption. However, Poll was the first to estimate successfully the strength of the CIA (4–0) band in Jupiter's spectrum. Improved values of these matrix elements have since been published by Hunt, Poll, and Wolniewicz.[30]

Higher resolution spectra of these bands were obtained by Belton and Spinrad[4] for all of the giant planets. Neptune's spectrum was too noisy to analyze in detail and they found the CIA in Jupiter and Saturn to be very weak. Uranus' spectrum provided the best CIA spectrum for analysis. They modeled the CIA H_2 (3–0) overtone band complex but fitted only the $S_3(0)$ band to analyze the Uranian haze structure. The CIA $S_3(1)$ band was too hidden in strong CH_4 absorption to be fitted. They found that the Uranian atmosphere cannot be completely clear; a haze or a cloud deck had to be present. This contradicted the earlier results of Belton, McElroy and Price[3] which were obtained before the availability of the matrix elements of the polarizability. Still higher resolution spectra of these bands for Uranus were obtained by Trafton[55] who modeled and fitted both the CIA (3–0) and (4–0) overtone bands and the (3–0) H_2 quadrupole lines. He showed that Uranus has a visible cloud deck and negligible aerosol scattering haze throughout the radiative zone. He also demonstrated the dominance of H_2 in Uranus' atmosphere. A more comprehensive modeling of Uranus' entire visual and near-IR spectrum by Baines and Bergstralh,[2] who additionally included the extinction from CH_4 and haze, further refined the derived haze distribution. Trafton[57] extended his earlier analysis to include his newly acquired high resolution spectra of the (3–0) H_2 quadrupole profiles. Both confirmed the necessity for a lower cloud deck and an elevated stratospheric haze; and they both found near-equilibrium values of the ortho–para H_2 ratio, in agreement with the later Voyager results.

Intermediate-resolution planetary spectroscopy of the lower overtone and fundamental band of H_2 became feasible only after the development of sensitive infrared (1–4 μm) detectors in the 1980's. However, strong blending of the (2–0) H_2 overtone with the CH_4 spectrum have discouraged analysis of this band except for estimates of upper limits on the H_2 column abundance for Jupiter and Saturn.[22]

In contrast to the (2–0) overtone band of H_2, the (1–0) fundamental band, especially $S_1(1)$, is situated in atmospheric windows relatively unobstructed by absorption from other gases in the atmospheres of the Earth, Jupiter, or Saturn. For Uranus and Neptune, however, the CH_4 absorption is strong enough to obscure much of the (1–0) band. The first spectrum of the fundamental H_2 band in planetary atmospheres was obtained for Jupiter by the balloon Stratoscope II.[21] Since the spectral resolution was only $\lambda/\delta\lambda=40$, analysis was severely limited by blended nearby CH_4 bands. Obtaining image tube spectra of both Jupiter and Saturn, Martin *et al.*[37] analyzed this band using the simple reflecting layer model. For each planet, they derived both the H_2 abundance above an assumed reflecting cloud layer and a mean temperature along the absorbing path.

The first theoretical modeling of the translational CIA band profile was done for H_2 and its enhancement by He by Trafton.[51] Semi-empirical modeling of the ro-vib CIA band profiles had been done for numerous molecules but the translational profile was needed to complete the thermal opacity for the giant planets, especially for the colder ones. Trafton[52,53] applied this CIA opacity to non-gray, radiative-convective models of the vertical atmospheric structure of the giant planets. He introduced the effective temperature[b] and He/H_2 ratio as free parameters to be determined by comparison with observed spectra when they became available. The models showed how the tropospheres of Jupiter and Saturn could become warm enough to explain the concentration of NH_3 vapor implied by spectra. They also showed that Jovian models omitting the CIA enhancement of He could not fit the observations, but models including it could, indicating that He must be present in Jupiter's atmosphere. Trafton's models also indicated the presence of an internal source of heat for Jupiter. All of the giant planets except Uranus have since proved to have internal heat sources.

During the first flybys of Jupiter, done by the Pioneer 10 and 11 spacecraft, the infrared radiometers probed Jupiter's thermal spectrum in two wavelength channels centered at 20.0 μm and 45.5 μm with equivalent widths of 11.6 μm and 22.7 μm, respectively, with spatial resolution varying from 1/30 to 1/200 of Jupiter's diameter.[15] Two important objectives were Jupiter's global He/H_2 ratio and the effective temperature, which would provide the magnitude of the internal heat source. Ingersoll *et al.*[31] derived effective temperature of 125$\pm$3 K, implying an internal heat source 0.9 times the absorbed solar radiation. However, significant uncertainty remained in the values of the the He/H_2

[b]This is the temperature of a black body radiating the same total heat flux as the planet.

ratio and internal heat source, largely as a consequence of the two-channel sampling of the spectrum.

3 CIA Enhancement of the Greenhouse Effect

The greenhouse effect occurs when an atmosphere is more transparent to sunlight than it is to thermal radiation. Sunlight absorbed in the atmosphere or on the surface is radiated as heat. Most planetary atmospheres are relatively transparent to sunlight because their bulk gases, CO_2, N_2, O_2, H_2, He, and CH_4 only weakly absorb near the peak of the solar spectrum; i.e., over the visual and near IR spectrum. On the other hand, these gases have strong CIA bands which overlap the thermal spectra of these bodies. CIA of H_2 is enhanced the in the giant planets by CH_4 and He, and by CH_4 and N_2 in Titan. Consequently, sunlight is converted to heat deep in these atmospheres and the heat is partially trapped by the thermal opacity. In order to balance the absorbed energy with that radiated to space, the energy density must build up progressively with increasing depth in the atmosphere. This results in an increase of temperature with depth. On Venus, this greenhouse heating is enhanced by the deep atmosphere and high pressure. Also, this effect is enhanced in the deep atmospheres of the giant planets because their internal heat flux contributes to the absorbed solar radiation to produce the thermal spectrum. It causes the temperature to increase down to the solid core, instead of becoming isothermal below the depth sunlight can penetrate.

Formation of the giant planets involved the accretion of planetesimals followed by the capture of H_2 gas in the protosolar nebula to form their envelopes. The capture of gas was assisted by CIA absorption which provided the opacity needed for the extended envelope to radiate away the heat generated by the molecular capture impacts. Although the density was very low in these protoplanetary envelopes, the path length was very large, leading to significant optical depth. As the gas in the envelope cooled, the envelope contracted, allowing further capture. Capture of gas ended with the depletion of the protosolar nebula but cooling and contraction continued as these bodies continued to radiate their primordial heat, aided by the allowed dipole absorptions from molecules formed during the process.

When the temperature gradient exceeds a critical value, an atmosphere becomes unstable to convection and heat is transferred efficiently by convection rather than by radiation below this level. CIA still plays a role, however, by maintaining the thermal opacity high enough that the radiative temperature gradient continues to exceed this value at deeper levels. Otherwise, convection would halt in those layers and radiative transfer would be the dominant heat transport mechanism. In other words, the CIA still causes the temperature to increase with depth in the convective parts of the atmosphere but the lapse rate is not as steep as for a region in radiative equilibrium.

The top of a convecting region often coincides with a cloud deck if this part of the atmosphere contains a saturated volatile gas. Convection adiabatically reduces the temperature of an upward moving parcel of air, causing its saturated volatiles to condense. The region of the atmosphere above the convecting zone is usually clear, or at worst, hazy. It is no accident that sunlight is absorbed primarily below the cloud deck, if one occurs, where it is converted to heat and sufficiently trapped by the thermal opacity to induce convection. Jupiter's upper cloud layer and Saturn's cloud deck are formed by condensing NH_3 while condensing CH_4 is primarily responsible for the cloud decks of colder Uranus and Neptune.

On Titan, CH_4 and hydrocarbons form an extensive stratospheric haze layer which absorbs part of the insolation, and which would cool the surface by 9 K in the absence of any thermal opacity. But the CIA greenhouse effect overcomes this loss to raise the temperature of Titan's surface 12 K warmer than the 82 K effective temperature.[39] In Titan's atmosphere, N_2 is the dominant gas while the mixing ratio is only $\simeq 0.1\%$ for H_2, 1.7–4.5% for CH_4, and less than 6% for argon.[20] Nevertheless, the primary thermal opacity is provided by the CIA of H_2, enhanced by collisions with N_2 and CH_4, because of its overlap with Titan's thermal spectrum. If a condensation cloud is present, Courtin *et al.*[20] find that it must have an optical depth less than 0.7 and be located significantly higher than the expected condensation level.

On Venus, the thermal opacity arises primarily from the allowed transitions of CO_2. The greenhouse warming is determined by the mean transmission of these bands, which is governed by the shape and strength of the distant wings of the lines since the cores of the lines are often dark. However, CIA can augment the thermal opacity of CO_2 by inducing absorption in the far wings of normally forbidden bands of the molecule. On the Earth, permanent dipole absorptions of H_2O and CO_2 provide the main thermal opacity.

Despite the wide variety of size and temperature among planetary atmospheres, the upper cloud deck, where it exists, lies at a pressure on the order of an atmosphere. For the giant planets, this reflects the steep pressure dependence and broad continuum of CIA relative to permanent dipole absorption. These typically cause the atmosphere to become optically thick to thermal radiation and to break down to convection at comparable pressure levels. One exception is the planet Uranus. For the other giant planets, the magnitude of the internal heating is on the order of the absorbed insolation, but for Uranus the internal heating is weak or non existent. In addition, CH_4 is frozen out of its upper atmosphere but not out of the lower troposphere, where it is unsaturated with a mixing ratio of about 2%. Consequently, in the presence of a low internal heat flux, the mean molecular weight of the atmosphere increases with depth enough to suppress the onset of convection above a pressure level of several bars. Hence, the cloud deck and the visible atmosphere are very deep, as is the radiative zone wherein CIA governs the transfer of heat. As is evidenced by its greater cloudiness, Neptune's atmosphere is more convective. The higher internal heat flux more than offsets the lower insolation flux and atmospheric gradient in the mean molecular weight.

The contribution of CH_4 to the thermal opacity in the atmospheres of Uranus and Neptune may be neglected because of the low concentration of CH_4 imposed by the saturation vapor pressure in the radiative equilibrium regions of these atmospheres.[54]

Since CH_4 is also saturated in Titan's atmosphere, its concentration is dependent on the temperature. Because CH_4 contributes to Titan's greenhouse warming, there is a positive feedback between the CH_4 concentration and temperature. Raising the surface temperature increases the saturated CH_4 concentration, which increases the thermal opacity through H_2–CH_4 collisions, which raises the temperature further. This process is limited by the decreasing overlap between the thermal spectrum and the CIA and by the reduced insolation of the surface with increasing CH_4 concentration.

4 Voyager Observations of the CIA Rotational-Translational Band of H_2–He

Just as for the Pioneer mission to Jupiter, the determination of the He/H_2 ratio and the effective temperature were important objectives for Voyager during each giant planet flyby. The thermal spectra were observed with the Infrared Interferometric Spectrometer (IRIS) experiment [28] on board Voyager. The instrument provided adequate spectral resolution and a spectral range that covered most of the thermal spectrum, especially for the warmer planets. Its narrow field of view provided much better spatial resolution of the CIA thermal spectrum than Pioneer obtained, allowing center-to-limb studies of the emission.

4.1 The He/H_2 Ratio

The primordial He/H_2 ratio of the Sun is related to the He/H_2 ratio in the proto-solar nebula during formation of the solar system, and so to the values of the ratio in the atmospheres of the giant planets. The He/H_2 ratio at the

Table 1: Helium mass fractions determined by the Voyager IRIS experiment.

Planet	Mass fraction He	Reference
Jupiter	0.18 ± 0.04	Gautier *et al.*[25]
Saturn	0.06 ± 0.05	Conrath *et al.*[17]
Uranus	0.26 ± 0.05	Conrath *et al.*[18]
Neptune	0.32 ± 0.05	Conrath *et al.*[19]
(Solar Neighborhood)	0.28 ± 0.02	Pagel[46]

time the universe was forming during the Big Bang can be deduced from the primordial solar ratio with the help of models of the evolution of galaxies.[18] Such models quantify how He, formed during nucleosynthesis in stars and released to space through supernovae and stellar winds, enrichens younger stars formed out of this released material.

Determination of the He/H_2 ratio from the thermal spectrum is possible because the enhancement to the CIA of H_2 by He results in a translational absorption coefficient much bigger in relation to the ro-vib component than for pure H_2. Hence, He enhances the thermal opacity at the longer wavelengths. This occurs because unlike for H_2 collisions, H_2–He collisions are not homopolar. Consequently, a large angle-independent overlap component in the induced dipole moment results, which is not modulated by rotation.

The values of the He/H_2 ratio determined for each planet by fitting synthetic spectra to the IRIS observations are listed in Table 1. These values were refined using the independent determination of the depth dependence of the temperature obtained from Voyager's Radio Occultation experiment.[23] The Table shows that the He/H_2 ratio varies significantly among the giant planets and that only Uranus and Neptune have values representative of stars in the solar neighborhood.

Diffusive separation of He from H is likely to occur in the high-pressure (> 3 Mbar) immiscible metallic phase of H as the planets cool while radiating their primordial heat.[50] This gravitational separation is thought to power their internal heating because they should have already radiated away their primordial heat. The solid cores of Uranus and Neptune may be big enough to inhibit this separation, leading to the indicated lack of evolution of their

He/H_2 ratios. Both Jupiter and Saturn are thought to have small cores and high central pressures so that diffusive separation is likely to be taking place in them. Because Saturn is smaller, it is likely to have lost its primordial heat sooner than Jupiter, so that diffusive separation of its He has occurred for a longer time, as evidenced by its lower He/H_2 ratio.

4.2 The Ortho–Para H_2 Ratio

In the giant planets most of the H_2 molecules radiating heat to space not only are in their ground electronic and vibrational states, but also populate either the J=0 or 1 rotational levels. This clustering around low-lying levels due to the low temperature makes the thermal spectrum sensitive to the fraction of para–H_2. The time scale for ortho–para H_2 to come to thermal equilibration is longer than either the convective time scale or the time scale for vertical mixing over a scale height. Departures of this ratio from its equilibrium value corresponding to the local temperature is diagnostic of the large-scale dynamical processes in the atmosphere.

Values of the ortho–para H_2 ratio were derived by fitting synthetic spectra to the IRIS observations with this ratio as a free parameter, along with the He/H_2 ratio, constraining the depth dependence of the temperature to be equal to the independent determination obtained from Voyager's Radio Occultation experiment.[23] For Jupiter, significant deviations from equilibrium were found, with values of the ratio lying in between the normal and equilibrium values. The greatest deviations were found near the equator, and the least near 60° latitude.[17] No evidence was found that cloud, winds, or color catalyzed the ortho–para H_2 conversion. Essentially equilibrium values were determined for the other giant planets, even for Neptune, where haze was found to affect the spectrum.

5 Post-Voyager Results

The most reliable determinations of the He/H_2 ratio and effective temperature for Jupiter, as well as for the other giant planets, are based on data obtained during the Voyager flybys using the IRIS spectrometer. In the near future, new results for Jupiter are expected from the Galileo mission. In the first decade of the next century, the Cassini mission to Saturn will provide an in-depth examination of the thermal spectra of both Saturn and Titan.

In spite of the impressive results from the Voyager mission, ground based spectroscopy of the planets over the past decade continued to make progress owing to advances in instrumentation. In addition, analysis of both space-based and ground-based planetary thermal spectra has benefited from the recent generation, particularly by A. Borysow, of quick but accurate algorithms for expressing the CIA ro-vib-trans profiles of various gases found in planetary atmospheres, and their binary mixtures, as a function of temperature and pressure.[7,8,12,10,11,62] These codes are based on fitting functions to both laboratory CIA profiles and extensive new ab initio calculations of CIA by Meyer and colleagues.[40,41,42]

5.1 *Ground Based Observations of the H_2 Fundamental Band*

The Voyager IRIS spectrometer did not measure radiation having wavelengths low enough to include the fundamental bands of H_2 or N_2. Our present knowledge of these bands on other planets rests on ground-based spectroscopic observations. However, since the range of the NIMS imaging spectrograph on board Galileo includes these wavelengths, it may soon provide additional information in the case of Jupiter.

Jovian spectra of the fundamental H_2 band having sufficient resolution ($\lambda/\delta\lambda$=200) to reveal the $S_1(0)$ band profile, resolving it from neighboring CH_4 absorption, were first obtained by Trafton *et al.*[56] using a cooled grating spectrometer. Fitting models of both the H_2 band profile and the CH_4 absorption to these spectra, Trafton[58] showed that Jupiter's north polar region was covered by a 'polar hood' of haze particles at the 30 mbar level having a clear transmission of 66%. Kim *et al.*[32] obtained higher resolution interferometric observations of Jupiter's polar haze and similarly concluded that an optically thin polar haze would be located mainly between the 70 and 5 mbar pressure levels; and the top of an opaque haze layer would lie near the 15 mbar level. Such a haze might result partially from the byproducts of auroral chemistry. Trafton also found good agreement with the equatorial haze distribution determined by analysis of the Voyager IRIS spectra.[58]

5.2 *Venus and CIA Night Sky Emission from CO_2*

The allowed dipole absorption of CO_2 and haze extinction are the most evident sources of near infrared and thermal opacity in the dense Venus atmosphere. In the near infrared, there are spectral windows between the strong CO_2 bands where the opacity is higher than can be explained by extrapolating the far wings of the CO_2 lines from nearby bands, modified in shape as indicated by laboratory data closer to the band cores. One such region is the 2.3 μm window.[5] Thermal emission from this window was detected on the nightside and radiates from levels below the planet-wide sulfuric acid cloud decks ($\simeq$8 bars).

Room temperature laboratory experiments [12] have yielded the absorption coefficient of CO_2 in this spectral region, confirming the excess emission and supporting Bezard *et al.*'s suggestion [5] that the excess is the CIA of CO_2. However, the Venus atmosphere at the source levels is hotter than room temperature so one cannot rule out the possibility that part of the excess opacity may be due to numerous weak "hot" bands of CO_2.

5.3 A Massive Early Atmosphere on Triton?

Triton has a retrograde orbit about Neptune which suggests that it was captured. Its initial orbit was then likely to have been eccentric, resulting in substantial heating of the surface through tidal evolution towards the present circular orbit. During this time, volatiles would have outgassed rapidly from Triton and may have formed a dense atmosphere. Lunine and Nolan [33] have modeled the growth and decay of such a massive atmosphere during tidal heating, using initially the CIA of CH_4, N_2, CO (species all presently observed in Triton's surface ice deposits), and trace amounts of H_2 ($\geq 0.01\%$) as the thermal opacity. They find that massive atmospheres could indeed have formed, and under certain conditions, could have lasted well beyond the time tidal heating ceased. They report that the lifetime of such an atmosphere is limited by the rate of escape of infrared-active gases and could have exceeded a billion years. A thick atmosphere could explain the youthful surface of Triton by shielding the surface from impactors up to a km in diameter.

5.4 The 2.15 μm N_2 Ice Band and the Atmospheres of Triton and Pluto

The atmospheres of Triton and Pluto are supported by the saturation vapor pressure of deposits of N_2-rich ice on their surfaces. [61,48] Consequently, the density of these atmospheres is very sensitive to the temperature of the ices, tripling for each 2 K increase in temperature. The ice temperature is determined by the diurnally averaged insolation and varies with the extreme seasons on these bodies, which arise primarily from the inclination of Triton's precessing orbit relative to Neptune's tilted equator and from Pluto's steep obliquity. Hence, the atmospheric mass undergoes dramatic and complex seasonal changes on these bodies, and volatile ices are transported seasonally between their northern and southern hemispheres. Further complicating the modeling of the seasonal changes is the proximity of the ice temperature to the 35.7 K temperature for the exothermic transition from the hexagonal hcp beta phase of solid N_2 to the denser cubic fcc alpha phase. This indroduces another phase into the thermodynamic system during the seasonal changes.

Grundy and Fink[27] have measured the temperature dependence and integrated absorption coefficient of the liquid N_2 (2–0) band to model the ice grain size. The integrated absorption is smaller than that of the gas but the half width of the profile exhibits the square root of the temperature dependence characteristic of CIA. They extrapolated the liquid absorption profile to Triton's nominal ice temperature to analyze the ice profile. They derived a mean N_2 ice grain size between 0.7 and 3 cm.

Temperature is a key parameter for modeling the state of these multiphase systems. Owing to the fortuitous proximity of the temperature to the phase transition temperature from beta to alpha N_2 ice, the ice temperature can be determined quite accurately from the shape and width of the (2–0) overtone band of N_2 ice observed at 2.15 μm in ground based spectra of these bodies. Over several degrees K, the band profile in laboratory spectra changes from a broad band for beta N_2, characteristic of CIA, to a very narrow band characteristic of the higher density alpha phase. This behavior provides a sensitive indicator of the temperature in this interval. This method has been used to derive the present values of the ice temperature of 38 ± 1 K and 40 ± 2 K for Triton and Pluto, respectively.[59,60] In addition, it has further constrained the multiple scattering properties of the ice, suggesting a grain size near one cm.

6 Extrasolar Planets

As of this writing, the first unambiguous detection of a cool brown dwarf has been reported, with a mass of 30–55 times that of Jupiter and apparently orbiting an M1 type star.[43] The spectrum of this object, Gliese 229B, shows absorption bands of CH_4 and H_2O.[45,26] Methane has not been observed in the spectrum of any star, and its presence in Gliese 229B indicates an effective temperature less than 1000 K. The near infrared spectrum is similar to that of Jupiter and Titan.[1,26] Preliminary nongray model atmospheres of this brown dwarf have been constructed which confirm the CH_4 and H_2O absorption and the object's substellar nature.[1,36] Neither grains nor clouds appear to play a significant role in its infrared spectrum. Models imply that the effective temperature is 960 ± 70 K with a surface gravity between 800 and 2200 m s^{-2}.[35] Convection is likely, even though no cloud deck has been detected. Models include the CIA opacity of the H_2 (1–0) fundamental;[1] however, the role of CIA appears to be quite subdued relative to the opacity of H_2O and CH_4, which strongly overlap the Planck function at this elevated temperature. All models indicate flux enhancement in the relatively clear window between 4 and 5 μm, which seems to be a feature of models ranging from Jupiter to brown dwarfs. It is therefore possible that the CIA of the N_2 fundamental band could play

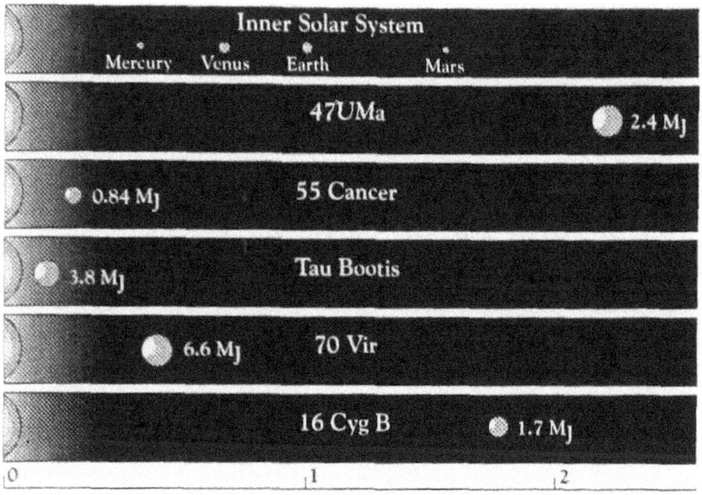

Figure 1: Representative planets detected around Sun-like stars. The planet sizes are magnified but the relative scale is uncertain owing to unknown densities and orbital inclinations. The distance scale from the star is distorted below 0.2 AU to accomodate the magnified star image. One AU is the Earth's distance from the Sun.

a role in limiting this flux in hot bodies, where the nitrogen is not tied up in NH_3.

Evidence is mounting that several planetlike bodies have been detected orbiting other stars.[38,34,13,24,16,14,44] A significant fraction of the candiate planet bodies appears to include large, Jupiter-like bodies, including bodies very close to the star; i.e., less than Mercury's distance to the Sun (see Fig. 1). The relative number of these detections may be atypically high since such bodies are easier to detect. It is known that a hydrogen atmosphere can be stable on such a body so close to the primary, once emplaced there. The remaining objects include smaller bodies at more conventional distances from their primary, and Jupiter-like bodies farther away.

Spectra for these close binary objects are not available but model atmospheres have been constructed for parameters spanning the range of observed masses and luminosities for the gaint bodies among them, which have estimated effective temperatures less than 800 K. As for the brown dwarf models, the reflected visible and near infrared thermal spectrum is dominated by H_2O, CH_4, and NH_3 absorption. Preliminary results suggest that H_2O clouds are present in some, but not all, of these objects.[36] Some may be cloud free, allowing a view to the deep atmosphere in the 4–5 μm window. For objects close to their primary, thermochemical equilibrium favors N_2 and CO rather than NH_3 and CH_4. The CIA opacity of N_2 may play a role in this window for such bodies, but is otherwise likely to be dominated by the allowed absorptions.

For giant extrasolar planets close to their primaries, heating by the primary may retard the radiation of their primordial heat and internal cooling. In this case, diffusive separation of He from metallic H may not have occurred in such bodies. Then the He/H_2 ratio would not have evolved and so would be the same as its value at the time the body was formed. Moreover, it would reflect the He/H_2 ratio of the material out of which the body formed, providing clues about the formation of such planetary systems.

As further observations of the spectra and heat fluxes of giant extrasolar planets become available, it should be possible to deduce their atmospheric structure and energy balance. For the cooler, Jovian-like giants, CIA is expected to play a dominant role in the thermal opacity, just as for such bodies in the solar system. This should not only help to characterize their individual atmospheres but also to provide clues as to their origin and evolution. We can expect a comparative study of independent planetary systems to follow.

Acknowledgements

The author thanks W. D. Cochran for helpful information on extrasolar planets. Support was provided by NASA grant NAGW-2314.

References

1. F. Allard, P. H. Hauschildt, I. Baraffe, and G. Chabrier, *Astrophys. J.* **465**, L123 (1996).
2. K. H. Baines and J. T. Bergstralh, *Icarus* **65**, 406 (1986).
3. M. J. S. Belton, M. B. McElroy, and M. J. Price, *Astrophys. J.* **164**, 191 (1971).
4. M. J. S. Belton and H. Spinrad, *Astrophys. J.* **185**, 363 (1973).
5. B. Bezard, C. de Bergh, D. Crisp, and J.-P. Maillard, *Nature* **345**, 508 (1990).
6. A. Birnbaum and J. D. Poll, *J. Atmos. Sci.* **26**, 943 (1969).
7. J. Borysow, L. Trafton, L. Frommhold, and G. Birnbaum, *Astrophys. J.* **296**, 644 (1985).
8. A. Borysow, L. Frommhold, and M. Moraldi, *Astrophys. J.* **336**, 495 (1989).

9. A. Borysow, *Icarus* **92**, 273 (1991).
10. A. Borysow, *Icarus* **96**, 169 (1992).
11. A. Borysow and C. Tang, *Icarus* **105**, 175 (1993).
12. C. Brodbeck, N. Van-Thanh, J.-P. Bouanich, C. Boulet, and A. Jean-Louis, *J. Geophys. Res.* **96**, 17497 (1991).
13. R. P. Butler and G. W. Marcy, *Astrophys. J.* **464**, L153 (1996).
14. R. P. Butler, G. W. Marcy, E. Williams, H. Hauser, and P. Shirts, *Astrophys. J.* **474**, L115 (1997).
15. S. C. Chase, R. D. Ruiz, G. Munch, G. Neugebauer, M. Schroeder, and L. M. Trafton, *Science* **183**, 315 (1974).
16. W. D. Cochran, A. P. Hatzes, R. P. Butler, and G. W. Marcy, *Astrophys. J.* **483**, 457 (1997).
17. B. J. Conrath, D. Gautier, R. A. Hanel, and J. S. Hornstein, *Astrophys. J.* **282**, 807 (1984).
18. B. J. Conrath, D. Gautier, R. A. Hanel, G. Lindal, and A. Marten, *J. Geophys. Res.* **92**, 15003 (1987).
19. B. J. Conrath, D. Gautier, G. F. Lindal, R. E. Samuelson, and W. A. Shaffer, *J. Geophys. Res.* **96**, 18907 (1991).
20. R. Courtin, D. Gautier, and C. P. McKay, *Icarus* **114**, 144 (1995).
21. R. E. Danielson, *Astrophys. J.* **143**, 949 (1966).
22. C. de Bergh, J. Lecacheux, and M. Combs, *Astron. Astrophys.* **35**, 333 (1974).
23. V. R. Eshleman, G. L. Tyler, G. E. Wood, G. F. Lindal, J. D. Anderson, G. S. Levy, and T. A. Croft, *Science* **204**, 976 (1979).
24. G. Gatewood, *Bull. Amer. Astron. Soc.* **28**, 885 (1996).
25. D. Gautier, B. Conrath, M. Flaser, R. A. Hanel, V. Kunde, A. Chedin, and N. Scott, *J. Geophys. Res.* **86**, 8713 (1981).
26. T. R. Geballe, S. R. Kulkarni, C. E. Woodward, and G. C. Sloan, *Astrophys. J.* **467**, L101 (1996).
27. W. M. Grundy and U. Fink, *Icarus* **93**, 169 (1991).
28. R. A. Hanel *et al.*, *Science* **204**, 972 (1979).
29. G. Herzberg, *Astrophys. J.* **115**, 337 (1952).
30. H. L. Hunt, J. D. Poll, and L. Wolniewicz, *Can. J. Phys.* **62**, 1719 (1984).
31. A. P. Ingersoll, G. Munch, G. Neugebauer, D. Diner, G. S. Orton, B. Schupler, M. Schroeder, S. C. Chase, R. D. Ruiz, and L. M. Trafton, *Science* **188**, 472 (1975).
32. S. J. Kim, P. Drossart, J. Caldwell, J.-P. Maillard, D. Goorvitch, A. Moorwood, A. Moneti, and J. Lecacheux, *Icarus* **91**, 145 (1991).
33. J. I. Lunine and M. C. Nolan *Icarus* **100**, 221 (1992).

34. G. W. Marcy and R. P. Butler, *Astrophys. J.* **464**, L147 (1996).
35. M. S. Marley, D. Saumon, T. Guillot, R. S. Freedman, W. B. Hubbard, A. Burrows, and J. I. Lunine, *Science* **272**, 1919 (1996).
36. M. S. Marley, T. Guillot, D. Saumon, and R. S. Freedman, *Bull. Amer. Astron. Soc.* **28**, 1114 (1996).
37. T. Z. Martin, D. P. Cruikshank, C. P. Pilcher, and W. M. Sinton, *Icarus* **27**, 391 (1976).
38. M. Mayor and D. Queloz, *Nature* **378**, 355 (1995).
39. C. P. McKay, J. B. Pollack, and R. Courtin, *Science* **253**, 1118 (1991).
40. W. Meyer, L. Frommhold, and Birnbaum, *Phys. Rev. A* **39**, 2434 (1989).
41. W. Meyer, A. Borysow, and L. Frommhold, *Phys. Rev. A* **40**, 6931 (1989).
42. W. Meyer, A. Borysow, and L. Frommhold, *Phys. Rev. A* **47**, 4065 (1993).
43. T. Nakajima, B. R. Oppenheimer, S. R. Kulkarni, D. A. Golimowski, K. Matthews, and S. T. Durrance, *Nature* **378**, 463 (1995).
44. R. W. Noyes, S. Jha, S. G Korzennik, M. Krockenberger, and P. Nisenson, *Astrophys. J.* **483**, L111 (1997).
45. B. R. Oppenheimer, S. R. Kulkarni, K. Matthews, and T. Nakajima, *Science* **270**, 1478 (1995).
46. B. E. J. Pagel, *Philos. Trans. R. Soc. London, Ser. A* **307**, 19 (1982).
47. J. D. Poll in *IAU Symp. No. 40: Planetary Atmospheres*, ed. C. Sagan, T. C. Owen, and H. J. Smith (Dordrecht, Reidel, 1971), p. 384.
48. J. R. Spencer, J. A. Stansberry, L. M. Trafton, E. F. Young, R. P. Binzel, and S. K. Croft in *Pluto and Charon*, ed. D. Tholen and S.A. Stern (University of Arizona Press, Tucson, 1997).
49. H. Spinrad, *Astrophys. J.* **138**, 1242 (1963).
50. D. J. Stevenson, *Annu. Rev. Earth Planet. Sci.* **30**, 755 (1982).
51. L. Trafton, *Astrophys. J.* **146**, 558 (1966).
52. L. Trafton, *Astrophys. J.* **147**, 765 (1967).
53. L. Trafton, and G. Munch, *J. Atmos. Sci.* **26**, 813 (1969).
54. L. Trafton, *Astrophys. J.* **172**, L117 (1972).
55. L. Trafton, *Astrophys. J.* **207**, 1007 (1976).
56. L. Trafton, D. Lester, J. Carr, and P. Harvey, *Bull. Amer. Astron. Soc.* **18**, 772 (1986).
57. L. Trafton, *Icarus* **70**, 13 (1987).
58. L. Trafton in *Spectral Line Shapes* **5**, ed. J. Szudy, (Ossolineum, Warsaw, 1988), p. 755.
59. K. A. Tryka, R. H. Brown, D. P. Cruikshank, and T. C. Owen, *Science* **261**, 751 (1993).

60. K. A. Tryka, R. H. Brown, D. P. Cruikshank, T. C. Owen, T. R. Geballe, and C. deBergh, *Icarus* **112**, 513 (1994).
61. R. V. Yelle, J.I. Lunine, J.B. Pollack, and R.H. Brown in *Neptune and Triton*, ed. D.P. Cruikshank (University of Arizona Press, Tucson, 1995).
62. C. Zheng and A. Borysow, *Icarus* **113**, 84 (1995).

CHEMICAL AND OPTICAL PROPERTIES OF MOLECULAR COMPLEXES USING MATRIX ISOLATION SPECTROSCOPY

L. SCHRIVER-MAZZUOLI

Laboratoire de Physique Moléculaire et Applications,
Unité propre du CNRS, Université Pierre et Marie Curie,
Tour 13, case 76, 4 place Jussieu,
75252 Paris Cedex 05, France

The application of the matrix isolation technique to molecular complexes study is described. Following a short description of the experimental technique three areas are reviewed in which matrix experiments are particularly well suited to determine the spectroscopic photochemical and photophysical properties of molecular complexes. The first one involves the IR spectroscopic properties of molecular complexes and is divided into three sections: charge transfer complexes, hydrogen bonded complexes, structural arrangement of weak complexes. The second one is devoted to infrared photodissociation of hydrogen bonded complexes. The third part is related to visible and UV photochemistry of some molecular complexes. Examples including mainly atmospheric species are chosen to evidence the power of the method for stabilizing a wide range of molecular complexes from those between highly reactive reagents to weakly bound complexes often postulated as reaction or catalytic intermediates at room temperature.

1 Introduction

The matrix isolation technique provides a powerful tool to simulate the low temperature and isolated conditions in planetary atmospheres. In particular since the atmospheres are mixtures of molecules, spectroscopic studies of both strong and weak complexes are important and can help the identification of astronomical infrared data.

The matrix isolation technique is based on the isolation of the species to be studied in an inert solid or matrix kept at cryogenic temperatures (usually rare gas or nitrogen). It can be used in conjunction with a wide variety of spectroscopic methods, IR, UV/Visible, X-rays absorption, ESR, Raman, Mössbauer and other methods such as secondary ion mass spectroscopy and NMR. This technique of trapping at low temperatures was originally developped by G. Pimentel in 1954[1] as a mean of stabilizing and studying unstable molecules or transient species as radicals, ions and reaction intermediates which, in the gas phase, have short life times because of the occurrence of bimolecular reactions which in the matrix can be restricted. Later the advantages to be gained in applying matrix isolation to the vibrational spectroscopy study of stable molecules and complexes were recognized and this technique has since been widely used. Isolation of monomeric solute molecules at high density in an inert environment reduces intermolecular interactions compared with other condensed phases. Furthermore with the exception of a few small hydrogenated molecules (water, hydrogen chloride, ammonia, methane) rotation does not occur in matrices and absorption lines for single vibrations are observed with a frequency weakly shifted from that in gas phase. Isolation of monomeric species is quite easily achieved at low concentration but at high concentration dimers and larger multimers can be also trapped. The association process can be controlled and followed by allowing the matrix to warm up slightly. Thus, in spite of the development of supersonic beam techniques for the studies of weakly bonded molecules in gas phase, matrix isolation remains a valuable tool for investigating all types of molecular complexes. For strong interactions it is the only method for characterizing one to one complexes. Moreover infrared and visible–UV photochemical behavior of complexes can be easily observed by irradiation of the matrix with appropriate sources emitting in the corresponding wavelength ranges.

The purpose of this Chapter is to demonstrate that the matrix technique coupled to the spectroscopy is a particular suited technique to study molecular complexes restricted to non metallic and atom species. The paper is divided in four sections. In the first section the experimental technique is described. The second section is devoted to the vibrational spectroscopy of complexes. It allows to understand the mechanism of formation, the structure and the bonding properties of molecular complexes. In most cases it has been combined with *ab initio* calculations to elucidate the nature of intermolecular mode and to determine the vibrational contribution to the thermodynamic properties of the complexed molecules. The third and fourth sections are devoted to the photochemistry of complexes by infrared irradiation and visible–ultraviolet irradiation, respectively.

2 Experimental Techniques

The matrix isolation technique (MIT) has been extensively discussed in the literature[2–8] and only a brief description is given here.

2.1 Closed-cycle Refrigeration

Development of closed cycled refrigerators has allowed new approaches in the matrix isolation research field. They operate on the Gifford-McMahon refrigerator cycle using helium gas from a helium compressor to produce low temperature upon a sample holder.

The Displex system's, a two steps closed refrigerator, provides 10 K–450 K temperature range and can be used in any orientation. Refrigeration is achieved as compressed helium gas is expanded in the Displex expander, the expander being supplied with the helium gas through a flexible hose. Following expansion, refrigeration is created at the expander sample tip. Through a second hose, the helium gas returns to the compressor to complete the closed loop, so that no helium is consumed in the process.

A three steps closed cycle systems Heliplex has been recently developped by ADP Cryogenics Inc. It allows one to reach 4.2 K temperature and replaces the liquid helium cryostat, for which circulation of liquid helium through a heat exchanger is necessary to attain such a temperature. This type of cryostat is needed for neon matrix isolation and photoluminescence studies. The Heliplex HS-4 refrigeration system comprises dual compressors and a refrigeration unit. Two water-cooled compressors supply helium gas to the refrigeration unit. One stream is cooled at progressively colder stages in a Displex regenerative expander and return to the first (expander) compressor unit. A split stream, which supplies the Joule-Thomson circuit, is successively cooled by the two external heat stations of the Displex expander and by its return stream after Joule-Thomson expansion at the 4 K cooling station. This stream then returns to the second (J-T) compressor unit where it is boosted to the suction pressure of the first compressor unit.

Figure 1 shows the schematic general design of a two steps closed cycle refrigerator. Cooldown of the expander to 11 K takes less than one and a half hour.

Before operating, the whole system is pumped down, to keep contamination to a minimum, pumping being maintained throughout the operation. Although oil diffusion pumps backed by a mechanical high vacuum pump are the ones mostly used for this operation, turbomolecular pumps which can be mounted in any orientation allowing minimization of the system configuration constraints are preferable. Antivibration materials to filter vibrations reaching the sample holder are necessary to prevent pollution of measurements by mechanical noise. It is possible with a special device to translate, rotate the vacuum shroud and rotate the sample holder under vacuum and cold (10^{-7} mm Hg, 10 K).

2.2 Sample Preparation

Mixtures of the solute-gas with a large excess of matrix gas are prepared with standard manometric techniques. Figure 2 shows a schematic representation of a vacuum line. This consists of generally a glass line with a great degree

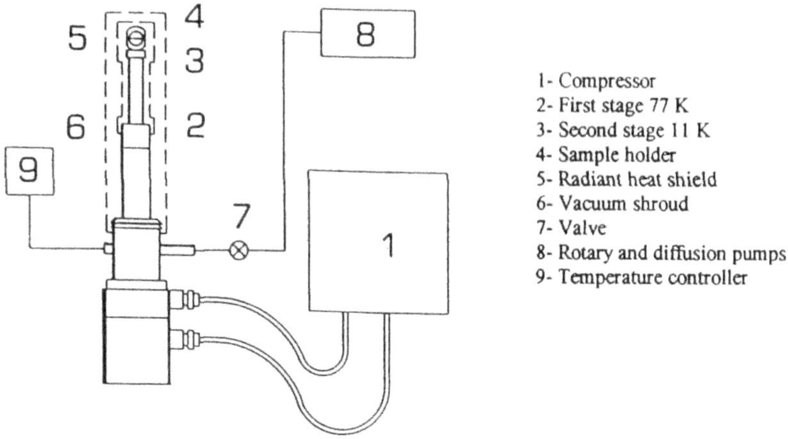

Figure 1: Schematic general design of a two steps closed cycle refrigerator.

of versatility. Mercury, oil (sometimes with perfluorinated oil) or a Barocel manometers are used to control the pressure range.

The matrix ratio (M/R) between inert molecules and impurity species should vary in a range 1000 to 10000 so that impurities are rigorously isolated in the matrix so that interactions between these species can be neglected. For compounds having a high vapor pressure, a bulk divided into two partitions with a volume ratio of 200 can be used. Successive dilutions are performed until the desired concentration is reached. For non-volatile compounds, a Knudsen furnace can be attached at the bottom of the cryostat with the inlet for the matrix gas next to the inlet from the Knudsen cell. Atoms, ions and substances which are unstable at room temperature can be generated either in the matrix gas phase by microwave discharge, flash photolysis or *in situ* after deposition of the matrix by photolysis of an appropriate precursor or by atoms or molecules reactions or by proton beam irradiation.[9] Recently new sources of transient species for matrix isolation studies have been devised: i) a cold window radial discharge which rely on an electrical treatment of gases proceeding to the cold surface,[10] ii) a source combining dc discharge with pulsed supersonic expansion,[11] iii) an apparatus to couple high voltage, low power

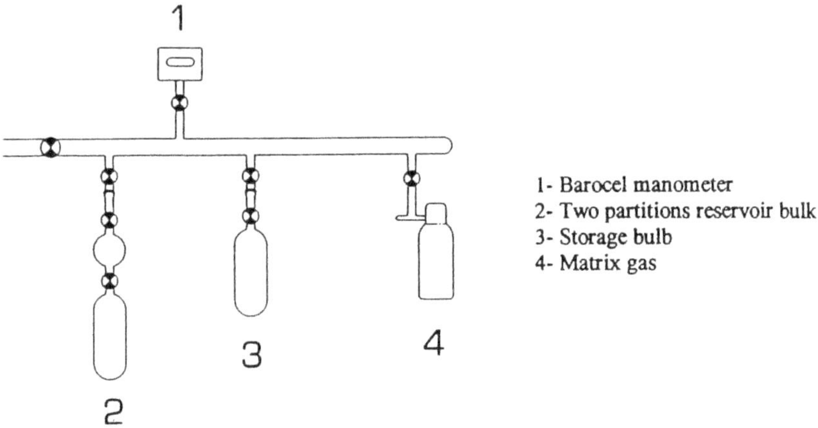

Figure 2: Schematic representation of a vacuum line for sample preparation.

corona discharge.[12]

They are three main methods to prepare molecular complexes for experimental studies. When the interaction is weak the solutes may be premixed in the gas phase prior to deposition but for solutes which react with one another the simultaneous codeposition technique is required. An other way to produce strictly one to one complex in considerably higher concentration can be obtained by a photolysis of a suitable precursor. Due to the so called cage effect, complexes between the photoproducts can be observed. For example identification of the $CH_2 = NH...H_2O$ complex was recently made by photoinduced dehydrogenated reaction of CH_3NH_2 by NO_2 in a cryogenic matrix.[13] The complex between atomic oxygen and ozone was produced by irradiation of an oxygen matrix.[14] Complexes of hydrogen oxides, carbon monoxide and carbon dioxide were formed by photooxidation of isotopic formaldehydes and glyoxal in solid oxygen.[15] Recently $NH : N_2$ and $NH : CO$ van der Waals complexes were prepared by *in situ* photolysis of the precursor molecules HN_3 and HNCO and by *in situ* photolysis of NH_3 in rare gas matrices doped with N_2 and CO respectively.[16]

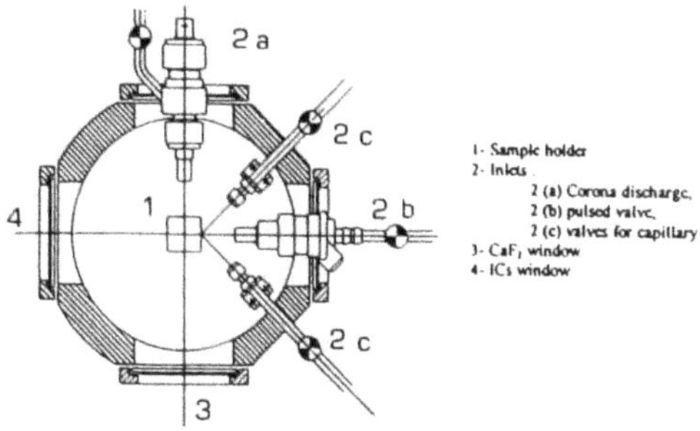

Figure 3: Section of a cell shroud.

2.3 Sample Deposition

Deposition of the matrix gas mixture is generally realized by slow spraying (few millimoles per hour) onto a cooled window, mirror or other substrate at a temperature such that the gas flow instantaneously freezes. Condensation of the sample can be also realized by controlled pulsed deposition, each pulse of mixture condensing before the next one arrives at the surface.[17] Inlets are generally placed in front of the sample holder at about 2 cm of distance and at 45° as illustrated on Fig. 3 which represents a section of the cell shroud at the level of the sample holder. The rate flow of sample mixture to the cryostat can be accurately regulated either with micrometer needle valves on the sample outlet line or using capillary of various lengths. In the latter case the debit of gas is given by the Poiseuille law:

$$D = \frac{P^2 \pi r^4}{8LRT\eta}. \tag{1}$$

where r and L are the radius and the length of the capillary tube; P is the pressure of the gas, η is the viscosity coefficient of the gas; T and R are the temperature and the universal gas constant.

The deposition temperature is an important parameter and depends on the gas nature as showed in Table 1. It is chosen in order to get: i) a rigid lattice preventing migration of the chemical species within the cryostat and its surface during deposition, ii) a size of microcrystals as bigger as possible to minimize light loss by multiple reflexion scattering. Annealing temperatures which allow a non random migration of the dopants by releasing matrix rigidity are also indicated in Table 1.

Table 1: Thermodynamic data of some matrices

	Critical point, K	Triple point, K	Sublimation temperature, K at P~ 10^{-6} Torr	Deposition temperature, K	Annealing temperature, K
Ne	44.4	24.6	9	5	9
Ar	150.7	83.8	31	20	35
Kr	209.5	115.8	42	30	45
Xe	289.7	161.4	58	35	60
N_2	126.2	63.15		17	27
O_2	154.36	54.4		20	40

During cooldown, deposition and annealing temperatures can be monitored using a silicon diode or a thermocouple coupled with an heater. The depth of deposited film, ideally in the range 100 to 250 μm can be measured from the intensities of interference fringes of a small He-Ne laser produced by the growing matrix layer.[18]

2.4 Matrix Properties

It is generally assumed that the rare gas cage of an isolated molecule has the structure of the *fcc* equilibrium phase of solid rare gas. However numerous experimental and theoretical studies of the structure of condensed inert phase show stacking fault density and polymorphism conversion.[19-22] Accordingly, three possible types of cavities exist in the face centered cubic crystals: two types of interstitial holes (tetrahedrical, equidistant from four noble gas atoms and octahedrical, equidistant from six gas atoms) and substitutional holes created by removing gas atoms from their lattice site. In the case of a single substitutional site, the center is equidistant from twelve lattice sites. The radii of such cavities together with other data concerning the noble gases are given in Table 2.[8,23]

Table 2: Structural and molecular properties of some matrices

	Ne(cfc)	Ar(cfc)	Kr(cfc)	Xe(cfc)	$N_2(\beta)$	$O_2(\alpha)$
Lattice constant, Å	4.47	5.31	5.61	6.13	5.66	a=5.403 b=3.429 c=5.086
Polarizability α, Å^3	0.39	1.63	2.46	4.02	1.76	1.6
Quadrupole moment Q, 10^{26} esu cm^2					-1.4	-0.4
Substitutional hole, Å	3.16	3.75	3.99	4.34	3.99	3.64
Octahedral hole, Å	1.31	1.56	1.65	1.8		
Tetrahedral hole, Å	0.71	0.85	0.90	0.97		

Some matrix materials have several solid phases but the only phase change considered in matrix studies concerns the $\alpha \rightarrow \beta$ phase change of oxygen which occurs to 23.8 K and which can complicate any interpretation of diffusion studies. Dynamics of matrix trapping involving distortion of the host material have been also the subject of theoretical[24,25] and experimental studies.[26–29] For a *fcc* lattice the statistical probability to get pair of two molecules A trapped in nearest position is given by the following formula[18]

$$P = 12x_A(1 - x_A), \tag{2}$$

x_A being the molecular fraction of A in gas phase. When two molecules A and B are trapped the probability to form AB pairs is given by:[30,31]

$$P = 24x_Ax_B(1 - x_A - x_B). \tag{3}$$

2.5 IR Spectroscopy in Matrices

Infrared spectroscopy has been the most readily applied method for identification of trapped species and the development of modern Fourier transform IR spectrometers gave rise to a resurgence of infrared spectroscopy. There are several advantages of FTIR over classical dispersive infrared spectroscopy. In particular all the frequencies are detected simultaneously (Feldgett's advantage), the energy throughput is only function of the size of the mirrors (Jacquinot's advantage) and a laser realizes the internal frequency calibration of the interferometer (Connes's advantage). The digital data acquisition and computerization permits multiple scanning and it is possible to record higher signal-to-noise spectra. The most sophisticated instruments allow one to record spectra from 10 cm^{-1} to 40 000 cm^{-1} with a resolution of 0.001 cm^{-1}.

Matrix isolation spectroscopy has a great advantage because the rigid cage prevents molecular rotation. Due to the low temperature and weak forces between the host crystal and the solute, infrared bands are sharp and weakly shifted relatively to gas phase. However, matrix isolated species can give multiplet spectra features which are not due to rotation, aggregation, existence of conformers. This trend has been explained in terms of multiple trapping sites. This is principally observed in argon matrices due to hexagonal close packing besides cubic close packing. In most cases, annealing of the matrix removes the less stable trapping sites.

3 Vibrational Spectroscopy

Molecular complexes formed by electron donor acceptor interaction ('charge transfer' complexes) and those formed by hydrogen bond will be considered successively.

3.1 Charge Transfer Complexes

The theory of 'charge transfer' interaction was first formulated by Mulliken[32] as a result of the observation of an absorption band in the ultraviolet and visible spectrum of many molecular complexes. This characteristic 'charge transfer' transition band does not appear in the spectrum of either component alone. The position of the band is often related to the ionization of the donor, the electron affinity of the acceptor and the charge separation. However, further theoretical and experimental studies have shown that according to the nature of the two partners, electrostatic forces play a dominant role in the determination of the properties of electron donor–acceptor complexes in their electronic ground state.

Interaction occurs generally by transfer of a lone electron pair from an electron donor or base into an unfilled molecular orbital of an electron acceptor or Lewis acid. For the n-v or n-σ^* type complexes, strong interaction mainly electrostatic in nature is formed when the partners are a polar acceptor and a polar donor. Intermediate strength is generally observed between a polar and a non polar molecule and there is a comparable contribution to the stabilization between electrostatic forces (mainly dipole-quadrupole interaction) and charge transfer. Between non polar molecules the stability is mainly due to dispersion forces and charge transfer. For the n-π^* type complex, the dipole of the donor is perpendicular to the π dipole moment of the acceptor and intermediate strength occurs with electrostatic contribution. In the other π complexes types $(\pi - \sigma^*)$ and $(\pi - \pi^*)$ there is a contribution of both dispersion, electrostatic forces and charge transfer. In fact, the Highest Occupied Molecular Orbital (HOMO) of the donor matches or mismatches with the Lowest Unoccupied Molecular Orbital (LUMO) of the acceptor and the exact strength of interaction depends on the difference in energy of the two concerned orbitals. The concept of hard and soft acids and bases (HSAB principle) first developped by Pearson [33] then by Douglas and Kollman [34] is attractive in determining the strength of the interaction in molecular complexes. Hard bases (nucleophile) is characterized by a low value for the energy of the HOMO frontier orbital, a soft base by a higher value. A hard acid on the contrary is characterized by a high value for the energy of the empty frontier orbital and its hardness decreases with the decrease of LUMO orbital energy. When the two frontier orbitals are nearly the same the interaction is predominant and a strong electron transfer occurs. For hard-hard interactions very little electron transfer occurs and interaction depends principally on the ionic interaction of the reagents.

Relative intensity of the vibrational stretching absorptions can be strongly changed by charge transfer. One of the most attractive example of this observation concerns the 1:1 complex between H_2O and the lithium atom. [35] The vibrational spectrum of the H_2O–Li complex recorded in krypton matrix is characterized by small frequency shifts but strong intensity changes of the water bands. The OH stretching modes ν_1 and ν_3 display a remarkable intensity inversion with respect to free water due to strong charge transfer from water towards Li.

For weakly bounded complexes interaction between the solutes and the quadrupolar nitrogen matrix can cause a change in the structure of the complex with respect of its structure in a rare gas matrix or in gas phase. An example of this has been observed for the $(SO_2)_2$ dimer as reported below.

Halogen–Base Complexes

Numerous studies in matrix have been carried out upon complexes involving halogen molecules as Lewis acid and various bases. Most of them have been reviewed and described by Ault.[36] Codeposition of Cl_2 and ClF with dimethylether,[37] methanol,[37] sulphide and phosphorus bases[38] give rise to very intensive absorptions assigned to the perturbed ClF (which interacts by chlorine atom) and Cl_2 stretching modes. The Cl_2 molecule is infrared inactive but the spectroscopic constants of chlorine isolated in argon have been determined by laser induced fluorescence and Raman spectroscopy. The ω_e values were found at 554.5 cm^{-1}, 547.2 and 539.3 cm^{-1} for $^{35-35}Cl_2$,$^{35-37}Cl_2$, and $^{37-37}Cl_2$.[39] In a complex, the chlorine stretching is IR allowed by symmetry. Similar shifts of about 50–100 cm^{-1} were observed for ClF and Cl_2 in the complexes with the oxygen bases studied. With trimethylsulfur and trimethyl phosphorus, the ClF stretching mode shifts to 300 cm^{-1} and 450 cm^{-1} respectively with an intense charge transfer absorption and a substantial intensification of the halogen stretching indicating a very strong interaction and large degree of electron distribution. Complexes of Cl_2 and ClF with alkenes and alkines have been also reported.[40] The structure appears to be T–shaped with the halogen perpendicular to the multiple bond. The magnitude of the halogen stretching mode is less than in the complexes with lone electron pair donors but nevertheless very substantial. Interaction for alkene complexes are considerably stronger than their analogous alkine complexes and addition of methyl groups leads to a stepwise decrease in the halogen stretching frequency.

Recently very weak complexes between halogenated carbonyls (COF_2, COFCl, $COCl_2$) and Cl_2 were studied.[41] Only a shift of about 8 cm^{-1} was observed for the Cl_2 stretching mode. The ν_{CO} and ν_{C-X} (X = Cl, F) stretching modes of the bases were also perturbed and allowed a qualitative measure of the interaction strength after removing of the Fermi resonance observed in ν_{CO} region for the COF_2 : Cl_2 and COFCl : Cl_2 complexes. Relative shifts of the ν_{CO} stretching mode were found to be nearly similar for COFCl (0.68%) and $COCl_2$ (0.52%) but weaker for COF_2 (0.26%). This observation suggests the attachment of Cl_2 to oxygen and seems to exclude a π complex. The strong electronegativity of fluorine atoms in COF_2 draw electron density from the C=O bond. Thus, the charge transfer from oxygen atom into the unfilled chlorine σ^* antibonding molecular orbital is weaker than in other complexes $COCl_2$: Cl_2 and COFCl : Cl_2. *Ab initio* calculations at the MP2 level were in good agreement with the experimental results. A correlation between the complexation energy and the intermolecular distance was evidenced. The three complexes have a plane equilibrium structure and the chlorine molecule is practically oriented with the oxygen lone pair.

Ozone–Base Complexes

A wide range of molecular complexes involving ozone has been recently investigated and their photochemistry has been studied.[42–56] They are summarized in Table 3.

Table 3: Molecular Ozone Base complexes

System	Matrix	ν_3, cm^{-1}	$\Delta\nu_3$, cm^{-1}	Reference
O_3 monomer	Ar	1039.7		55
	N_2	1043.0		56
$O_3 \ldots SO_2$	Ar	1039.9	0.2	42
$O_3 \ldots NO$	N_2	1042.0	1.0	43
$O_3 \ldots N_2$	Ar	1040.9	1.2	44
$O_3 \ldots CO$	Ar	1041.4	1.7	45
$O_3 \ldots PH_3(PCl)_3$	Ar	1037.2	-2.5	46
$O_3 \ldots C_2H_4$	N_2	1040.2	-2.8	47
$O_3 \ldots H_2CO$	N_2	1047.2	4.2	47
$O_3 \ldots AsH_3$	Ar	1035.5	-4.2	48
$O_3 \ldots C_2H_5I$	Ar	1033.7	-6.0	49
$O_3 \ldots Cl_2$	Ar	1032.1	-7.6	50
$O_3 \ldots CF_3I$	Ar	1032.0	-7.7	51
$O_3 \ldots NH_3$	Ar	1049.1	9.3	52
		1047.6	9.3	52
$O_3 \ldots Br_2$	Ar	1030.0	-9.7	50
$O_3 \ldots BrCl$	Ar	1028.0	-11.7	50
$O_3 \ldots ClF$	Ar	1026.0	-13.7	53
$O_3 \ldots ICl$	Ar	1019.5	-20.2	54
	N_2	1023.0	-20.0	54

Some of them (ozone complexes with ClF, ICl, CH_3I, CF_3I, Br_2, BrCl) show a transfer charge character which explains that visible irradiation, which does not photodissociate isolated ozone, can induce photodecomposition of the complex as developped in the Sec. 5. The strength of interaction has been mainly characterized by the shift of the antisymmetrical vibrational ν_3 mode of ozone. As seen in the Table 2 the strength of the interaction ranges from the extremely weak complex $O_3...SO_2$ complex ($|\Delta\nu| = 0.2$ cm^{-1}) to the stronger $O_3...ICl$ complex ($|\Delta\nu| = 20.0$ cm^{-1}). Red shift of the ν_3 mode is generally observed but for N_2, SO_2, NO, CO, H_2CO and NH_3 in interaction with ozone, a weak blue shift is measured, suggesting a different bond interaction. Note that a blue shift was also observed for the $O_3 : H_2O$[57] and $O_3 : CH_3OH$[58] complexes which present a weak hydrogen bonding character, in opposition with the red shift of 13 cm^{-1} observed in O_3 : HF complex.[59]

As previously demonstrated by Andrews,[60] information upon the structure of ozone complexes can be obtained from scrambled isotopic 16–18 ozone, (^{16}O /^{18}O = 1) since the type of multiplet and relative intensities of its components indicate the number of equivalent oxygen atoms in the complex. In the case of symmetric ozone complexes, the ozone submolecule in the complex has the same symmetry as ozone itself. Consequently, one observe in the ozone ν_3 region a sextet with relative intensities 1/2/1/1/2/1 which matches the intensity distribution of the isotopic ozone bands. In the case of asymmetric ozone complexes due to a terminal oxygen attachment, a multiplet with eight component of equal intensity is expected in the ν_3 ozone region. With respect to interchange of terminal oxygen bond, two bands characterize the 16-16-18 and the 16-18-18 complexed species and one band characterizes the corresponding symmetric submolecule ozone species.

SO_2 and SO_3 -Base Complexes

Sulfur dioxide can act as a weak base (lone pair electron donor) or as acid (π electron acceptor) for the sulfur atom. Some complexes with NH_3 and methylamine have been reported[61] as well as with water.[62] Elucidation of the SO_2 dimer structure has also received attention because it can be a part of the mechanism of photooxidation of atmospheric sulfur dioxide.[63]

The earliest identification of $(SO_2)_2$ came from infrared spectroscopy of SO_2 trapped in nitrogen matrix.[64] It was concluded that two subunits in the dimer were not equivalent with nearly orthogonal orientation of the transition dipole moments and that SO...S bond was formed. No infrared study in the gas phase has been published to date and only microwave studies have been performed. Recently Fourier Transform spectroscopy has been used to study various isotopic species of sulfur dioxide isolated at different concentrations in solid argon.[65] The most stable $(SO_2)_2$ dimer possesses a structure with the transition dipole moments of both subunits parallel or antiparallel one to another with the four SO oscillators not equivalent by pair. This suggested structure (C_i symmetry) which is consistent with *ab initio* calculations and not in opposition with microwave study, is different from that observed in nitrogen matrix. Such a difference can be explained by a strong anisotropic solute-matrix potential in nitrogen which tends to distort the highly symmetric structure of isolated aggregates. Rare gas matrices preserve generally the structure as in the gas phase.

Infrared spectra of 1:1, 1:2 and 2:1 complexes between water and sulfur dioxide in argon and nitrogen have been reported using isotopic substitution for water and sulfur dioxide.[62] In the 1:1 complex the interaction is of charge transfer type, the oxygen atom playing the role of electron donor. This interaction is not comparable to that deduced from microwave spectroscopy for the $SH_2 : SO_2$ complex in which weak O...H hydrogen bonds have been evidenced.[66] The greater basicity of the oxygen atom can explain this difference. The $(SO_2)_2 : H_2O$ complex has probably an asymmetrical structure in which one SO_2 molecule forms a charge transfer complex with water through its sulfur atom, the second one being H bonded through one of its oxygen. The spectrum of the $(SO_2) : (H_2O)_2$ complex is interpreted as occurring from a water dimer engaged in a cyclic structure through formation of one hydrogen bond and one charge transfer interaction O...S with SO_2. The $SO_2 : NH_3$ complex as the complex $SO_2 : H_2O$ shows weak sulfur-nitrogen coordinate bond with a formally S(IV) center.[61]

SO_3 molecule is a strong Lewis acid but it can also oxide the partners rather than form complexes and the matrix isolation appears as a good tool for studying these complexes. Strong interaction were evidenced between SO_3 and ammonia,[67] methylamine[67] and various oxygen bases as methanol, ethylene oxide, formaldehyde, furan, THF, dimethylether.[67] For all the complexes there is coordination of either a nitrogen or oxygen atom to the sulfur atom. In most of the complexes the antisymmetric stretch of the SO_3 subunit appeared splitted in two components as a consequence of the lowering of symmetry in the complex and the forbidden symmetric stretch was activated. Surprisingly the complex between SO_3 and water postulated as the intermediate in the formation of H_2SO_4 in the atmosphere is not yet accurately identified in spite of some papers on the subject.[68,69] In the most recent study a weak $H_2O : SO_3$ complex (S...O interaction) was only identified in nitrogen matrix and not in argon matrix.[70]

Oxygen–Base Complexes

Due to a possible role of the oxygen dimer in oxygen photolysis in the Herzberg continuum, several studies have been devoted to its identification in gas phase with different methods. Results showed that the weakly bonded $(O_2)_2$ is metastable. Its spectrum in the visible and UV was recorded by Long and

Ewing.[71]

The first study in matrix on this subject has been performed by Pimentel and Cairns thirty years ago.[72] None of the spectra of solid oxygen including isotopic mixtures suggest the presence of O_4 molecules. The spectrum of solid oxygen codeposited at 15 K or 20 K consists of a sharp absorption centered at 1549 cm^{-1} (half width 4 cm^{-1}) and a more intense broad absorption (half width 45 cm^{-1}) with well defined peaks at 1591 and 1617 cm^{-1}. The broad band is due to combination involving translational lattice modes; the sharp feature is apparently the O_2 fundamental appearing because of the lattice imperfection.

More recently high resolution electronic spectroscopy study of the O_2-O_2 dimer has been made by Goodman and Brus in a solid neon matrix at 4.2 K.[73] The spectra were found consistent with a parallel D_{2h} geometry for the dimer.

In matrices little is known about heteromolecular complexes involving O_2 molecule. Recently a very weak complex between nitrous oxide and oxygen was identified in nitrogen matrix by the shift of about 0.5 cm^{-1} of the ν_1 and ν_3 modes of N_2O.[74] However, in solid oxygen containing various alkenes, Hashimoto and Akimoto observed prominent bands of contact charge transfer bands well separated from the intrinsic absorptions of isolated alkenes.[75] Their maxima were measured at 218, 243, 234, 235, 263 and 287 nm for propene 2, methylpropene, *cis* 2 butene, *trans* 2 butene, 2 methyl 2 butene and 2,3-dimethyl-2-butene respectively. Excitation within these bands was studied.

Hydrogen–Base Complexes

Prominent shifted H–H stretching vibration has been observed in argon matrices containing dihydrogen and Lewis bases suggesting formation of weak van der Waals complexes in which hydrogen interacts with the base.[76] The relatively weak magnitude of the red frequency shifts of the ortho–H_2 band (10 cm^{-1} with H_2O, 14 cm^{-1} with NH_3, 20 cm^{-1} with $(CH_3)_3N$) correlates with the base strength suggesting that the interaction is based on hydrogen acting as a Lewis acid. When dihydrogen is codeposited with alkali halides in argon matrices, the H–H stretching vibration is red shifted by as much as 114 cm^{-1} from the position of the Q branch vibration of ortho–H_2. The observations suggest that it is the amount of the alkali halide ion pair which is responsible for the perturbation of H_2.[77]

3.2 Hydrogen Bonded Complexes

Hydrogen bonding is formed between a proton donor group AH where A is an electronegative atom (O, N, S, Cl, F, Br, C,...) and a proton acceptor

B which is a lone electron pair of an electronegative atom or a π orbital. A profound similarity exists between hydrogen bonded and donor–acceptor complexes. The hydrogen bond is strongly electrostatic in nature with small but significant contribution from charge transfer. Hydrogen bonding is more localized and hence it is a distinctly directional and specific interaction. The enthalpy of H bonding falls in the range of less 4 to 60 $kJ\,mol^{-1}$.

Infrared and Raman spectroscopy have been principally used in the studies of hydrogen bonding. Several vibrations of the complex AH...B can be characterized:

- the fundamental A–H stretching which is shifted to lower frequencies than the corresponding band of the isolated AH with a broadening and increasing of the integrated intensity by a factor of up to ten or more due to the rate of change of the electrostatic dipole moment,

- the AH deformations (bending) which are shifted to higher frequencies and which do not show any substantial band broadening or intensity change,

- new vibrational stretching or deformation modes H...B which appear in the far IR region,

- the vibrational modes of the H acceptor B which are weakly shifted to longer or shorter wavelengths.

The perturbed AH fundamental stretching has been more widely investigated in H bonding studies than any other experimental effect. The relative frequency shift $(\nu_0 - \nu_s)/\nu_0$ where ν_0 and ν_s are the frequencies of the isolated and hydrogen bonded AH oscillators, respectively, is a measure of the strength of the hydrogen bond. It varies with the relative proton affinity of the two bases A^- and B. Ault and Pimentel proposed a vibrational correlation diagram to compare the AH frequency shifts with a normalized proton affinity difference $D = [PA(B) \quad PA(A^-)] \ / \ [PA(B) + PA(B^-)]$ and distinguish three classes of hydrogen bonds.[78] Class 1 corresponds to weak and medium interactions and is characterized by a decrease of the AH stretching frequency of at most 20 to 30%. Class 2 corresponds to a proton sharing between the two proton acceptors A^- and B identified through the stretching mode of the proton along the B–A axis which is always found below 1500 cm^{-1}. Class 3 pictures an ion pair structure $A^-...BH^+$ oscillator perturbed by an anion A^-. For weak and medium hydrogen bonded complexes the electrostatic interaction is the dominant term in the intermolecular potential, while for strong hydrogen bonded complexes the polarisability of the proton donor bond plays an important role.

This trend is well illustrated in the Fig. 4 which compares the relative frequency shifts for some HX...B complexes (X = Cl, Br, I) plotted *versus* the proton affinities of the various bases B. With respect to HCl and HBr peculiar properties are observed for HI, tied on the one hand to the low value of its dipole moment and, on the other hand, to its lower stability, which makes this acid the strongest within the HX series. As a consequence, in the case of weak interactions which are mainly electrostatic in nature, the vibrational perturbations are weak because the dipolar contributions are small; conversely, in the case of strong interactions, the ionic dissociation of HI is more easily achieved, leading to proton sharing and even to proton transfer of the type $BH^+ \ldots I^-$.

The frequency shifts are sensitive to the matrix nature. W.A.P. Luck *et al.* found a linear dependence between perturbed ν_{OH} of alcohols and the square of the critical temperature of the rare gas materials.[80 a]

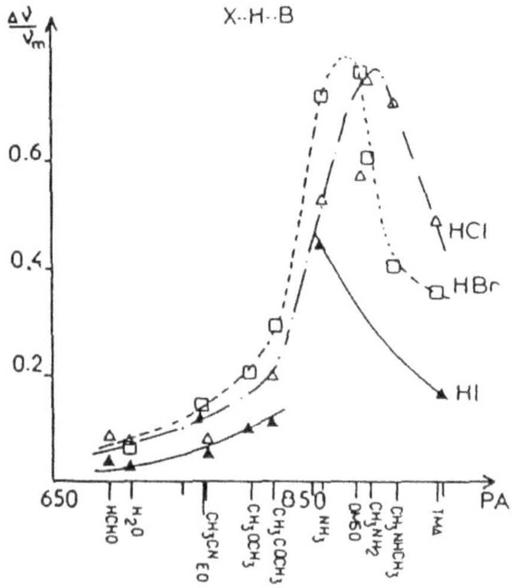

Figure 4: Correlation between relative frequency shift and proton affinity (kJ mol^{-1}) for some XH...B complexes (X=Cl, Br ot I) ($\Delta\nu = \nu_m - \nu_s$). ν_m is the HX monomer frequency.[79]

The broadening of the A–H stretching band in a H bonded complex and its structure have been the subject of numerous studies in liquid and gas phases. In the theories of hydrogen bonding, it is generally assumed that the main band

[a] About this correlation see also Chapter 8, Sec. 4.4 of the present book. (A.V.)

shaping factors are the mechanical and electrical coupling of the internal and external modes of the hydrogen bonds.[81,82] Broadening of the ν_s bands have been explained by stochastic effects in liquids[83,84,85] and rovibrational structure in the gas phase.[86] However, in matrices usually a substantial narrowing of the absorption bands is observed due to the extreme low temperatures. This has been found for weak hydrogen bonding ($\Delta\nu/\nu < 0.1$), where the ν_s band is very narrow (FWHM < 10 cm^{-1}). In the case of a medium strong hydrogen bonding ($\Delta\nu/\nu$ between 0.1 and 0.2) two situations were encountered: i) either narrow bands as for HF...NH_3[87] and HCl...$S(CH_3)_2$[88] systems, or ii) broad bands with a strong narrowing on deuterium substitution as for HCl...DME[89] or HCl...ethylene oxide[90] complexes.

Far infrared provides direct information on the intermolecular interactions. As a matter of fact two 1:1 complexes between CH_3CN and HCN in solid argon were observed: a predominantly antiparallel structure giving rise to four IR bands and a more stable linear hydrogen bonded structure exhibiting only two far IR bands.[91]

Weak and Medium Hydrogen Bonded Complexes

A wide variety of complexes have been described in the literature and complexes to be discussed here include mainly the most recent investigations and systems of interest for atmospheric chemistry. They are presented in regard to the different proton donors.

Hydrogen Halides

HX dimers and polymers (X = F, Cl, Br, I) trapped in rare gas and nitrogen have been extensively studied as well in the near and the far IR as in Raman (see[92,93,94] and references therein). The dimers have an open structure while trimers and tetramers have a cyclic structure of high symmetry. In the case of HF the cyclic polymers have been identified till n = 6[95] and some open polymeric structures have been identified. Several base...HF complexes have been described by Andrews[96] and base...hydrogen halides complexes have been reviewed by Barnes.[97] Among them, comparison of the spectral properties of molecular complexes between acetonitrile and hydrogen halides (HCl, HBr, HI) is of interest. Results suggest different possibilities of attack on the cyanogroup by HX molecules.[98] In the case of X = Cl, this attack is mainly axial; for X = Br, the presence of two groups of bands suggests the existence of two structures: one axial and the other T–shaped. This interpretation could explain that in gas phase the CH_3CN/ HBr mixture is unstable because the double attack of the cyano group giving rise to the addition/ionization process and formation of $(CH_3C{=}NH_2)^+Br^-$ particles are not observed with HCl.

Recently infrared study and *ab initio* calculations of NO and ClNO complexes with HCl have been investigated in argon matrices. In the NO : HCl complex[99] the NO stretch strongly increases in intensity and shifts to higher energy with respect to the monomer NO stretch. *Ab initio* calculations show that the most stable structure HCl/NO is linear with 0.002 Å shortening of the NO bond distance and an increase of the HCl bond length with an identical amount. In the ClH : ONCl complex,[100] interaction of the Cl, N and O atoms of ClNO with the H atom of HCl was evidenced. Comparison with the vibrational spectra for the HCl/ClNO and HCl/NO systems is of interest. The magnitude of $\Delta\nu/\nu$ ($\Delta\nu = \nu_{free} - \nu_{complexed}$) calculated for HCl in the HCl : NO and HCl : ClNO complexes $\Delta\nu/\nu = 1.5 \cdot 10^{-2}$ and $4.5 \cdot 10^{-2}$, respectively, shows that HCl forms a substantially stronger complex with ClNO than with NO. In parallel the relative blue shift $\Delta\nu$ observed for ν_{NO} in the HCl : ClNO complex ($1.7 \cdot 10^{-2}$) is stronger than that in the HCl : NO complex ($3 \cdot 10^{-3}$). Such a blue shift has been explained by charge redistribution in perturbed NO stretching. If we take into account, as indicated by Lucas *et al.*,[101] the frequencies of NO^+ (2240 cm^{-1}) and NO^- (1350 cm^{-1}) as measurements of antibonding molecular orbital occupancy of NO, the shifts observed towards the high frequencies indicate a 0.07 electron and 0.02 charge removal from the NO group in the ClNO : HCl and NO : HCl complexes, respectively.

Water: Self Association

The infrared spectrum of the water dimers in matrices has been the subject of many investigations in rare gas and nitrogen matrices.[102–105] The dimer has an open linear structure with one water molecule acting as a proton donor and the other as proton acceptor. Isotopic studies show that the D bonded form is preferred to the H bonded form in the mixed dimer and for $(HDO)_2$ in argon only the D bonded form is observed. Thermal equilibrium data between the two forms indicate that the D bonded dimers are approximately 60 cm^{-1} more stable than the H bonded dimers.[106] Intensity evolution of intermolecular vibrations of the water dimer in terms of the corresponding intensities for the monomer have been theoretically explained by Zilles and Pearson[107] taking into account the difference in atomic polar tensors from monomer to dimer. The water polymers have been studied far less than the water dimers. However, it now seems established that in matrices the trimer is cyclic with three equivalent water molecules.[108]

Water: Heteromolecular Complexes

In heteromolecular complexes the water molecule can act either as proton donor or as proton acceptor. For weak complexes characterized by weak frequency shifts of the water vibrations it is sometimes difficult to choose between these two possibilities. Isotopic studies are needed. In the complex of hydrogen bonding the water submolecule does not keep the C_{2v} symmetry and the coupling of the two OH and OD oscillators is not symmetrical. As a matter of fact it was found that water acts as proton donor in the $H_2O : O_3$,[57] $H_2O : I$,[109] $H_2O : CO$,[110] $H_2O : C_2H_4$[111] complexes but as lone pair donor in the $H_2O : SO_2$,[62] $H_2O : CO_2$,[110] $H_2O : C_2H_2$[112] and $H_2O : X_2$ (X= Cl, Br)[109] complexes.

A representative selection of water–hydrogen bonded complexes is presented in Table 4 with indication of the water bands in cm^{-1} observed in the infrared spectra. Bases have been classified by their proton affinity.[113] It can be seen that the relative frequency shifts are not the same for the three modes and in some cases the frequency shift does not increase with proton affinity. Comparison of the water–water frequencies in the $(H_2O)_2$ dimer and in the $H_2O : CH_3CN$ complex is of interest.[114] Frequencies of the ν_1 and ν_2 modes are similar for both complex whereas the ν_3 frequency is higher with acetonitrile. With dimethylether and acetonitrile the structure of complexes involving two base molecules has been reported. The 2:1 dimethylether–water complex has a symmetrical structure in which the water molecule keeps its C_{2v} symmetry[115] whenever the 2:1 acetonitrile–water complex is asymmetrical with a structure H–O–H. . .$O(CH_3CN)_2$ in which the hydrogen bond is in perpendicular plane to that of the dimer.[114]

Water Ices

Due to the importance of vapor-deposited water ice in our solar system and interstellar space, several studies have been performed upon the low temperature, low pressure structures of ices using techniques such as thermal analysis,[121,122] electron diffraction,[123] X-rays diffraction,[124] inelastic neutron scattering spectra,[125,126] and infrared spectroscopy.[127] Water ice which is formed by vapor deposition at low pressure shows several phases during warm up from 10 to 188 K. Vapor deposition at 10 K produces a high density (1.07 $g\,cm^{-3}$) amorphous form (I_{ah}) which persists until 38 K which then undergoes a transition into a low density (0.94 $g\,cm^{-3}$) amorphous I_{al} between 38 and 68 K. At 131 K the ice transforms into a third amorphous form (I_{ar}) which precedes the crystallization of cubic ice (I_c) and coexists metastably with I_c from 148 K until at least 188 K. At about 210 K, the cubic ice transforms irreversibly into

Table 4: Hydrogen bonded complexes between water and various bases. Proton affinities are given in column 1 in parentheses; the frequency shift $10^3\Delta\nu/\nu = (\nu_{complex} - \nu_{free})/\nu_{free}$ is shown in columns 3 to 5 in parentheses. Asterisks label wavenumbers of the major component.

Base	Matrix	ν_1, cm^{-1}	ν_2, cm^{-1}	ν_3, cm^{-1}	Ref.
Reference	Ar	3638.0	1589.1	3734.3	118
H_2O	N_2	3634.9	1597.5	3727.5	118
monomer	O_2	3636.7	1614.6	3732.3	110
frequencies			1612.8		
			1591.6		
CO	O_2	3635.8 (-0.5)	1594.0 (-11.6)	3727.5 (-2.4)	110
(140)					
C_2H_4	Ar	3613.0 (-6.9)	1594.7 (3.5)	3717.5 (-4.5)	111
(158)					
H_2CO	Ar	3585.1 (-14.5)	1610.7 (13.6)	3709.3 (-6.7)	116
(170)		3580.4	1620.6 (14.5)	3703.5 (-6.4)	
	N_2	3577.6 (-15.8)		3701.3	
		3572.8			
H_2O	Ar	3574.0 (-17.6)	1611.0 (13.8)	3709.0 (-6.8)	103
(170)	N_2	3550.0 (-23.4)	1618.5 (14.5)	3699.0(-7.6)	104
CH_3OH	N_2	3627.2* (-2.1)	1603.1 (3.5)	3713.2 (-3.8)	117
(171)					
C_6H_6	Ar	3615.9* (-6.0)			118
(183)	N_2	3610.6 (-6.7)	1604.0 (4.1)	3709.6 (-4.8)	
		3609.0	1600.1	3708.6	
CH_3CN	Ar	3581.0 (-15.7)	1613.5 (15.4)	3719.5 (-4.0)	114
(187)		3575.5	1609.5	3707.5 (-5.4)	
	N_2	3568.0 (-18.4)	1633.0 (22.2)		
		3565.0	1631.0		
CH_3OCH_3	Ar	3506.5 (-36.1)	1615.3 (16.5)	3706.3 (-7.5)	115
(190)	N_2	~ 3510 (-34.3)	1622.0 (15.3)	3699.0 (-7.6)	
NH_3	Ar	3434.9 (-55.8)		3702.2 (-8.6)	119
(205)	N_2	3414.0 (-60.8)		3697.0 (-8.2)	
		3374.0			
$(CH_3)_2NH$	Ar	3193.4 (-122)		3702.2 (-8.6)	120

hexagonal crystalline ice (I_h). However the specific temperature at which the different transitions occur depend strongly on heating rate, deposition condition and annealing time.

Vibrational spectrum of solid $H_2O(D_2O)$ is principally characterized by a strong and broad band at about 3250 cm^{-1} due to modes of OH stretching vibrations [127] and a weaker absorption at 1660 cm^{-1}. Temperature dependence of the 3250 cm^{-1} band (band intensity, peak frequency, band width) have been observed and related to structural differences between amorphous and crystalline ices. Recently optical constants and integrated absorbances have been measured at different temperatures from 4000 to 50 cm^{-1}. [128] Recent spectroscopic studies of amorphous ice revealed a weak feature on the high frequency side of the intense OH (OD) broad band. [129] This feature measured at 3693 cm^{-1} (H_2O) and at 2728 cm^{-1} (D_2O) has been assigned to a stretching mode of water molecules (at the surfaces) whose hydrogen bonds are not saturated. During the built up of the ice film the "dangling band" grows to a saturation value of 0.02 absorbance units. Adsorption of species as CCl_4, $CFCl_3$, CH_3COCH_3 show that these compounds interact with the ice surface via the OH(D) bond dangling so that the frequency shifts towards lower frequency. [130]

Infrared spectra of water ices containing various species have been also described as for example ice containing methanol, [131] carbon oxide, [132] carbon dioxide. [133]

Alcohols

Hydrogen bonding between alcohols of the type R_3COH (R= H, CH_3, CH_2F, CF_3) as proton donors and various bases: N_2,[134] CO,[134] acetone,[135] nitromethane, [135] acetonitrile, [136] dimethylether, [137] water, [117] as proton acceptor have been studied.

In the IR spectra, the OH stretching and the large amplitude τ_{OH} (torsion) motion of alcohols are highly sensitive to complexation effect. The largest values of the frequency shifts have been observed for the perfluoro-t-butyl alcohol (PFTB) for which the proton acceptor character is very weak so that the extent of self association is relatively small and the acidity properties are strong with a pK_a value of 5.2 at 25°C close to that measured for the carboxylic acids. As for the PFTB monomer, the PFTB : base complexes band are very sensitive to matrix nature. A red shift of 46 cm^{-1} was measured for the ν_{OH} of the PFTB : CH_3CN complex from argon to nitrogen matrix. [136]

Ammonia, Hydroxylamine

Infrared spectra of ammonia clusters have been studied in neon, argon and nitrogen matrices.[120] The NH_3 dimer spectrum show three non linear intermolecular hydrogen bonds, one N–H bond from one submolecule and two N–H bonds from the other submolecule forming an asymmetric cyclic dimer structure with two unequivalent weakly bonded N–H submolecules.

Infrared spectra of 1:1 complexes of NH_3 with mono-, di-, trimethylamine in solid argon have been described.[120] Absorptions assigned to the NH_3 submolecule in the complexes exhibit red shifts nearly similar to the shifts observed for the more strongly perturbed submolecule in the $(NH_3)_2$ dimer. There is no correlation of NH_3 submolecule shifts with base proton affinities which suggests similar electrostatic interaction and similar structures in $(NH_3)_2$ and ammonia–methylamine complexes.

Hydrogen bonded complexes involving hydroxylamine as proton donor has received little attention. IR spectra of the hydroxylamine dimer has been interpreted in terms of a cyclic doubly hydrogen bonded dimer of C_{2h} symmetry consistent with that predicted theoretically.[139] A complex between HF and $HONH_2$ has been spectroscopically identified in argon matrix. Results suggest the formation of a primary hydrogen bond to the amine and a secondary chelating hydrogen bond between the fluorine and the hydroxyl hydrogen.[140]

Nitric and Nitrous Acid

Recently nitric acid complexes with N_2, H_2O, CO and NH_3 have been identified by their OH stretching mode[141] and two detailed studies of the interaction of nitric acid with methane and ethylene were reported.[142,143] The spectra of a $HNO_3/CH_4/Ar$ matrices prove the formation of a well defined $CH_4 : HNO_3$ complex with a relatively strong perturbation of OH(OD) group vibration (43 cm^{-1} red shift) which indicates the complex geometry with the OH group direct towards a face of the methane tetrahedron.[142] When methane is substituted by ethylene the OH stretching fundamental is 300 cm^{-1} red shifted whereas the OH in plane bending and the OH torsion modes are 100, 150 cm^{-1} blue shifted respectively. The most perturbed ethylene mode is the ν_7 CH_2 out of plane wagging fundamental which is blue shifted by 26 cm^{-1} from the ν_7 mode of C_2H_4 monomer. The most stable structure is probably the structure in which the plane of HNO_3 molecule is perpendicular to the C_2H_4 plane and to the C=C bond of the ethylene molecule.[143]

Matrix infrared spectra and *ab initio* calculations of the *trans* and *cis* nitrous acid isomers complexes with N_2, CO and NH_3 have been carried out recently.[144,145] The perturbation of the OH group vibration shows that the six complexes are hydrogen bonded. The blue shift of the CO vibration in the *trans* and *cis* HONO complexes indicates a OC...HONO structure. The OH stretching vibration in the H_3N...HONO *trans* complex is 800 cm^{-1} red shifted with respect to the *trans* HONO monomer indicating a very strong hydrogen bond.

Carboxylic Acids

Matrix isolation FTIR and *ab initio* study of complexes between formic acid with carbon monoxide and nitrogen have been recently reported.[146] The most stable complexes are formed via a nearly linear hydrogen bond yielding a substantial perturbation of the OH stretching and torsional vibrations of the formic acid.

Proton Transfer

The proton transfer process between two interacting molecules is one of the most interesting problems in the hydrogen bonding phenomena. Several experimental studies based on the matrix isolation technique have been devoted to the problem of proton transfer within bimolecular species involving hydrogen halides[78] or carboxylic species[147,148] and ammonia or volatile organic amines. In rare gas matrices (argon) complete transfer does not occur even in the pair involving the best proton donor candidate (HI) and strong bases as ammonia or amines.[149] As shown by theoretical approaches of the proton transfer, effects of the environment on such complexes via the introduction of a reaction field[150,151,152] play a crucial role. Spectroscopic studies of the 1-1 complexes of HI and either NH_3 or $N(CH_3)_3$ (TMA) trapped in four different matrices (Ar, O_2, N_2, C_2H_4) allowed to show that the rate of proton transfer is strongly matrix dependent.[149] All the steps of proton transfer from a perfect sharing between the two proton acceptor I^- and NH_3 to a complete proton transfer were evidenced. The complexes were characterized by the stretching vibration (ν_s) of the proton along the I–N axis. Figure 5 gathers the position and the structure of ν_s for both complexes in the different matrices. For IH–NH_3, ν_s was measured to 1255 cm^{-1} (Ar), 1310 cm^{-1} (O_2), 1847–1967 cm^{-1} (N_2) and 2315 cm^{-1} (C_2H_4). In ethylene matrix, the frequency is higher than that of the isolated HI monomer (2170 cm^{-1}), a unique situation. The replacement of NH_3 by TMA which is more basic increases the proton transfer but a common limit (2315 cm^{-1}) corresponding to the ion pair I^- ^+HB is reached for both HI–NH_3 and HI–TMA complexes in ethylene, the most perturbing material. In molecular matrices anisotropic interactions which are of electrostatic nature (dipole-quadrupole) occur (only induction type in argon). This electrostatic contribution is responsible for ν_s frequency increase from Ar to N_2. The ν_s frequency increase from N_2 to C_2H_4 both of which have same absolute value of quadrupole may be understood as arising from larger value of the polarizability of ethylene.

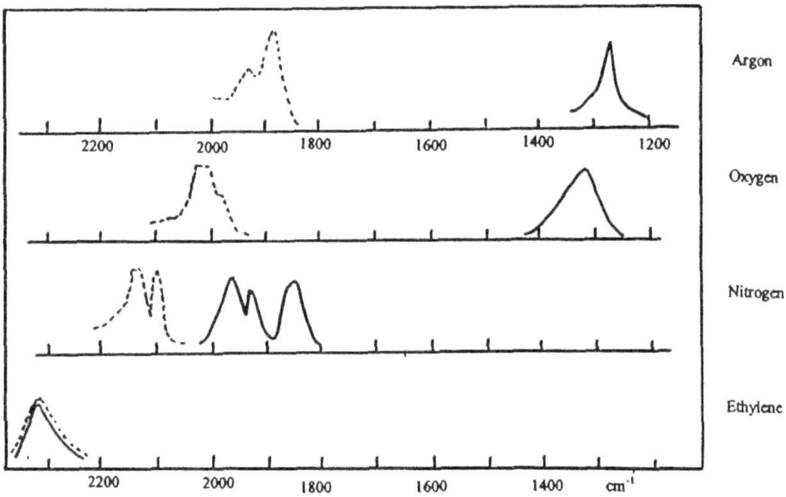

Figure 5: Behavior of the ν_s band of HI–NH_3 (full line) and HI–TMA (dotted line) (1:1) complexes as a function of the matrix.[149]

In aggregates larger than the one to one complexes previously described, the proton transfer leading to the formation of cation and anion can occur due to cooperative effects c.a. to the proton solvation energy of a B_n cluster. Thus, by increase of the NH_3 concentration, typical bands below 1000 cm^{-1} were observed in the different matrices (680–737 cm^{-1} in Ar, 675 cm^{-1} in O_2 and 670–830 cm^{-1} in N_2). They were assigned to the centrosymmetric cation $(H_3N\text{-}H\text{-}NH_3)^+$ I^-. The same behavior was also observed for other systems. The infrared spectra of the complexes between dimethyl ether (DME) and hydrogen iodide at high concentrations revealed the formation of a symmetrical cation $(DME)_2$ H^+ and an asymmetrical cation $(DME)_2$ H^+ in which the hydrogen atom is more tipped towards one of the two oxygen than in the former cation.[153,154] In water rich $HCl/H_2O/Ar$ matrices proton transfer occurred in $(HCl)(H_2O)_4$ aggregate and the entity $(H_3O^+)(H_2O)_3$ Cl^- was characterized by a broad intense feature at 1250 cm^{-1} and by two less intense absorptions at 3170 and 1720 cm^{-1}. Replacement of HCl by HBr showed as expected little effects.[155]

The reaction proton transfer can also take place in the electronically excited state of the neutral clusters. In particular it is well known that there is a large difference between the acidities of hydroxy aromatics in their S_0, S_1 and T_1 states. A recent study of the phenol-ammonia (ϕ-OH/NH_3) system in argon showed that the ionic form (ϕ-O^-H $(NH_3)_n^+$) corresponding to the energy minimum in $n \geq 3$ clusters in the S_1 state was formed.[156]

3.3 Trapping of Two Structural Arrangements of Weak Complexes

In some complexes competition between different bonding interactions can occur due to the nature of bonded species. Matrix isolation technique can trap two structural forms whereas in gas phase one form, the lower energy structure is generally observed. Matrix cage quenches internal energy and hence prevents rearrangement of the less stable to the more stable structural form. One of first experimental evidence for such weak complexes was observed with HF and HCN.[157] Owing to the dual acid–base nature of the submolecule, two forms were identified: a form I, HF...HC=N, and a form II, HC=N...HF. Upon warming the matrix above 20 K the concentration of I decreased while the amount of II increased indicating that II is the more stable complex. An other interesting example is the CO/HF complex[158] which in principe is able to form three different isomeric species. The forms I [OC...HF] and II [CO...HF] were produced in argon matrix by photodissociation of matrix isolated HC(O)F. In complex I the frequency ν_{CO} is increased (2162.1 cm^{-1}) and in complex II decreased (2129.3 cm^{-1}) compared to the value for the matrix CO molecule (2138.3 cm^{-1}). Annealing experiments changed the ratio of concentration I/II indicating that the form II was slightly less stable than form I. An enthalpy difference between the isomers at 20 K was estimated to 0.3 kJ mol^{-1}. Recently, two structural forms of the CO/CO_2 complex were evidenced in argon matrix.[159,160] The form (LF) which appears at a frequency lower than that of isolated CO converts spontaneously upon temperature increase in a form (HF) which appears at a frequency higher than that of isolated CO. This effect is reversible at low temperature and is time dependent as illustrated in Fig. 6.

The nature of the matrix plays a non negligible role for stabilizing the less stable isomer as evidenced for the nitric oxide dimer which has a planar structure in gas phase as in inert matrix (Ar and Xe) but which presents both *cis* and *trans* planar structures in nitrogen matrices.[161]

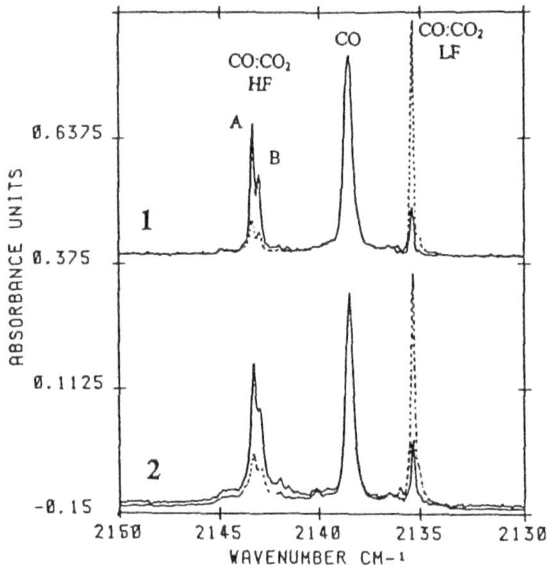

Figure 6: IR absorption of the CO:CO_2 complex in the CO range. (1) CO/CO_2/Ar = 1/3/2000; $l \sim 33$ μm; T = 5.4 K, — t = 120 s, - - - t = 1200 s. (2) CO/CO_2/Ar= 4/3/2000; $l \sim 9$ μm; T = 5.4 K, — t = 120 s, - - - t = 1200 s. Reproduced with permission from Ref. 160. ©The American Institute of Physics.

4 Infrared Photodissociation of Hydrogen Bonded Complexes

With the advent of free molecular beam sources the infrared photodissociation of weakly bound van der Waals and hydrogen bonded clusters prepared in the gas phase is a very active field of research.[162,163] The role of the frequency and fluence of the laser radiation for the dissociation of size selected clusters has been investigated. In all cases a single vibrational quantum of excitation much larger than the binding energy of the van der Waals bond is sufficient to break the bond. However the vibrational mode selectivity in gas phase is difficult to observe due to fast intramolecular energy randomization, whereas cryogenic matrices offer a unique environment for studying quantum efficiencies and product branching upon selective infrared excitation. The weak coupling between the guest and the solid lattice implies that intramolecular vibrational relaxation within the guest molecule degrees of freedom can be much faster than the vibrational energy redistribution in the lattice. In a matrix the rotations are prevented and IR excitation is normally into a zero phonon transition opening the possibility of vibrational selectivity study. The first works of infrared conversion in matrices have focussed upon unimolecular rotational isomerization as the interconversion of *cis* and *trans* HONO,[164] the conformer changes of halogeno and amino ethanols and unsaturated alcohols and amines (see Refs.[165–170] and references therein). More recently the infrared photodissociation of hydrogen bonded complexes involving HI as proton donor and weak to medium strength bases B (B = cyanomethane,[171,172] water,[173] acetone,[174] ethylene oxide,[175] methanol,[176] dimethylsulfoxide,[177] N–N dimethylacetamide,[178]) was studied as well as the photodissociation of dimers and trimers of methanol,[179,180,181] and water.[182]

4.1 Hydrogen Bonded Complexes IH...B

The $CH_3CN/HI/N_2$ system was chronologically the first to be examined.[183] As illustrated in Fig. 7 after simultaneous deposition at 20 K of two beams CH_3CN/N_2 and HI/N_2 , the IR spectrum at the same temperature displays, in the HI stretching region, the monomer HI band at 2237 cm^{-1} and the shifted HI absorption at 2142 cm^{-1} assigned to the hydrogen bonded form of the CH_3CN one to one complex. A second spectrum of the same sample cooled at 10 K and recorded after some minutes exposure in the beam of spectrophotometer or in the dark during 30 mn shows that the band at 2142 cm^{-1} disappears with a smaller increase in the intensity of the 2137 cm^{-1} absorption. In a third step, the sample is heated to 25 K in the dark for 10 mn and the spectrum recorded at this temperature is identical to the first one indicating that the hydrogen bonded complex has been regenerated through a thermal process. Similar observation were made for the other systems. In all cases the most stable hydrogen bonded form of the complex C photodissociates upon absorption of IR photons yielding a non hydrogen bonded pair U with HI frequency close to that of free HI. The reverse transformation U $\longrightarrow$ C is achieved by increasing the temperature of the sample. Extensive kinetic studies in various experimental conditions showed that the IR process was first order and allowed to obtain quantitative informations upon the C $\longrightarrow$ U and U $\longrightarrow$ C conversion.

C $\longrightarrow$ U Conversion

The most efficient vibrational excitation is that of the HI stretching mode but in the case of the HI/H_2O[173] and HI/CH_3OH system,[176] use of an infrasil filter which absorbs more than 95 per cent of the light below 2100 cm^{-1} prove also the selective efficiency of the ν_{OH} of H_2O and of the ν_{OH} or ν_{CH} modes of CH_3OH. Quantum yields for the C $\longrightarrow$ U photoconversion of the HI/H_2O complex were estimated of 0.75 and 0.41 respectively for vibrational excitation

HI and HO stretching modes. The high yield for ν_{HI} confirms that this mode is at the origin of the photoprocess and suggests the following internal conversion ν_{HI} (v = 1) + ν_l (v_l = 0) $\longrightarrow$ ν_{HI} (v = 0) + ν_l (v_l = n) where ν_l is the librational mode of HI induces the reorientation of HI within the pair which corresponds to the formation of U species. The relatively high yield (0.40) for ν_{OH} for the H_2O : HI pair suggests a two step process involving firstly an intramolecular v-v transfer.

For some systems changing HI to DI, the photoconversion B : DI was not observed, indicating that the vibrational energy of DI transferred to librational levels is not high enough to induce the C $\longrightarrow$ U conversion. A remarkable situation was encountered for $(CH_3)_2CO$: DI trapped in solid argon, the hydrogen bonded form of which exits in two forms, one photosensitive (ν_{DI} = 1442 cm^{-1}) and the other insensitive to the irradiation (ν_{DI} = 1502 cm^{-1}). Thus the C $\longrightarrow$ U barrier height lies in the range 1450–1500 cm^{-1}. In spite of a barrier height lower than 1700 cm^{-1} the ν_{CO} vibration of complexed acetone (as ν_{CO} of methanol) was inefficient for inducing the C $\longrightarrow$ U conversion.[174]

C $\longleftarrow$ U Back Conversion

The reverse process C $\longleftarrow$ U is easily obtained upon temperature increase. From kinetic studies as a function of temperature the activation energy for inducing the C $\longleftarrow$ U conversion was calculated in the range of 1.5–4 kJ according to the strength of the base. In the case of CH_3OH/HI system,[176] the C $\longleftarrow$ U conversion was also observed upon excitation of the CO stretching mode of U by the P(36) emission line of a cw low power CO_2 laser. However this process is not very efficient due to weak absorption cross section of the HI stretching mode of the U species, explaining the predominance of the photodissociation of the hydrogen bonded complex in polychromatic IR irradiation experiments.

Due to the small dipole moment and large polarizability of hydrogen iodide compared with the other hydrogen halides, the C $\longleftrightarrow$ U interconversion can be understood as a head to tail motion of the HI molecule, a conclusion confirmed by *ab initio* calculations.[184] However, a similar infrared induced conformational change was also observed for the two strong hydrogen bonded complexes 2-1 between dimethylether and hydrogen iodide identified as ion pairs $(DME)_2$ H^+ I^-, the C $\longleftarrow$ U conversion occurring by irradiation in the range 2600–1500 cm^{-1} whereas the reverse process was observed by irradiation below 1160 cm^{-1} suggesting that both hindered rotations of HI and base B cannot be ruled out.[154]

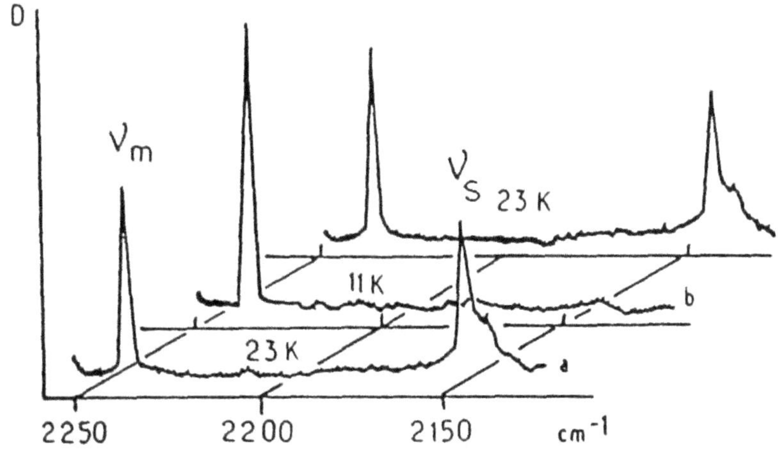

Figure 7: Spectral changes for a $CH_3CN/HI/N_2$ = 2/1/800 mixture in the range 2250–2110 cm^{-1} (a) Spectrum recorded at 23 K after deposition at 17 K in the dark. (b) Spectrum recorded after 30 mn at 11 K in the dark. (c) Spectrum (b) followed by warming to 23 K for 10 mn in the dark. ν_m, HI monomer; ν_s, HI stretching mode of $CH_3CN \cdot HI$ hydrogen-bonded complex.[171]

4.2 Infrared Photoisomerization of the ROH Aggregates (R = CH_3,H)

Methanol Dimer

It was first recognized for IR spectroscopy of methanol in nitrogen that the dimer exists as an open chain structure, one proton playing the role of proton acceptor and the other that of proton donor. Four forms which can interconvert upon temperature change or infrared irradiation were evidenced.[179] *Ab initio* calculations[185] as well as monochromatic irradiation in the ν_{OH} region with a tunable laser[186] confirmed the existence of several minima in the potential surface corresponding to different values of the torsional angles. Recently, reinvestigations of methanol dimers trapped in solid nitrogen using temperature effects and selective photoconversion upon excitation of CO stretching modes with CO_2 laser lines allowed to classify the four dimeric species into two pairs of closely relative forms (A, B and C, D) and to identify a transient species E.[180] The barrier heights involved in the conversion between the two groups of conformers were estimated and are reported in the Figure 8.

Methanol Trimer

IR photoconversion of the methanol trimer in nitrogen matrix, recently reported[187] shows that the difficult identification of oligomers can be reached by selective photoconversion/photodissociation processes upon vibrational excitation and temperature effects. After deposition a cyclic trimer was identified by its ν_{OH} IR active mode located at 3462 cm^{-1} (3574 cm^{-1} in gas phase) and by its ν_{CO} mode at 1036 cm^{-1}. Upon irradiation in the mid infrared the cyclic trimer is photoconverted into an open chain structure and the back conversion is easily obtained by increasing the temperature.

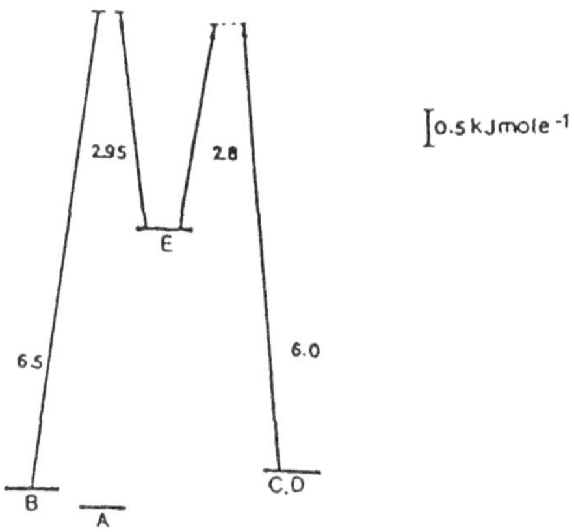

Figure 8: Energy diagram of the five dimer species of methanol trapped in solid nitrogen. Reproduced with permission from Ref. 180. ©The American Institute of Physics.

Water Dimer

For the water dimer and other water base complexes the HDO molecule in the complex prefers to form a D bond than an H bond as previously reported. Recently it was shown that when the perturbed OD (OH) fundamental of HDO submolecule is irradiated with a monochromatic tunable source, HDO changes from D to H bonding (or from H to D). This has been used to identify the out of plane shear vibration of HDO, H- and D- bonded to another molecule in the far infrared.[188]

5 UV and Visible Photochemistry

Photochemistry of molecules in the gas phase has been extensively studied. Less is known about the photochemistry of molecules included in solids or adsorbed to surfaces. Photochemistry processes in rare gas solids provide a model system for exploring reaction in condensed phases. Although it should be emphasized that the primary photochemical step in both gas and matrices is the same, results show that chemical pathways in solid rare gas are frequently different from homogeneous gas phase mechanisms. Many phenomena have to be taken into account:

- the 'cage effect' which can inhibit the separation or the migration of photofragments and which can lead to recombination of the precursor molecules,
- quenching of excited species which can be enhanced by multiple collisions with the host atoms,
- spin forbidden processes which can become efficient and arise from crossing of the singlet and triplet surfaces mixed by spin orbit interactions,
- reactions which can occur with the cage walls if the matrix is reactive.

Photodissociation of H_2S is an attractive example of both cage exit dynamics and cage induced bimolecular reaction.[189] Upon excitation to the first absorption continuum, H_2S photodissociates via several direct channels which are identified as: i) permanent dissociation in H + SH with cage exit of H atom, ii) permanent dissociation into S + H_2 with cage exit of S atom, iii) in-cage, reaction between H and SH photofragments. In some cases, cage recombination can lead not to the precursor but to an isomer species of parent molecules and hence offers an opportunity to identify species difficult to prepare in gas phase. Two recent published examples illustrates this behavior. Photodissociation of nitric acid in argon matrix produces *cis* and *trans* peroxynitrous acid from the recombination of the primary OH and NO_2 fragments.[190,191] Photodissociation of SO_3 in argon matrix has allowed to identify for the first time the OSOO isomer formed by recombination of O and SO_2 fragments.[192]

The exit of fragments from the cage requires crossing of an energy barrier. It depends strongly on the mass ratio of the fragment to the lattice atom. For a high ratio, exit is not observed. The exit probability can depend also in some cases on the initial orientation of the molecule in the cage or on the formation of new intermolecular ionic state due to charge transfer from the matrix to the guest as shown for the F_2 dissociation process.[193]

Photochemical behavior of specific molecular complexes in matrices is still an emerging field of research and only a few systems have been studied in spite of their considerable interest for heterogeneous and homogeneous chemistry. As a matter of fact, the photochemical of charge transfer molecular complexes is different from that of no bound species and photochemistry occurs with light of wavelength longer than the absorption light of the partners and lead to the formation of species which depend strongly of the complex structure. Mechanisms have been established to proceed via selective access to low lying reaction states which are charge transfer in nature. Several examples of such behavior have been reported. They concern mainly complexes of ozone ICl,[54] PH_3,[46] PCl_3,[194] Br_2,[50] BrCl,[50] SiH_3,[195] CH_3I,[196] where light absorption is followed by oxygen transfer, complexes of CH_3SCH_3, CH_3SH with Cl_2,[197] complexes of N_2O_4 with various alkenes where oxygen transfer from NO_2 to dienes leads to the oxirane biradical as primary product.[198] In some cases it has been observed that the reaction of an oxygen atom with an other species occurs only through the complex formation between the atom precursor and the related species. Thus the monomeric sulfur dioxide was found to be photochemically inactive in oxygen matrix and only dimeric species $(SO_2)_2$ was found to be readily photooxidized to sulfur trioxide.[63] In the same way, reaction of $O(^1D)$ produced by photodissociation of ozone with SO_2 [70] and H_2O [57] were not observed; only photoabsorption of the $O_3 : SO_2$ and $O_3 : H_2O$ complexes formed SO_3 and H_2O_2, respectively.

Photochemistry of some hydrogen bonded complexes has been also reported as hydrogen halide–acetylene complexes[199,200] and more recently NO : HCl complex,[99] ClNO : H_2O complex,[201] $HNO_3 : C_2H_4$ complex.[143] Photochemistry of iodide acetylene complex[199] shows a distinct wavelength dependence of product branching; at longer wavelengths only iodoacetylene is formed, whereas at shorter wavelengths acetylene isotopic exchange (H/D) is observed along with formation of vinyl iodide. Photolytic production of iodoacetylene involves hydrogen (deuterium) atom abstraction from acetylene by hydrogen iodide. At 266 nm wavelength, the photodissociation of the HCl : NO, HCl : $(NO)_2$ complexes which leads to ClNO and ClNO : HNO respectively, involves hydrogen atom abstraction from HCl followed by attack of NO by the outer chlorine atom.

5.1 *Case Study: Photodissociation of the O_3 : Br_2 and O_3 : BrCl Complexes in Argon Matrix at 532, 633, and 870 nm*[202]

Mixing bromine vapor with chlorine gas can produce BrCl if the mixture is allowed to equilibrate. When O_3 /Ar (1/1000) and Br_2 /Cl_2 /Ar (1/1/200) were condensed together, three new features in the ν_3 ozone region were observed with relative intensity of 1/1/2. Absorptions at 1032.1 and at 1030.1 cm^{-1} characterize the perturbed ν_3 mode of ozone submolecule in the O_3 : Cl_2 and O_3 : Br_2 complexes as illustrated Fig. 9 whereas the absorption at 1028.0 cm^{-1} characterizes a complex between O_3 and BrCl, namely O_3 : BrCl or O_3 : ClBr. Photolysis of the matrix at various wavelengths allowed to characterize for the first time new species ClBrO, BrClO and BrOCl, to determine the binding between O_3 and BrCl and to confirm the enhancement of the photo cross-section of ozone in the Chappuis band in condensed medium at low temperature.

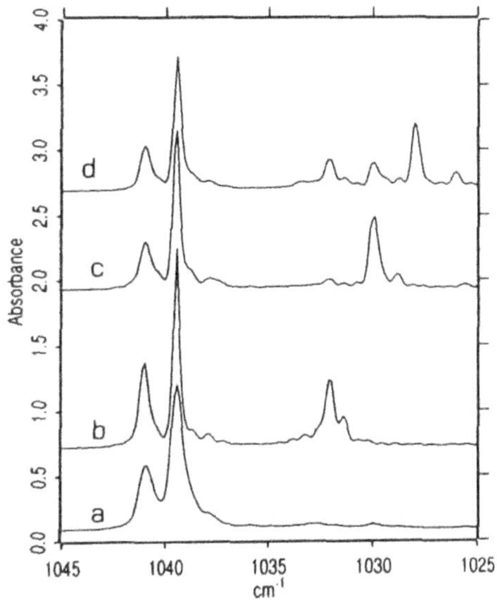

Figure 9: Comparison between IR spectra after deposition in the ozone ν_3 spectral region of: (a) O_3/Ar 1/200 mixture; (b) O_3/Cl_2/Ar 1/1/400 mixture; (c) O_3/Cl_2/Ar 1/1/400 mixture; (d) O_3/Cl_2+Br_2/Ar 1/1-1/400.

Irradiation at 10 K with the 532 nm laser line of a matrix containing O_3, Br_2 and Cl_2 diluted in argon (O_3 / Br_2-Cl_2 /Ar = 1/5/1000) led to the disappearance of the O_3 : (BrCl) and O_3 : Br_2 complexes and to the appearance, in the 1000–800 cm^{-1} region, of absorptions characterizing ClBrO and BrBrO. Thus, the photolysis of the O_3 : (BrCl) complex at this wavelength led only to one photoproduct, namely ClBrO, identified by a doublet with equal components at 819.6 and at 817.9 cm^{-1} corresponding to the BrO stretching mode of $Cl^{79}BrO$ and $Cl^{81}BrO$ species, respectively. This result supports the existence of only one stable complex between ozone and bromine monochloride and suggests that ozone is bound to BrCl in the complex via the bromine atom. Under irradiation at 532 nm, there is a transfer of the terminal oxygen atom of ozone to BrCl and the final product contains BrO but not ClO bonds.

Irradiation of a $Br_2/Cl_2/O_3/Ar$ mixture with the 633 nm laser line led to different results and allowed to understand why irradiation of a similar mixture with the filtered xenon lamp produced both ClBrO and BrClO.[203]

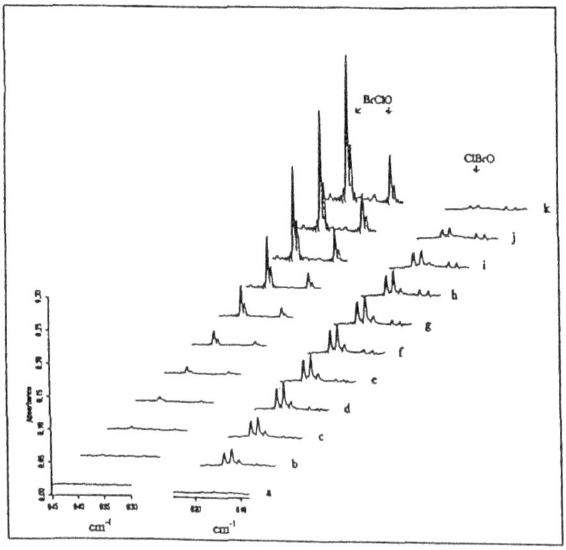

Figure 10: Complex O_3 Br_2 decay and product band growth during laser photolysis at 10 K at 532 nm laser line of $O_3/Br_2/Ar$ 1/5/2000 mixture in the ν_3 ozone. Photolysis time a) 0 mn, b) 5 mn, c) 10 mn, d) 20 mn, e) 25 mn, f) 35 mn, g) 45 mn, h) 55 mn, i) 110 mn.[202]

Figure 10 shows in the 1000–800 cm^{-1} region the spectra obtained after various irradiation times with the 633 nm laser line. At the beginning of the irradiation, only the ClBrO absorption is observed. With longer irradiation times, it increases, reaches nearly stationary state, then decreases in intensity upon prolonged photolysis after the total depletion of the O_3 : BrCl complex. In parallel after a 30 mn irradiation time, the BrClO absorption at 940.5–932.5 cm^{-1} appears, then increases strongly and continues to grow upon prolonged photolysis. The time evolution of the two species (ClBrO and BrClO) under irradiation at 633 nm laser clearly indicates that BrClO is a secondary product arising from the isomerization of ClBrO at 633 nm.

Series of photolysis experiments were also performed on a $Br_2/Cl_2/O_3/Ar$ mixture with the 870 nm laser line. The O_3 : Br_2 and the O_3 : BrCl complexes diminished very slowly and reached 35% of their initial IR intensity after 300 mn of irradiation time. As expected, in the 900–800 cm^{-1} region the doublet at 819.6–817.9 cm^{-1} characterizing the BrO stretching mode of ClBrO species was observed but the corresponding absorption of the BrBrO species was surprisingly not observed. With a longer irradiation time the doublet at 819.6–817.9 cm^{-1} decreased strongly. Concurrently a new set of bands at 675.9, 623, 559.5 and 525.6 cm^{-1} appeared and grew upon irradiation. Bands at 623 and 525.6 cm^{-1} were assigned in [204] as ν_1 and ν_3 fundamentals of monomeric BrOBr. By analogy the absorptions at 675.9 cm^{-1} and 559.5 cm^{-1} were assigned to the BrOCl species which was identified for the first time. Subsequent *ab initio* calculations confirmed this assignment.[205,206]

On the basis that the primary pathway of the photolysis of O_3 : Br–Br and O_3 : BrCl at 870 nm is the formation of BrBrO and ClBrO species, these results indicate that the BrOBr and ClOBr species are secondary products formed from the isomerization of BrBrO and ClBrO. The BrBrO at this wavelength is more photosensitive than ClBrO and it is destroyed as soon as it is produced.

Finally the photolysis of O_3 : Br_2 and O_3 : BrCl complexes can be described as follows:

$$O_3 : Br_2 \xrightarrow{532-870\text{nm}} \text{BrBrO} \xrightarrow{870\text{nm}} \text{BrOBr}$$

$$O_3 : \text{BrCl} \xrightarrow{532-870nm} \text{ClBrO} \begin{cases} \xrightarrow{870nm} \text{BrOCl} \\ \xrightarrow{633nm} \text{BrClO.} \end{cases}$$

The photodissociation of the O_3 : Br_2 complex at 532 nm is faster than that of the O_3 : BrCl complex whereas at 870 and 633 nm, a similar rate is observed for the depletion of the two complexes.

The concentration of the complex at initial time (subscript 0) and at the time t is proportional to the integrated absorption intensity of the ν_3 vibrational mode of complexed ozone. Plots of $\ln([O_3 : X_2]^0/[O_3 : X_2]^t)$ versus time for the O_3 : Br_2 and O_3 : BrCl complexes photolyzed at 633 nm and 532 nm with a fluence of $5.6\cdot10^{16}$ photons cm^{-2} sec^{-1} and $4.0\cdot10^{16}$ photons cm^{-2} sec^{-1}, respectively, were linear and the kinetic rate constant which depends on the quantum yield, the ozone cross-section and the photons flux allowed to discuss the values of the cross-section of ozone submolecule in each complex in regard to the cross-section of monomer ozone in gas phase.

References

1. E. Whitle, D. A. Dows, and G. C. Pimentel, *J. Chem. Phys.* **22**, 1943 (1954).
2. *Low Temperature Spectroscopy*, ed. B. Meyer (Elsevier, 1972).
3. *Vibrational Spectroscopy of Trapped Species*, ed. H. E. Hallam (Wiley, 1973).
4. *Vibrational Spectra and Structure, vol.4*, ed. J. R. Durig (Elsevier, 1975).
5. *Cryochemistry*, ed. M. Moskovits and G. A. Ozin (John Wiley, 1976).
6. A. J. Barnes, W. J. Orville-Thomas, A. Muller and R. Gaufres, *Matrix Isolation Spectroscopy* (Reidel, Dordrecht, 1981).
7. *Adv. in Spectroscopy, vol. 17*, ed. R. J. H. Clark and R. E. Hester (Wiley, 1989).
8. L. Andrews and M. Moskovits, *Chemistry and Physics of Matrix Isolated Species* (North Holland, Amsterdam, 1989).
9. S. Cradock and A. J. Hinchliffe, *Matrix Isolation: A Technique for the Study of Reactive Inorganic Species* (Cambridge University Press, 1975).
10. R. Kolos, *Chem.Phys. Letters* **247**, 289 (1995).
11. R. Schlachta, G. Lask, and V. E. Bondybey, *Chem. Phys. Letters* **180**, 275 (1991).
12. H. Bai and B. S. Ault, *Chem. Phys. Letters* **188**, 126 (1992).
13. N. Tanaka, J. Oike, Y. Kajii, K. Shibuya, and M. Nakata, *Chem. Phys. Letters* **232**, 109 (1995).
14. L. Schriver-Mazzuoli, A. de Saxce, C. Lugez, C. Camy-Peyret, and A. Schriver, *J. Phys. Chem.* **102**, 690 (1995).
15. T. L. Tso, and E. K. C. Lee, *J. Phys. Chem.* **89**, 1618 (1985).
16. M. Maas, C. H. Gross, and U. Schurath, *Chem. Phys.* **189**, 217 (1994).
17. M. M. Rockind, *Anal. Chem.* **39**, 567 (1967); **40**, 762 (1968).
18. P. Groner, I. Stocklin, and H. S. Günthard, *J. of Physics E: Scientific Instruments* **6**, 122 (1973).
19. J. A. Venables and D. J. Ball, *J. of Crystal Growth* **4**, 180 (1968).
20. S. I. Kovalenko and N.N. Bagrov, *Soviet Solid State* **11**, 2207 (1970).
21. W. Langel, W. Schuller, E. Knözinger, H. W. Fleger, and H. J. Lauter, *J. Chem. Phys.* **89**, 1741 (1988).
22. J. Crutz and G. E. Lopez, *J. Chem. Phys.* **104**, 4294 (1996).
23. C. Cabana, G. B. Savitsky, and D. F. Hornig, *J. Chem. Phys.* **39**, 2942 (1963).

24. R. Fraenkel and Y. Hass, *J. Chem. Phys.* **100**, 4324 (1994).
25. J. Crutz and G. E. Lopez, *J. Chem. Phys.* **104**, 4301 (1996).
26. E. Knözinger, and R. Wittenberg, *J. Am. Chem. Soc.* **105**, 2154 (1983).
27. E. Knözinger, W. S. Schuller, and W. Langel, *Faraday Discuss. Chem. Soc.* **86**, 285 (1988).
28. C. Girardet and D. Maillard, *J. Chem. Phys.* **77**, 5923 (1982).
29. B. Laraoui, J. P. Perchard, and C. Girardet, *J. Chem. Phys.* **97**, 2347 (1992).
30. K. Harvey, H. Shurvell, and J. Henderson, *Can. J. Chem.* **42**, 911 (1964).
31. M. M. Kreitman and D. L. Barnet, *J. Phys. Chem.* **43**, 364 (1963).
32. R. S. Mulliken, *J. Am. Chem. Soc.* **64**, 811 (1952).
33. R. G. Pearson, *J. Chem. Educ.* **45**, 643 (1968).
34. J. Douglas and P. Kollman, *J. Phys. Chem.* **85**, 2717 (1981).
35. L. Manceron, A. Loutellier, and J. P. Perchard, *Chem. Phys.* **92**, 75 (1985).
36. B. S. Ault, *Reviews of Chemical Intermediates* **9**, 233 (1988).
37. N. P. Machara and B. S. Ault, *Inorg. Chem.* **24**, 4251 (1985).
38. N. P. Machara and B. S. Ault, *J. Phys. Chem.* **91**, 2046 (1987).
39. B. S. Ault, W. F. Howard, Jr., and L. Andrews, *J. Mol. Struct.* **55**, 217 (1975).
40. V. Branchadell, A. Oliva, and J. Bertran, *Chem. Phys. Letters* **113**, 197 (1975).
41. L. Schriver-Mazzuoli, O. Abdelaoui, and A. Schriver, *J. Phys. Chem.* **96**, 8069 (1992).
42. L. Schriver-Mazzuoli and A. Schriver, unpublished results.
43. H. Frei and G. C. Pimentel, *J. Phys. Chem.* **85**, 3355 (1981).
44. M. Bahou, L. Schriver-Mazzuoli, C. Camy-Peyret, and A. Schriver, to be published.
45. V. Raducu, D. Jasmin, R. Dahoo, P. Brosset, B. Gauthier-Roy, and L. Abouaf-Marguin, *J. Chem. Phys.* **101**, 1873 (1994).
46. R. Withnall, M. Hawkins, and L. Andrews, *J. Phys. Chem.* **90**, 575 (1986).
47. L. Nord, *J. Mol. Struct.* **96**, 37 (1982).
48. L. Andrews, R. Withnall, and B. W. Moores, *J. Phys. Chem.* **93**, 1279 (1989).
49. R. J. H. Clark and J. R. Dann, *J. Phys. Chem.* **100**, 532 (1996).
50. M. Bahou, L. Schriver-Mazzuoli, A. Schriver, and P. Chaquin, *Chem. Phys* **216**, 105 (1997).
51. L. Andrews, M. Hawkins, and R. Withnall, *Inorg. Chem.* **24**, 4234 (1985).

52. R. Withnall and L. Andrews, *J. Phys. Chem.* **92**, 2155 (1988).
53. L. Andrews, K. Chi, and A. Arkell, *J. Am. Chem. Soc.* **96**, 1997 (1974).
54. M. Hawkins, L. Andrews, A. J. Downs, and D. J. Drury, *J. Am. Chem. Soc.* **106**, 3076 (1984).
55. P. Brosset, R. Dahoo, B. Gauthier-Roy, L. Abouaf-Marguin, and L. Lakhlifi, *Chem. Phys.* **172**, 3515 (1993).
56. L. Schriver-Mazzuoli, A. Schriver, C. Lugez, A. Perrin, C. Camy-Peyret, and J. M. Flaud, *J. Mol. Spectrosc.* **176**, 85 (1996).
57. L. Schriver-Mazzuoli, C. Barreau, and A. Schriver, *Chem. Phys.* **140**, 429 (1990).
58. C. Lugez, A. Schriver, R. Levant, and L. Schriver-Mazzuoli, *Chem. Phys.* **181**, 129 (1994).
59. L. Andrews, R. Withnall, and R. D. Hunt, *J. Phys. Chem.* **92**, 78 (1988).
60. B. W. Moores and L. Andrews, *J. Phys. Chem.* **93**, 1902 (1989).
61. C. S. Sass and B. S. Ault, *J. Phys. Chem.* **88**, 432 (1984).
62. L. Schriver-Mazzuoli, A. Schriver, and J. P. Perchard, *J. Mol. Spectrosc.* **127**, 125 (1988).
63. J. R. Sodeau and E. K. C. Lee, *J. Phys. Chem.* **84**, 3358 (1980).
64. L. Nord, *J. Mol. Struct.* **96**, 19 (1982).
65. L. Schriver-Mazzuoli, A. Schriver, and M. Wierzejewska-Hnat, *Chem. Phys.* **199**, 227 (1995).
66. D. J. Pauley, R. E. Bumgarner, and S. G. Kukolich, *Chem. Phys. Letters* **132**, 67 (1986).
67. C. S. Sass and B. S. Ault, *J. Phys. Chem.* **90**, 4533 (1986).
68. T. L. Tso and E. K. C. Lee, *J. Phys. Chem.* **88**, 2776 (1984).
69. V. E. Bondybey and J. H. English, *J. Mol. Spectrosc.* **107**, 221 (1985).
70. L. Schriver-Mazzuoli, D. Carrere, A. Schriver, and K. Jaeger, *Chem. Phys. Letters* **181**, 505 (1991).
71. C. A. Long and G. E. Ewing, *J. Chem. Phys.* **58**, 4824 (1973).
72. B. R. Cairns and G. C. Pimentel, *J. Chem. Phys.* **43**, 3432 (1965).
73. J. Coodman and L. E. Brus, *J. Chem. Phys.* **67**, 4398 (1977).
74. M. Bahou, L. Schriver-Mazzuoli, A. Schriver, T. Chiavassa, and J. P. Aycard, *Chem. Phys. Letters* **265**, 145 (1997).
75. S. Hashimoto and H. Akimoto, *J. Phys. Chem.* **91**, 1347 (1987).
76. A. Moroz, R. L. Sweany, and S. L. Whittenburg, *J. Phys. Chem.* **94**, 1352 (1990).
77. J. S. Ogden, A. J. Rest, and R. L. Sweany, *J. Phys. Chem.* **99**, 8485 (1995).
78. B. S. Ault, E. Steinback, and G. C. Pimentel, *J. Phys. Chem.* **79**, 615 (1975).

79. L. Schriver-Mazzuoli, A. Schriver, and J. P. Perchard, *J. Mol. Struct.* **222**, 141 (1990).
80. A. Behrens-Griesenbach, W. A. P. Luck, and O. Schrems, *J. Chem. Soc. Faraday* **2**, 80 (579)1984.
81. G. N. Robertson, *Phil. Trans. R. Soc. London* **A 286**, 25 (1977).
82. Y. Boutellier and Y. Guissani, *Mol. Phys.* **38**, 617 (1979).
83. S. Bratos, *J. Chem. Phys.* **63**, 3499 (1975).
84. V. P. Sakun, *Chem. Phys.* **99**, 457 (1985).
85. W. G. Johnson and D. W. Oxtoby, *J. Chem. Phys.* **87**, 781 (1987).
86. J. Lascombe and J. C. Lassegues, *Mol. Phys.* **40**, 969 (1980).
87. L. Andrews, *J. Mol. Struct.* **100**, 281 (1983).
88. G. Maes and M. Graindourze, *J. Mol. Spectrosc.* **113**, 410 (1985).
89. L. Schriver-Mazzuoli, A. Loutellier, A. Burneau, and J. P. Perchard, *J. Mol. Struct.* **95**, 37 (1982).
90. P. Bernadet and L. Schriver-Mazzuoli, *J. Mol. Struct.* **130**, 193 (1985).
91. E. Knözinger and R. Wittenbeck, *J. Chem. Phys.* **80**, 5979 (1984).
92. D. Maillard, A. Schriver, J. P. Perchard, and C. Girardet, *J. Chem. Phys.* **71**, 505 (1979).
93. J. P. Perchard, D. Maillard, A. Schriver, and C. Girardet, *J. Raman. Spectrosc.* **11**, 406 (1981).
94. D. Maillard, A. Schriver, F. Fondere, J. Obriot, and C. Girardet, *J. Chem. Phys.* **75**, 1091 (1981).
95. L. Andrews and G. L. Johnson, *Chem. Phys. Letters* **96**, 133 (1983).
96. L. Andrews in *Chemistry and Physics of Matrix Isolated Species*, eds L. Andrews and M. Moskovits (Elsevier Science Publishers, BV, Chapter 2, 17, 1989).
97. A. J. Barnes, *J. Mol. Struct.* **100**, 259 (1983).
98. L. Schriver-Mazzuoli, A. Schriver, and J. P. Perchard, *J. Chem. Soc., Faraday Trans.,2* **81**, 1407 (1985).
99. A. de Saxce, N. Sanna, A. Schriver, and L. Schriver-Mazzuoli, *Chem. Phys.* **185**, 365 (1994).
100. N. Sanna, L. Schriver-Mazzuoli, A. Hallou, A. de Saxce, and A. Schriver, *J. Chem. Phys.* **103**, 6930 (1995).
101. D. Lucas, L. J. Allamandola, and G. C. Pimentel, *Croatica Chem. Acta* **55**, 121 (1982).
102. G.P. Ayers and A. D. E. Pullin, *Spectrochim. Acta* **32 A**, 1629 (1976).
103. L. Fredin, B. Nelander, and G. Ribbegård, *J. Chem. Phys.* **66**, 4065 (1977).
104. R. M. Bentwood, A. J. Barnes, and W. J. Orville-Thomas, *J. Mol Spectrosc.* **84**, 391 (1980).

105. W. J. Orville-Thomas and H. Ratajczak, *Croatica Chemica Acta* **55**, 191 (1982).
106. A. Engdahl and B. Nelander, *J.Chem. Phys.* **86**, 1819 (1987).
107. B. A. Zilles and W. B. Pearson, *J. Chem. Phys.* **79**, 65 (1983).
108. A. Engdahl and B. Nelander, *J.Chem. Phys.* **86**, 4831 (1987).
109. A. Engdahl and B. Nelander, *Chem. Phys.* **100**, 273 (1985).
110. T. L. Tso and E. K. C. Lee, *J. Phys. Chem.* **89**, 1612 (1985).
111. A. Engdahl and B. Nelander, *Chem. Phys.Letters* **113**, 49 (1985).
112. A. Engdahl and B. Nelander, *Chem. Phys.* **100**, 129 (1983).
113. R. A. Walder and J. L. Franklin, *Intern. J. of Mass Spectrosc. and Ion Physics* **36**, 85 (1980).
114. A. Burneau, L. Schriver-Mazzuoli, L. Manceron, and J. P. Perchard, *J. Chem. Phys.* **82**, 19 (1985).
115. L. Schriver-Mazzuoli, A. Burneau, and J. P. Perchard, *J. Chem. Phys.* **82**, 9 (1985).
116. B. Nelander, *J. Chem. Phys.* **72**, 77 (1980).
117. N. Bakkas, Y. Boutellier, A. Loutellier, J. P. Perchard, and S. Racine, *J. Chem. Phys.* **99**, 3335 (1993).
118. A. Engdahl and B. Nelander, *J. Chem. Phys.* **86**, 2253 (1982).
119. B. Nelander and L. Nord, *J.Chem. Phys.* **86**, 4375 (1982).
120. R. B. Bohne and L. Andrews, *J. Phys. Chem.* **95**, 9707 (1991).
121. E. Mayer and R. Pletzer, *Nature* **319**, 298 (1986); *J. Chem. Phys.* **80**, 2939 (1983).
122. B. Schmitt, J. Ocampo, and J. Klinger, *J. Phys.* **48**, C1-519 (1987).
123. P. Jenniskens and D. F. Blake, *Science* **265**, 754 (1994).
124. A. H. Narten, C. G. Venkatesch, and S. A. Rice, *J. Chem. Phys.* **64**, 1106 (1976).
125. M. R. Chowdhury, J. C. Dore, and D. G. Montague, *J. Phys. Chem.* **87**, 4037 (1983).
126. Jichen Li and D. K. Ross, *Nature* **365**, 327 (1981).
127. W. Hagen, A. G. G. M. Thielens, and J. M. Greenberg, *Chem. Phys.* **56**, 367 (1981).
128. D. M. Hudgins, S. A. Sandford, L. J. Allamandola, and A. G. G. M. Thielens, *Astrophys. J.* **86**, 713 (1993).
129. B. Rowland, M. Fisher, and J. P. Devlin, *J. Chem. Phys.* **95**, 1378 (1991).
130. J. E. Schaff and J. T. Roberts, *J. Phys. Chem.* **98**, 6900 (1994).
131. S. A. Sandford and L. J. Allamandola, *Astrophys. J.* **417**, 815 (1993).
132. S. A. Sandford, L. J. Allamandola, A. G. G. M. Thielens, and G. C. Valero, *Astrophys. J.* **329**, 510 (1988).

133. P. Ehrenfreund, P. A. Gerkines, W. A. Schutte, M. C. van Hemert, and E. F. van Dishoek, *Astronomy and Astrophysics* **296**, 810 (1995).
134. J. Murto and M. Ovaska, *Spectrochem. Acta* **39 A**, 149 (1983).
135. L. Schriver-Mazzuoli, A. Schriver, and O. Schrems, *Can. J. Chem.* **65**, 1520 (1991).
136. W. A. P. Luck, S. Peil, L. Schriver-Mazzuoli, and A. Schriver, *J. Mol. Struct.* **224**, 185 (1990).
137. O. Schrems, H. M. Oberhoffer, and W. A. P. Luck, *J. Phys. Chem.* **88**, 4335 (1984).
138. S. Süzer and L. Andrews, *J. Chem. Phys.* **87**, 5131 (1987).
139. G. A. Ayed and T. A. Ford, *J. Mol. Struct.* **217**, 307 (1990).
140. R. Lascola and L. Andrews, *J. Am. Chem. Soc.* **109**, 4765 (1987).
141. A. J. Barnes, E. Lasson, and C. Nielsen, *J. Mol. Struct.* **322**, 165 (1994).
142. Z. Mielke, K. G. Tokhadze, M. Hulkiewicz, L. Schriver-Mazzuoli, A. Schriver, and F. Roux, *J. Phys. Chem.* **99**, 10498 (1995).
143. Z. Mielke, L. Schriver-Mazzuoli, and A. Schriver, *J. Phys. Chem.* **A101**, 4560 (1997).
144. Z. Mielke, Z. Latajka, J. Kolodziej, and K. G. Tokhadze, *J. Phys. Chem.* **100**, 11610 (1996).
145. Z. Mielke, K. G. Tokhadze, and Z. Latajka, *J. Phys. Chem.* **100**, 539 (1996).
146. J. Lundell, M. Rasanen, and Z. Latajka, *Chem. Phys.* **189**, 245 (1994).
147. A. J. Barnes, K. Szczepaniak, and W. J. Orville-Thomas, *J. Mol. Struct.* **59**, 39 (1980).
148. M. Wierzejewska-Hnat, Z. Mielke, and H. Ratajzak, *J. Chem. Soc. Faraday Trans.,2* **76**, 834 (1980).
149. L. Schriver-Mazzuoli, A. Schriver, and J. P. Perchard, *J. Am. Chem. Soc.* **105**, 3843 (1983).
150. E. Clementi and J. N. Gayles, *J. Chem. Phys.* **47**, 3837 (1967).
151. O. Tapia and B. Silvi, *J. Phys. Chem.* **84**, 2676 (1980).
152. J. E. Sanhueza and O. Tapia, *J. Mol. Struct.* **89**, 131 (1982).
153. A. Loutellier, L. Schriver-Mazzuoli, A. Burneau, and J. P. Perchard, *J. Mol. Struct.* **82**, 165 (1982).
154. A. Schriver, S. Racine, O. Schrems, and L. Schriver-Mazzuoli, *Spectrochem. Acta* **50A**, 287 (1994).
155. C. Amirand and D. Maillard, *J. Mol. Struct.* **176**, 181 (1988).
156. C. Crepin and A. Tramer, *Chem. Phys.* **156**, 281 (1991).
157. G. L. Johnson and L. Andrews, *J. Am. Chem. Soc.* **105**, 163 (1983).
158. G. Schatte, H. Willner, D. Hoge, E. Knözinger, and O. Schrems, *J. Phys. Chem.* **93**, 6025 (1989).

159. V. Raducu, D. Jasmin, R. Dahoo, P. Brosset, B. Gauthier-Roy, and L. Abouaf-Marguin, *J. Chem. Phys.* **102**, 9235 (1995).
160. V. Raducu, B. Gauthier-Roy, R. Dahoo, L. Abouaf-Marguin, J. Langlet, J. Caillet, and M. Allavena, *J. Chem. Phys.* **105**, 10092 (1996).
161. F. Legay and N. Legay-Sommaire, *Chem. Phys. Letters* **211**, 516 (1993).
162. R. E. Miller, *J. Phys. Chem.* **90**, 3306 (1986)
163. F. Huisken in *Adv. Chem. Phys.*, Vol. LXXXI, ed. I. Prigogine, S.A. Rice (1992), p. 63.
164. R. T. Hall and G. Pimentel, *J. Chem. Phys.* **38**, 1889 (1993).
165. W. F. Hoffman and J. J. Shirk, *J. Mol. Struct.* **157**, 275 (1987).
166. T. Lotta, J. Murto, M. Rasanen, A. Aspiala, and P. Sarkka, *J. Chem. Phys.* **82**, 1363 (1985).
167. M. Nakata and M. Tasumi, *Spectrochem. Acta, Part A* **41**, 1015 (1985).
168. M. Rasanen and V. E. Bondybey, *J. Phys. Chem.* **90**, 5038 (1986).
169. A. Schriver, S. Racine, and L. Schriver-Mazzuoli, *Chem. Phys.* **100**, 377 (1985).
170. T. Beech, R. Gunde, P. Feloer, and H. Gunthard, *Spectrochem. Acta, Part A* **41**, 319 (1985).
171. L. Schriver-Mazzuoli, A. Schriver, and J. P. Perchard, *J. Chem. Phys.* **84**, 5553 (1986).
172. L. Schriver-Mazzuoli, A. Schriver, S. Racine, and J. P. Perchard, *Chem. Phys.* **119**, 95 (1988).
173. Y. Hannachi, L. Schriver-Mazzuoli, A. Schriver, and J.P. Perchard, *Chem. Phys.* **135**, 285 (1989).
174. L. Schriver-Mazzuoli, *J. Chem. Soc. Faraday Trans., 2* **85**, 607 (1989).
175. P. Bernadet, L. Schriver-Mazzuoli, A. Schriver, and J. P. Perchard, *J. Phys. Chem.* **92**, 7204 (1988).
176. N. Bakkas, A. Loutellier, S. Racine, and J.P. Perchard, *Chem. Phys.* **166**, 167 (1992).
177. A. J. Barnes, L. Schriver-Mazzuoli, A. Schriver, and J. P. Perchard, *J. Mol. Struct.* **166**, 167 (1990).
178. Z. Mielke and A. J. Barnes, *J. Chem. Soc. Faraday Trans., 2* **82**, 447 (1986).
179. L. Schriver-Mazzuoli, A. Burneau, and J. P. Perchard, *J. Chem. Phys.* **77**, 4926 (1982).
180. N. Bakkas, A. Loutellier, S. Racine, and J. P. Perchard, *J. Chim.* **90**, 1703 (1993).
181. S. Coussan, N. Bakkas, A. Loutellier, J. P. Perchard, and S. Racine, *Chem. Phys. Letters* **217**, 4073 (1994).
182. L. Fredin, N. Nelander, and G. Ribbegård, *J. Chem. Phys.* **66**, 4073

(1977).
183. L. Schriver-Mazzuoli in *Communication at the 5th International Conference on Spectroscopy of Matrix Isolated Species*, (Fontevraud, France, 1985).
184. Y. Hannachi, B. Silvi, and J. P. Perchard, *Chem. Phys.* **154**, 23 (1991).
185. J. Murto, M. Rasanen, A. Aspiala, and E. Kemppinen, *Acta Chem. Scand.* **A 37**, 323 (1983).
186. O. Schrems, *J. Mol. Struct.* **141**, 451 (1986).
187. S. Coussan, N. Bakkas, A. Loutellier, J. P. Perchard, and S. Racine, *Chem. Phys. Letters* **217**, 123 (1994).
188. A. Engdahl and B. Nelander, *Chem. Phys.* **213**, 333 (1966).
189. J. Zoual and V. A. Apkarian, *J. Phys. Chem.* **98**, 7945 (1994).
190. B. M. Cheng, J. W. Lee, and Y. P. Lee, *J. Phys. Chem.* **95**, 2814 (1991).
191. W. J. Chen, W. J. Lo, B. M. Cheng, and Y. P. Lee, *J. Phys. Chem.* **97**, 7167 (1992).
192. Shen-Horn Jou, Min-Yi Shen, Chin-Hui Yu, and Yuan-Pern Lee, *J. Chem. Phys.* **104**, 5745 (1996).
193. J. Feld, H. Kunttu, and V. A. Apkarian, *J. Chem. Phys.* **93**, 1009 (1990).
194. B. W. Moores and L. Andrews, *J. Phys. Chem.* **93**, 1902 (1989).
195. R. Wihtnall and L. Andrews, *J. Phys. Chem.* **89**, 3261 (1985).
196. M. Hawkins and L. Andrews, *Inorg. Chem.* **24**, 3285 (1985).
197. N. P. Machara and B. S. Ault, *J. Phys. Chem.* **91**, 2046 (1987).
198. D. J. Fitzmaurice and H. Frei, *J. Phys. Chem.* **96**, 10308 (1992).
199. S. A. Abrash and G. C. Pimentel, *J. Phys. Chem.* **93**, 5828 (1989).
200. S. A. Abrash, C. M. Carr, M. T. C. McMahon, and R. W. Zehner, *J. Phys. Chem.* **98**, 11109 (1994).
201. A. Hallou, L. Schriver-Mazzuoli, A. Schriver, N. Sanna, and A. Pieretti, *Asian J. Spectrosc.*, in press.
202. M. Bahou, L. Schriver-Mazzuoli, A. Schriver, and P. Chaquin, *Chem. Phys.* **216**, 105 (1997).
203. L. Schriver-Mazzuoli, O. Abdelaoui, C. Lugez, and A. Schriver, *Chem. Phys. Letters* **214**, 519 (1993).
204. S. D. Allen, M. Poliakoff, and J. J. Turner, *J. Mol. Struct.* **157**, 1 (1987).
205. P. Chaquin, M. Bahou, A. Schriver, and L. Schriver-Mazzuoli, *Chem. Phys. Letters* **256**, 609 (1996).
206. T. J. Lee, *J. Phys. Chem.* **99**, 15074 (1995).

INFRARED SPECTROSCOPY OF SIZE-SELECTED FREE AND ADSORBED WATER COMPLEXES

F. HUISKEN
Max-Planck-Institut für Strömungsforschung, Bunsenstrasse 10, D-37073 Göttingen, Germany

Infrared molecular beam depletion and fragment spectroscopy have been employed to study the absorption behavior of small water complexes [$(H_2O)_n$, $n = 2 - 5$] in the region of their stretching vibrations between 3300 and 3800 cm^{-1}. This spectral range was covered with an injection-seeded optical parametric oscillator. Size-specific information was obtained by dispersing the water clusters with a secondary helium beam and measuring the laser-induced depletion as a function of the scattering angle. Three absorption bands could be assigned to the water dimer $(H_2O)_2$, with the bonded O-H stretch being localized at 3601 cm^{-1}. For each of the larger water clusters ($n = 3 - 5$), which have cyclic structures, two absorption bands could be identified, one belonging to the free O-H stretch and the other being due to the excitation of the O-H ring vibration. The measurements on free gas phase water clusters were complemented by studies on small water complexes attached to large argon clusters or solvated in liquid helium droplets. Such host-guest cluster systems provide the possibility to investigate the molecular aggregates in an intermediate state between the free gas phase and the bulk matrix. The measured frequency shifts are compared with the available matrix data and discussed in terms of a correlation with the critical temperature of the gas providing the solvent. As an example for the interaction between a molecular chromophore and a large water cluster, we present first results on the system $CH_3OH{\cdot}(H_2O)_N$.

1 Introduction

The exploration of the structural and binding properties of small water complexes provides a key for the understanding of bulk water in its liquid and solid phase and for the understanding of solvation phenomena. Since several years it has been well accepted that small ice particles encountered in stratospheric clouds play an important role in the depletion of Antarctic ozone. [1] Although the details are not completely understood it is believed that chemical reactions taking place on the surface of the ice particles liberate molecular chlorine which, if dissociated by UV radiation, attacks and destroys the ozone. [2] In order to investigate how the properties of water vary as a function of the size of the system (from single isolated molecules to the condensed phase), considerable attention has been given to the theoretical and experimental study of small water complexes, with the emphasis being placed more and more on larger systems.

Early structural information on water polymers, $(H_2O)_n$, has been obtained from molecular beam electric deflection studies. [3] These experiments revealed the polar nature of the dimer, consistent with a linear hydrogen bond, and the nonpolar character of the higher polymers with $n = 3 - 6$, suggesting that the larger complexes have cyclic structures. The latter result has been confirmed by early [4] and more recent [5–7] *ab initio* calculations. Employing molecular beam electric resonance spectroscopy in the radio frequency and microwave regimes, Dyke and coworkers [8,9] obtained detailed structural information on the dimer.

Recent experimental [10–14] and theoretical [15,16] work has focused upon the tunneling motions in small water complexes which arise due to the existence of equivalent configurations in a floppy overall structure and which manifest themselves in a splitting of spectral lines and in the occurrence of new transitions. Saykally and coworkers [10] have developed the far-infrared vibration-rotation-tunneling (FIR-VRT) spectroscopy to study this phenomenon. Detailed information on the tunneling dynamics has already been obtained for the small water clusters up to the pentamer. [11–14] In combination with the theoretical investigations of Gregory and Clary [15,16] employing the diffusion Monte Carlo (DMC) technique, these studies yield valuable information with unprecedented accuracy on the multidimensional potential surfaces and the dynamics occurring on them.

While the FIR spectroscopy probes *inter*molecular vibrations, complementary results may be obtained by studying the *intra*molecular modes. The water monomer is characterized by three fundamental vibrations, the bending mode (ν_2) near 1595 cm^{-1}, the symmetric O-H stretch [17] (ν_1) at 3657.05 cm^{-1}, and the asymmetric O-H stretch [17] (ν_3) at 3755.97 cm^{-1}. In the water dimer which consists of two non-equivalent water molecules (the proton donor and proton acceptor) each vibration splits into two components. Correspondingly, the larger $(H_2O)_n$ polymers have n components for each intramolecular vibration. [5,7]

Due to the unfavorable wavelength around 6.3 μm needed to excite the ν_2 bending mode, this vibration has only been studied in matrix isolation experiments. [18–21] In order to investigate free gas phase water clusters in molecular beams, tunable lasers are highly desirable; but unfortunately suitable lasers are not available in this spectral range. In contrast, it is possible to access the 2.6 μm region where the symmetric and asymmetric O-H stretching vibrations can be excited. For this purpose one may use a color center or a diode laser if cw sources are desired, or it is possible to employ pulsed laser sources in combination with nonlinear frequency conversion techniques, for example in a Raman shifter, in a difference frequency mixing crystal, or in an optical parametric oscillator (OPO).

The first absorption studies devoted to the stretching vibrations of water polymers, were carried out by Lee and coworkers,[22] employing molecular beam vibrational predissociation spectroscopy with an optical parametric oscillator. Although the problem of fragmentation in the electron bombardment ionizer was realized and care was taken to minimize this effect, the broad spectra seemed to be strongly affected by the fragmentation of larger polymers. The following study, carried out by Page *et al.*[23] in the same laboratory, focused on the IR spectroscopy of the water dimer, $(H_2O)_2$. The four major absorption bands were assigned to the excitation of the symmetric (ν_1) and asymmetric (ν_3) stretching vibrations in the nonequivalent dimer constituents as discussed before. The most redshifted band at $\tilde{\nu}_0 = 3545$ cm^{-1} was attributed to the donor ν_1 vibration which, in later studies,[24] was identified with the stretching vibration of the O-H group engaged in the hydrogen bond and which is now referred to as the bonded O-H stretch. Shortly after that, Coker *et al.*[24] studied the evolution of water polymer spectra from the dimer to large clusters, employing their molecular beam color center laser apparatus with bolometric detection. At their mildest expansion conditions they observed four absorption bands, in agreement with the earlier study of Page *et al.*[23] Again the most redshifted band (now at 3532 cm^{-1}) was attributed to the bonded O-H stretch of the water dimer. The assignments of all four spectral features to the water dimer was strongly supported by quantum simulation calculations carried out within the same study.[24] The extremely good agreement was taken as evidence that (i) the assignment of the experimentally observed spectral features to dimer bands was correct and that (ii) the employed pair potential[25] was a rather accurate representation of the true water surface in the region of its minimum. In another molecular beam study, using coherent anti-Stokes Raman scattering (CARS), Wuelfert *et al.*[26] observed a strong Raman line at 3480 cm^{-1} which they attributed to the bonded O-H stretch of the water dimer.

The assignments of the molecular beam studies were in disagreement with various investigations carried out in cryogenic matrices.[18,19,27,28] Nelander[29] draw the attention to this conflict and proposed that the bands at 3532 and 3545 cm^{-1} could be due to the water trimer and that instead the strong band at 3600 cm^{-1} should be attributed to the bonded O-H stretch of the dimer. The recent high resolution study of Huang and Miller[30] yielded very detailed information on the asymmetric stretch of the proton acceptor molecule in the water dimer but was not able to resolve the conflict concerning the band position of the bonded O-H stretch and to locate the absorption bands of the larger water polymers.

In order to provide an unambiguous assignment of the absorption band near 3532 cm^{-1} to the size of the water complex (dimer or trimer) and to obtain size-specific information on the spectroscopy of the larger water polymers, we have combined the molecular beam depletion spectroscopy (MBDS) with the cluster selection technique introduced by Buck and Meyer. [31] This technique has already been proved useful for the investigation of weakly bound complexes absorbing CO_2 laser radiation near 10 μm. [32] In order to access the spectral region at 3 μm where O-H, H-F, and C-H stretches can be excited, we have developed an IR injection-seeded optical parametric oscillator (OPO) based on $LiNbO_3$ as the nonlinear medium. [33] The first experiments were devoted to the spectroscopy of small $(HF)_n$ and $(H_2O)_n$ complexes [34,35] where n was in the range between 2 and 6. Additional experiments were carried out to investigate the spectral shifts occurring when the polymers were attached to or embedded in large rare gas clusters. Employing the pickup technique of Scoles and coworkers, [36] we have produced and spectroscopically investigated small water complexes embedded in argon [35] and liquid helium [37] host clusters. Such composite clusters may be considered as mesoscopic systems between free gas phase polymers and molecular complexes trapped in macroscopic cryogenic rare gas matrices and therefore deserve special attention.

The aim of this Chapter is to collect the vibrational spectroscopy data assembled so far on small water polymers in the gas phase as well on host clusters and in cryogenic matrices and to compare them with each other and theoretical predictions. In addition, it will be shown that the whole set of data can be regarded as highly consistent if the frequency shifts are analyzed in terms of a correlation with the square root of the critical temperature of the gas constituting the matrix or host cluster. [38]

2 Experimental

2.1 Principle of Measurements

The measurements presented here have been carried out in a crossed molecular beam machine which has already been described in detail in several publications. [32,34,35,37] In the experiments with free water complexes, we made use of the size selection technique introduced by Buck and Meyer, [31] by employing a secondary helium beam to disperse the water clusters and using a rotatable mass spectrometer detector to select and identify their size. Water complexes embedded in rare gas host clusters were produced by first generating the rare gas clusters and then attaching water molecules as monomeric species with the so-called pickup method. [36]

At first we would like to discuss some general aspects of the various ex-

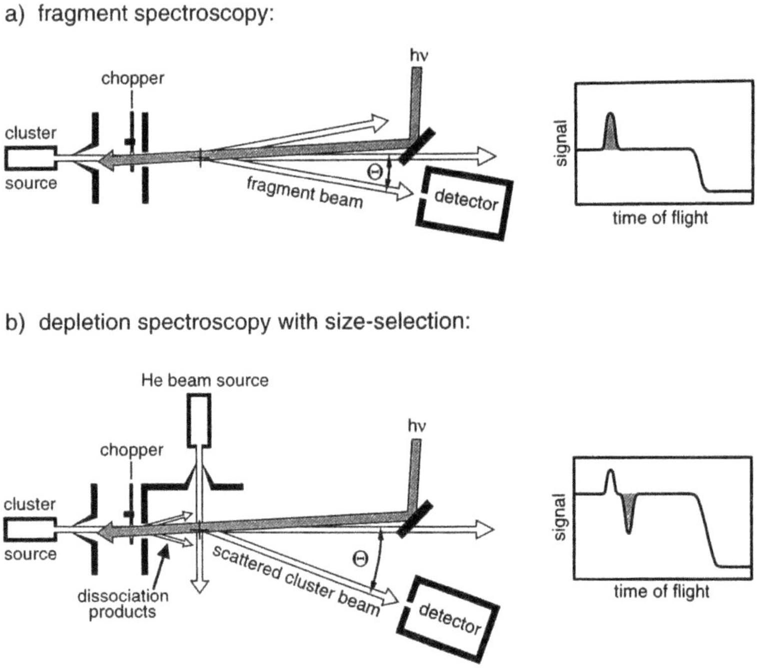

Figure 1: Schematic sketch of the experimental setups used for the IR spectroscopy of gas phase water clusters.

perimental setups employed in the present study. The different configurations applied for the spectroscopy of free gas phase water clusters are schematically shown in Fig. 1. The upper part (Fig. 1a) depicts the setup used for fragment spectroscopy. In this technique, the detector is positioned to a small angle (e.g. $\Theta = 3 - 5^{\circ}$) and the pulsed laser-induced fragments are detected as a transient increase in signal with the mass spectrometer tuned to a specific mass (see inset of Fig. 1a). By scanning the laser through the entire absorption region of the cluster beam, an *overview* absorption spectrum is obtained. Note that this spectrum is not size-selected and may contain contributions from various larger clusters. For example, if the mass spectrometer is tuned to the dimer mass, the smallest cluster size being probed is the trimer; but it is also possible that larger clusters contribute to the measured overview spectrum.

In order to obtain size-selective information and to determine the correlation between absorption band and cluster size, we employ the He secondary beam to deflect the clusters with different sizes into different angular ranges.[31,32] The corresponding setup is sketched in Fig. 1b. If the detector is positioned to a scattering angle characteristic for a specific cluster size and if the laser is tuned to dissociate this species, a laser-induced depletion will be detected in the time-of-flight (TOF) spectrum as is depicted in the inset of Fig. 1b. By measuring this depletion at fixed laser frequency as a function of the deflection angle, it is possible to determine, from the cutoff angle, the cluster size responsible for the absorption band. In this way, the various absorption bands measured in an overview spectrum without size selection can be assigned to specific cluster sizes.

2.2 Experimental Details

Water complexes embedded in large argon host clusters $[(H_2O)_n \cdot Ar_N]$ are generated by employing the pickup technique introduced by Gough *et al.*[36] The corresponding setup, which is described in more detail elsewhere,[35,39] is schematically shown in Fig. 2a. Briefly, argon clusters (Ar_N) with an estimated mean size of $\langle N \rangle = 50$ are produced by expanding argon gas at high pressure into the vacuum. Before entering the differential chamber through the skimmer, the argon cluster beam is crossed by a beam of H_2O molecules emanating from an effusive molecular beam source operated at stagnation pressures below 50 mbar. Upon the collisions, monomeric water molecules are attached to the argon clusters and carried by them to the detector. At larger water pressures, multiple adsorption occurs, and, by interacting through their large dipole moment, the water molecules aggregate and form small polymers within or on the surface of the host clusters. Due to the strong fragmentation of the composite clusters in the ionizer, the small water complexes can be detected on the typical fragmentation channels known from the mass spectrometry of gas phase water polymers.[35] Unfortunately, the kinematic size selection technique discussed before cannot be applied since the large argon clusters are too heavy to be sufficiently deflected by the helium beam.

When the embedded water complexes are excited by the IR radiation the vibrational energy is transferred to the host cluster which acts as a heat bath. The excess energy is rapidly released by evaporating around ten Ar atoms. This evaporation leads to a broadening of the angular width of the cluster beam and to a reduction of the ionization cross section. Both effects result in a depletion signal measured with the mass spectrometer as is shown in the time of flight spectrum of Fig. 2a.

For the production of water complexes embedded in large helium clusters $[(H_2O)_n \cdot He_N]$, the setup has been slightly changed[37] (see Fig. 2b). The helium cluster beam is generated by supersonic expansion of helium gas through a

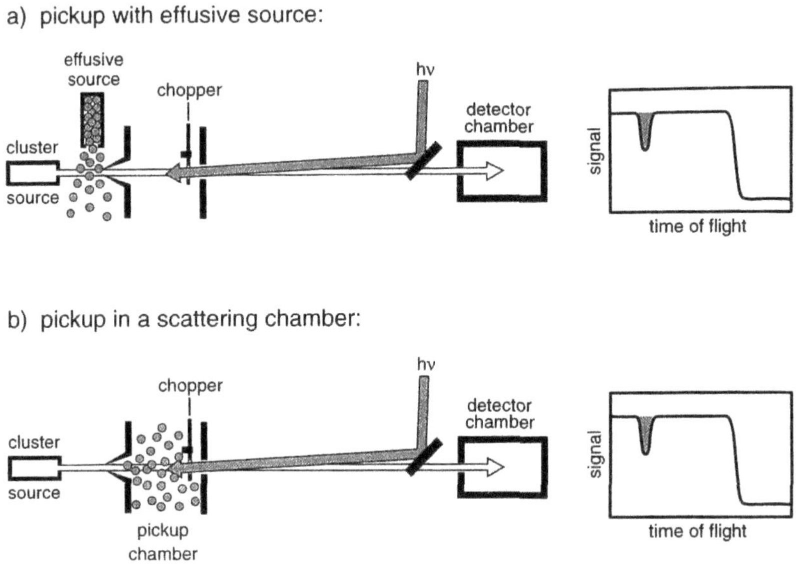

Figure 2: Schematic sketch of the experimental setups used for the IR spectroscopy of rare gas clusters doped with water complexes.

pinhole nozzle cooled with liquid helium to temperatures around $T_0 = 20$ K. At a stagnation pressure of $p_0 = 40$ bar, we produce helium clusters containing between 1000 and 6000 He atoms on the average. The helium clusters then pass through the differential chamber which serves as the pickup chamber. For this purpose it is filled with water vapor adjusted to the desired partial pressure around 1×10^{-5} mbar.

The free H_2O cluster beam was generated by expanding a mixture of 8% H_2O in helium at a total pressure of $p_0 = 2$ bar through a 0.1 mm diameter pinhole nozzle kept at a temperature of 100 °C.[35] After passing through the differential chamber, the skimmed and collimated water cluster beam could be crossed by a secondary helium beam if size-selective information was required. Finally the scattered cluster species were detected by a rotatable mass spectrometer detector equipped with electron bombardment ionizer ($E = 100$ eV, $I = 15$ mA), quadrupole mass filter, and electron multiplier. To produce the Ar (He) host clusters, we used a 50 (5) μm diameter pinhole nozzle, kept at 300 (18 - 24) K, and a backing pressure of 7.5 (40) bar.[35,37] These conditions resulted in mean cluster sizes of $\langle N \rangle = 50$ for Ar_N and $\langle N \rangle = 1000 - 5400$ for He_N.

The tunable near-infrared radiation was obtained from an injection-seeded optical parametric oscillator (OPO).[33] It consisted of a master oscillator containing a $LiNbO_3$ crystal, which was pumped by the fundamental of a Nd:YAG laser. This master oscillator was seeded by the narrow-bandwidth infrared radiation obtained by difference frequency mixing the output of a pulsed dye laser and the 532 nm radiation of the same Nd:YAG laser in a $LiIO_3$ crystal. The typical output energy for the low frequency component (idler) was ≥ 4 mJ per pulse in the entire spectral range covered in this study (3300 – 3800 cm^{-1}). The bandwidth of the infrared radiation, $\Gamma = 0.25$ or 2 cm^{-1}, which was determined by the bandwidth of the dye laser, was chosen to be 2 cm^{-1} in most experiments.

The interaction between cluster and laser beams was realized by using the quasicollinear configuration with counterpropagating laser beam, which intersected the cluster beam at a small angle.[34] The absorption of the IR photons by the clusters and the following dissociation was monitored by observing either the reduction of the signal in the direct or scattered beam (depletion spectra) or the enhancement of the signal on a fragment mass with the detector positioned to a small deflection angle (fragment spectra). In each case, the signal was monitored in real-time, using a computer-controlled time-of-flight analyzer[40] triggered by the laser. In order to determine the dissociation yields quantitatively, the measured TOF spectra were convoluted taking into account the clusters' time-of-flight distributions.[32]

An additional study on large water clusters doped with methanol as chromophore molecule was carried out in the configuration of Fig. 2a. In order to achieve enhanced cluster formation, the H_2O/He mixture was expanded through a conical nozzle of 0.09 mm diameter ($T_{bath} = 60$ °C, $T_0 = 100$ °C, $p_0 = 1.5$ bar). Under these conditions a $(H_2O)_N$ cluster distribution with an estimated mean size of $\langle N \rangle = 80$ was produced. Before leaving the source chamber through the skimmer, the water clusters were doped with methanol molecules emanating from the effusive molecular beam source ($p_0 = 30$ mbar) as is shown in Fig. 2a. The deposition of methanol molecules on the water host clusters was detected by monitoring the signals on $m = 31$ amu [CH_3O^+] and $m = 33$ amu [$(CH_3OH)H^+$].

Since, in the O-H stretching region, the absorption behavior of the composite clusters was governed by the O-H groups of the superabundant water molecules the first spectroscopic study was carried out in the region of the methanol C-O stretch at 10 μm where the water clusters absorb only weakly. For this purpose, the radiation of a line-tunable pulsed CO_2 laser (Urenco, model ML 104 S) was introduced into the differential chamber and crossed with the cluster beam at right angles. Details of the corresponding setup have been described in an earlier publication.[39]

3 Results

3.1 Free Water Clusters

In the configuration shown in Fig. 1a, we have measured absorption spectra of free gas phase water clusters by detecting the laser-induced dissociation products at small deflection angles with the mass spectrometer tuned to $m =$ 18 amu [H_2O^+] and $m = 19$ amu [$(H_2O)H^+$]. The results are summarized in Fig. 3. The left part of the spectrum, containing the two peaks designated with $n = 4$ and 5, was measured on mass $m = 19$ amu and $\Theta = 2.5°$ while the other part was recorded on $m = 18$ amu and $\Theta = 3.5°$. The gap at 3480 cm^{-1} could not be accessed, due to a serious power drop of the $LiNbO_3$ OPO. Since the fragment spectrum measured on mass $m = 19$ amu cannot contain contributions from dissociating water dimers (which must be monomers), the spectral features of the water dimer, as indicated in the figure, can be readily assigned by comparing the spectra measured on mass $m = 18$ and 19 amu. Note that the corresponding part of the $m = 19$ spectrum is not shown. For more details the reader is referred to the original literature.[35]

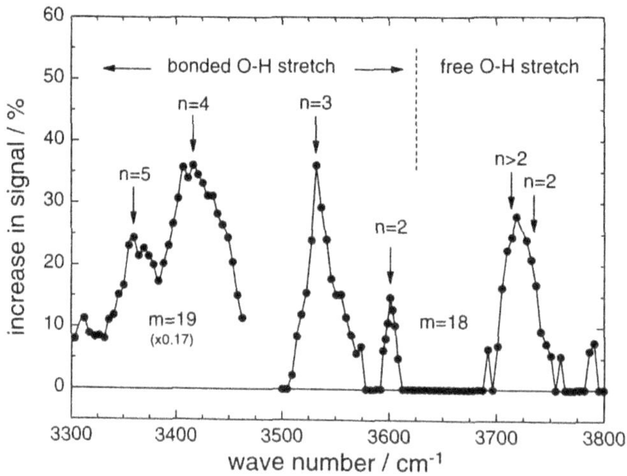

Figure 3: Overview absorption spectrum of small water clusters in the O-H stretching region. To record this spectrum, the laser-induced dissociation products were measured with the detector positioned to a small deflection angle and the mass spectrometer tuned to $m =$ 19 amu (left part) and $m = 18$ amu (right part), respectively. The assignments of the various absorption bands to cluster sizes have been made by employing the secondary helium beam and measuring the laser-induced depletion as a function of the scattering angle (see Fig. 4).

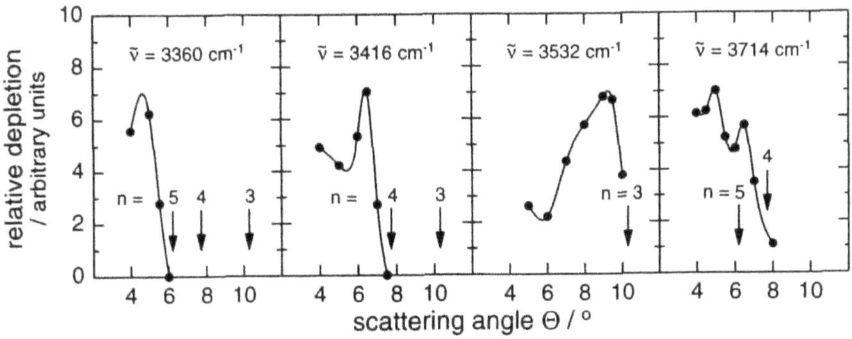

Figure 4: Relative depletion measured as a function of the scattering angle with the secondary helium beam turned on. For each curve the laser frequency $\tilde{\nu}$ has been kept fixed at the maximum of an absorption band labeled in Fig. 3. The positions marked by arrows indicate the maximum scattering angles as expected for the various cluster sizes n from the Newton diagram.

The other assignments given in the spectrum, which are much less obvious, were determined in a scattering experiment, employing the secondary helium beam (see Fig. 1b). As described before, the laser frequency was tuned to the maxima of the absorption bands marked by arrows in Fig. 3 (except those designated with $n = 2$) while the laser-induced depletion was measured as a function of the scattering angle. From the cutoff angle, the size of the cluster responsible for the absorption could be determined. The results are depicted in Fig. 4. In each plot the arrows indicate the maximum deflection angles of the water clusters of size n, as calculated from the Newton diagram.[32] The first angular depletion curve, measured at $\tilde{\nu} = 3360$ cm^{-1}, goes to zero just before the maximum deflection angle of the water pentamer ($n = 5$) is reached, indicating that the corresponding absorption band must be assigned to the pentamer. In contrast, the depletion signals measured at $\tilde{\nu} = 3416$ cm^{-1} and $\tilde{\nu} = 3532$ cm^{-1} can be observed up to the maximum tetramer and trimer deflection angles from which we conclude that tetramers and trimers are responsible for the respective absorption bands. The angular depletion curve measured for $\tilde{\nu} = 3714$ cm^{-1} shows two maxima and a cutoff below the tetramer angle. Thus, it appears that both tetramers and pentamers are contributing to the band at $\tilde{\nu} = 3714$ cm^{-1}. The latter frequency position is marked in Fig. 3 by the arrow labeled with $n > 2$. All band positions and assignments discussed so far are collected in Table 1.

Up to now we have not yet discussed the vibrational modes responsible for the various absorption bands. As is quite well established by theoretical calculations,[5-7] the smaller water complexes (from trimer on) have cyclic structures. Their vibrational stretching modes are identified as either bonded O-H stretches or free O-H stretches, each of which has n components for a complex containing n water molecules. While these n components are very close in frequency so that they may not be resolved in the experiment, the separation between the free and bonded O-H stretches is quite large (> 130 cm^{-1}), with the bonded O-H vibrations having significantly lower frequencies. Finally, the positions of the bonded O-H stretching bands are shown to depend much more sensitively on the cluster size than the frequencies of the free stretching vibrations which lie close together. With these theoretical results in mind, it is quite obvious that the measured absorption bands shown in Fig. 3 must be assigned, for $\tilde{\nu} < 3625$ cm^{-1}, to bonded O-H stretches and, in the unresolved peak at 3720 cm^{-1}, to free O-H stretches of various water cluster sizes.

3.2 Water Complexes Embedded in Argon Host Clusters

In order to provide a link between the free gas phase data just discussed and the large body of data available for water complexes trapped in macroscopic rare gas matrices, we have investigated the vibrational spectroscopy of water molecules embedded in large argon clusters.[35] Because of their nanoscopic dimensions these systems may be considered as *nano*matrices. Under the conditions employed in this experiment, the mean size of the argon clusters (before deposition of the water molecules) is estimated to be around $\langle N \rangle = 50$. Depletion spectra have been measured in the configuration of Fig. 2a for stagnation pressures of 30 and 50 mbar in the effusive water source and with the mass spectrometer tuned to $m = 19$ amu [$(H_2O)H^+$] and $m = 54$ amu [$(H_2O)_3^+$], respectively. The results are displayed in Fig. 5. The upper spectrum, measured on $m = 19$ amu, is dominated by three absorption bands located at 3516, 3576, and 3714 cm^{-1}, as indicated in the figure. Note that the ionization-induced fragmentation of larger water complexes to the protonated monomer ion [$(H_2O)H^+$] is still a serious problem, even if the water complexes are surrounded by argon atoms, and that the size selection cannot be applied. However, on the basis of the results obtained for free water clusters, the 3576 cm^{-1} band is readily assigned to the bonded O-H stretch of the water dimer while the 3714 cm^{-1} band is due to the excitation of the free O-H group in the same complex. The band at 3516 cm^{-1} is attributed to the trimer. The latter assignment becomes clear when the water pressure is raised to 50 mbar and the mass spectrometer tuned to $m = 54$ amu [$(H_2O)_3^+$] (see lower spectrum). The 3516 cm^{-1} peak, which is assigned to the bonded O-H stretch in the trimer ring, is now very pronounced. The other peak, observed

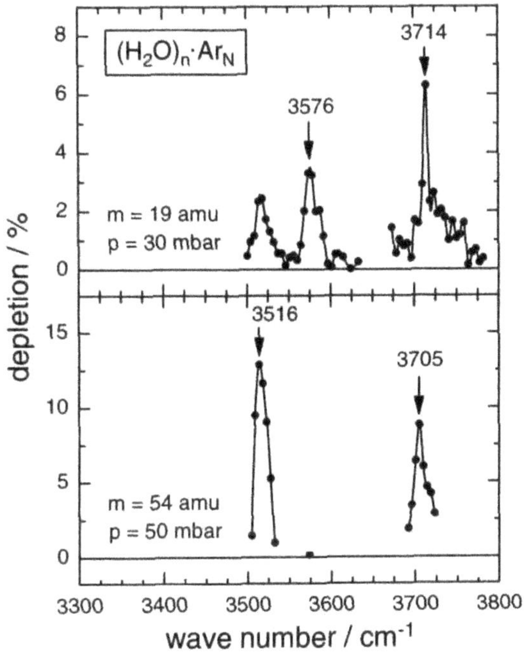

Figure 5: Depletion spectra of water dimers and trimers embedded in argon clusters. Due to the mass setting, water dimers are absent in the lower spectrum. Reprinted with permission from Ref. 35. ©1996 American Institute of Physics.

at 3705 cm^{-1}, is somewhat redshifted with respect to the dimer band. It is attributed to the free O-H stretch of the trimer. All line positions are included in Table 1.

As can be seen from the presence of the trimer peak at 3516 cm^{-1} in the upper $m = 19$ spectrum, the evaporation of a monomer unit following the ionization-induced protonation reaction is an important fragmentation channel even if the water trimer is embedded in the argon host cluster.[35] This indicates that the $(H_2O)_n \cdot Ar_N$ spectra might be affected by fragmentation artifacts and that care must be taken to interpret the data correctly. In this respect, the spectra of size-selected free water clusters are essential for a proper assignment of the observed features.

It is interesting to compare the $(H_2O)_n \cdot Ar_N$ results with the band positions observed in macroscopic argon matrices.[18,19] In these studies the band centers were determined for the bonded O-H stretch of the trimer at 3516 cm^{-1}, for the bonded O-H stretch of the dimer at 3574 cm^{-1}, for the trimer free O-H stretch at 3700 cm^{-1}, and for the dimer free O-H stretch at 3709 cm^{-1}. Comparing these values with the data summarized in Table 1, we state that the absorption frequencies of water dimers and trimers embedded in argon host clusters are very close to the values observed in conventional cryogenic argon matrices. A closer look reveals that they are slightly but consistently shifted to the gas phase data, in particular for the free O-H stretches. This indicates that the effective number of argon atoms interacting with the chromophore is smaller than in the bulk matrix. Regarding the question of whether the water complex adopts a surface or matrix site, we are tempted to conclude (because of the small shifts from the matrix data) that the water complexes are fully solvated. This conclusion is in agreement with recent structure calculations carried out by Liu *et al.*[41] for $H_2O{\cdot}Ar_N$ ($N = 1-14$) which show that the water *monomer* is fully solvated. The same behavior should be expected for the water dimer and trimer. The nice agreement between nanomatrix and bulk matrix data and the close resemblance of the spectra measured for solvated and free water complexes strongly support the assignments established earlier on the basis of the size-selective experiments, in particular for the bonded O-H stretch of the dimer.

3.3 Water Complexes Embedded in Helium Host Clusters

Recent experiments have shown that helium host clusters may be considered as ideal nanomatrices with minimum interaction between the chromophore and the environment.[42] Among all possible clusters, helium clusters are the only ones which are definitely liquid. They are characterized by a very low temperature of approximately 0.4 K, a value which was first theoretically determined[43] and later experimentally confirmed.[44] Larger helium clusters with $N > 64$ are predicted to be superfluid.[45,46] Because of these extraordinary properties, it seems extremely interesting to study the spectroscopy of water complexes dissolved in helium nanodroplets.

The water molecules have been embedded into the helium clusters, using the pickup technique in the configuration of Fig. 2b. At first we have studied the spectroscopy of water monomers trapped in helium nanodroplets.[37] The studies focused on the strongest IR-active mode, the ν_3 asymmetric stretch vibration. For the free water molecule the origin of this vibration[17] is located at 3755.97 cm^{-1}. The depletion spectrum measured with the mass spectrometer tuned to mass $m = 18$ amu (H_2O^+) is shown in Fig. 6. The spectrum is characterized by three distinct absorption bands at 3730, 3778, and 3799 cm^{-1} and two overlapping bands at 3751 and 3760 cm^{-1}. The positions have been determined by fitting Lorentzian line shape functions to the spectrum. By

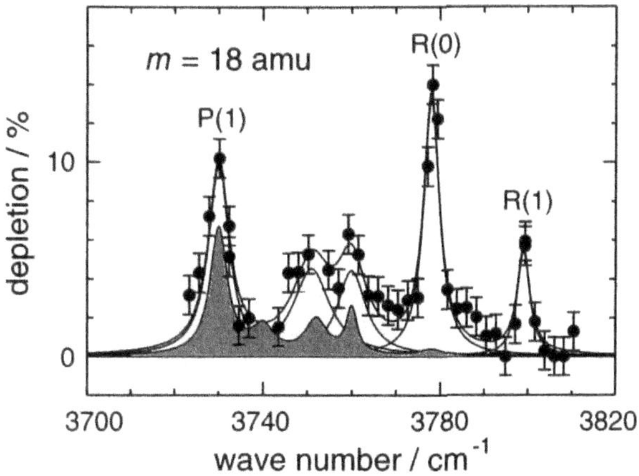

Figure 6: Absorption spectrum of the water monomer embedded in helium clusters ($H_2O{\cdot}He_N$). The depletion was measured on m = 18 amu (T_0 = 22 K, $\langle N \rangle$ = 1500, p_{H_2O} = 5 × 10^{-6} mbar). The estimated dimer contribution is indicated by the shaded area. Reprinted with permission from Ref. 37. ©1996 American Institute of Physics.

comparison with the free gas phase data of water,[17,47] the three most intense absorption lines can be readily attributed to the $P(1)$ (3730 cm^{-1}), $R(0)$ (3778 cm^{-1}), and $R(1)$ (3799 cm^{-1}) transitions which start from the lowest levels of *para* [$R(0)$] and *ortho* water [$P(1)$ and $R(1)$]. Compared to the free gas phase, the lines are redshifted by 1.5 – 2.5 cm^{-1}. The expected third *ortho* transition, starting from the ground state but ending in another K sublevel of the $J = 2$ state, could not be observed since it falls outside the tuning range of the employed laser system. According to the spectroscopic data on free water molecules,[17,47] this transition is expected just below 3863 cm^{-1}.

The assignment of the two weaker and somewhat broader lines is not as clear. According to their positions, they could be assigned to the $J_{K_aK_c} = 1_{11} \leftarrow 1_{10}$ and $J_{K_aK_c} = 1_{10} \leftarrow 1_{11}$ transitions. However, the corresponding levels are practically not populated at the low temperatures achieved in the present experiment. Another explanation could be that we have contributions from water dimers fragmenting to the monomer mass, m = 18 amu. This is rather plausible since, as we shall see later (in Fig. 7), the water dimer absorbs in the same spectral region. From gas phase spectroscopy[48] of the water monomer it is known that the intensity of the $P(1)$ line is 57% of that of the $R(1)$ line. Recalling that both transitions start from the same rotational level, this ratio should be the same even if the temperature is much lower. Assuming that the ratio of the transition probabilities is not affected by the helium cluster environment, we come to the conclusion that only 3.4% of the total 10.1% depletion at 3730 cm^{-1} is due to the monomer and that the residual signal must be attributed to the dimer which has a strong absorption at the same frequency, due to the excitation of the free O-H stretch (peak C in Fig. 7a). Having determined the dimer contribution in the $P(1)$ line, we are in the position to give the dimer contribution for the entire spectrum. This is visualized by the shaded area whose border curve is the same as the one fitted to the dimer spectrum of Fig. 7a, except that its maximum is normalized to a depletion of 6.7%. Focusing now our interest to the spectral region between 3740 and 3770 cm^{-1}, it is evident that the absorption bands localized at 3751 and 3760 cm^{-1} cannot entirely be explained by dimer absorptions. Some depletion observed in this region seems to originate from monomer absorptions. However, a plausible explanation for this observation is still missing.

With our highest laser resolution, $\Delta\tilde{\nu} = 0.25$ cm^{-1}, we have studied the strongest monomer line $R(0)$, which is practically free of dimer contributions, as a function of the helium cluster size.[37] Varying the nozzle temperature between $T_0 = 24$ and 20 K, under otherwise identical conditions, we obtained mean helium cluster sizes between $\langle N \rangle = 1000$ and 3200. As expected, the largest relative depletion signals were observed for the smallest host cluster size. Only careful inspection of the spectra revealed a very small increase in red shift (< 0.3 cm^{-1}) when the cluster size was changed from $\langle N \rangle = 1000$ to $\langle N \rangle = 3200$; but this effect might well be within the experimental error. All measured profiles could essentially be represented by the same Lorentzian line shape function peaked at $\tilde{\nu}_0 = 3778$ cm^{-1}. The linewidth was also constant with $\Gamma = 1.5$ cm^{-1}. Thus it appeared that, in the range investigated, the position and shape of the $R(0)$ line were independent of the helium cluster size.

Unfortunately, it was not possible to quantify the temperature of the helium clusters because only the ground states of the two nuclear spin modifications (*ortho* and *para* water) were populated. Compared to the free gas phase case, the rotational lines were only slightly shifted to the red by 1.5 - 2.5 cm^{-1}, indicating that the influence of the helium environment on the rotational energy levels of the water molecule was very weak and that the rotations were only slightly perturbed. In addition, it was found that the profile of the $R(0)$ line did not change when the size of the He_N clusters was varied. Summarizing the water monomer results, it is stated that the helium host clusters induce only minor matrix shifts and that they may be considered as ideal nanomatrices for studying trapped species at extremely low temperature.

At higher water vapor pressures in the pickup chamber, two or even more water molecules may be captured by the helium host cluster. The solvated water molecules interact through their large dipole moment and form dimers and eventually larger complexes. Thus, with this technique, it is possible to investigate the spectroscopy of small water polymers trapped in low temperature nanomatrices. In the present study we have concentrated on the bonded and free O-H stretches of the water dimer and trimer as well as on the ν_3 band of the proton acceptor molecule in the dimer. The spectra have been measured for different helium cluster sizes and for various water concentrations in the pickup chamber.

Figure 7a shows a dimer depletion spectrum that we have measured between 3700 and 3800 cm^{-1} with the mass spectrometer tuned to $m = 19$ amu [$(H_2O)H^+$]. The nozzle temperature was $T_0 = 24$ K, resulting in a mean helium cluster size of $\langle N \rangle = 1000$, and the pressure in the pickup chamber was chosen to be $p_{H_2O} = 1.8 \times 10^{-5}$ mbar. These conditions were found by optimizing the depletion signal on mass $m = 19$ amu in the main absorption band C at 3730 cm^{-1} and, on the other hand, minimizing the contribution from trimers on the basis of the count rate at $m = 37$ amu [$(H_2O)_2H^+$]. The spectrum of Fig. 7a could be reproduced by fitting five Lorentzian line shape functions to the measured data. The strongest band at $\tilde{\nu}_0 = 3730$ cm^{-1} coincides with the position observed in the high resolution study of Huang and Miller[30] for the free O-H stretch of the donor molecule (vibration C in their notation[30]) and is therefore assigned to the excitation of this vibration. The three bands, located at 3740, 3752, and 3778 cm^{-1} and designated with A, have also been observed by Huang and Miller.[30] They are assigned to various subbands of the ν_3 proton acceptor band involving transitions between different K_a levels. As has been pointed out by Huang and Miller,[30] the K_a levels split into two components, due to the internal-rotation-like tunneling motion involving a 180°-rotation of the proton acceptor molecule about its C_2 axis. The assignment of the fourth band at $\tilde{\nu}_0 = 3760$ cm^{-1}, designated with N, is not completely clear. Unfortunately, this spectral region was not investigated by Huang and Miller;[30] but it should be mentioned that the 3760 cm^{-1} peak was also observed in the earlier medium-resolution study of Page *et al.*[23] We tentatively assign this peak to the $K_a = 2 \leftarrow 1$ transition, involving the higher tunneling level (A_1^+) in the ground state and the lower one (A_1^-) in the vibrationally excited state.[37] Since the highest frequency transition at 3778 cm^{-1}, which involves different levels of the tunneling splitting, is observed at the same position as in the gas phase it follows that the tunneling motion is practically not hindered when the dimer is embedded in the helium cluster. This observation is in agreement with the almost free rotation found for the water monomer.

In the following experiment, we have studied the influence of the helium cluster size on the absorption spectrum of the water dimer. If the temperature of the helium cluster source was reduced to $T_0 = 20$ K the spectrum shown

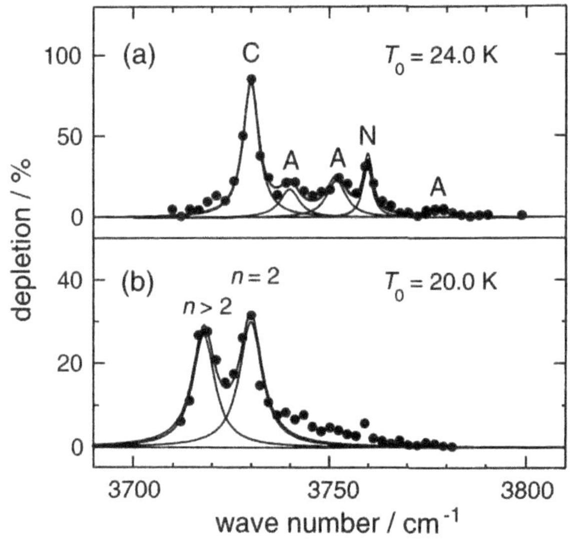

Figure 7: Variation of the absorption spectrum measured on m = 19 amu as a function of the size of the host cluster. The nozzle temperatures given in the spectra correspond to mean helium cluster sizes $\langle N \rangle$ = 1000 (a) and $\langle N \rangle$ = 3200 (b) prior to the deposition of the water molecules. The pressure in the pickup chamber was 1.8×10^{-5} in both experiments.

in Fig. 7b was measured. Going from spectrum (a) to (b), the helium cluster size increases from $\langle N \rangle = 1000$ to $\langle N \rangle = 3200$. Due to the larger capturing cross section, the probability of embedding more than two water molecules is increased, and as a result, the formation of water complexes larger than the dimer is considerably enhanced. Comparing the two spectra, the most striking difference is the appearance of a strong absorption band at $\tilde{\nu}_0 = 3719$ cm^{-1} in spectrum (b). In our study on small water clusters adsorbed on Ar_N host clusters (see Section 3.2) we have shown that the free O-H stretch of the water trimer is redshifted by 9 cm^{-1} from the corresponding dimer absorption. Therefore, the new band at $\tilde{\nu}_0 = 3719$ cm^{-1}, which is redshifted by 11 cm^{-1} from peak C, is assigned to the excitation of the free O-H stretch in water clusters larger than the dimer (presumably trimers). From this observation we conclude that the ionization-induced fragmentation of larger water clusters into $(H_2O)H^+$ ions ($m = 19$ amu) is an important reaction channel, even if the water complexes are embedded in helium host clusters. This is in contrast to what has been proposed earlier.[49]

In order to obtain more information about which cluster sizes contribute to the absorption band at $\tilde{\nu}_0 = 3719$ cm^{-1}, we have carried out depletion experiments with the mass spectrometer tuned to higher masses ($m = 37$ amu $[(H_2O)_2H^+]$ and 55 amu $[(H_2O)_3H^+]$. Since we found that the absorption band became broader but kept its maximum at $\tilde{\nu}_0 = 3719$ cm^{-1} we conclude that the larger water polymers with $n \geq 4$ also have their absorption maxima close to this frequency.[37]

As far as the bonded O-H stretches of helium-embedded water polymers are concerned, we have studied the water dimer and trimer. For the dimer we found a rather narrow absorption band at $\tilde{\nu}_0 = 3598$ cm^{-1}, redshifted by 3 cm^{-1} from the gas phase dimer absorption. A subsequent study of this band with higher resolution ($\Delta\tilde{\nu} = 0.25$ cm^{-1}) revealed a splitting, yielding one peak at $\tilde{\nu}_0 = 3596.6$ cm^{-1} and another one at $\tilde{\nu}_0 = 3598.15$ cm^{-1} each of which was 1 cm^{-1} wide. The positions of these peaks and their amplitude ratio practically did not change when the host cluster size was varied from $\langle N \rangle = 1500$ to 4200. For the trimer we found an absorption band at $\tilde{\nu}_0 =$ 3532 cm^{-1} with a distinct shoulder on its blue side (at 3545 cm^{-1}).[37]

The essential results of the $(H_2O)_n \cdot He_N$ experiment are summarized in Fig. 8 where we compare the absorption bands of free water dimers and trimers with those observed in helium host clusters. The free water cluster spectrum (a) is the same as the one shown in Fig. 3. It has already been discussed in Section 3.1. The results obtained in the helium cluster experiments are collected in form of an overview spectrum in the lower panel (b) of Fig. 8. In order to allow for a better comparison, we have combined three measurements which have not been carried out under identical conditions. The assignments given in the figure have already been discussed.

The most striking difference between the two spectra is the fact that all spectral features are considerably narrower for $(H_2O)_n \cdot He_N$. This is due to the very low temperature encountered in the liquid helium clusters. We expect this temperature to be as low as 0.4 K.[43,44] The temperature of the free water polymers, produced in the adiabatic expansion with helium as carrier gas, is not exactly known but is estimated to be between 20 and 30 K. Due to the reduced linewidths, the free O-H stretch of the dimer embedded in He_N can be clearly separated from the free O-H stretch of the larger water polymers. As far as the bonded O-H stretch of the water trimer is concerned, the present helium nanomatrix study clearly reveals a splitting of 13 cm^{-1}. Recent *ab initio* calculations[50–53] as well as absorption experiments in the far infrared[12] indicate that the water trimer has a nonplanar ring structure, giving rise to three IR-active bonded O-H stretches. According to a vibrational analysis,[7] the two strongest bands are separated by 9 cm^{-1} whereas the third band, which is redshifted by 60 cm^{-1}, is weak in IR spectroscopy but has the highest Raman activity. Although the theoretical intensity ratio of the two stronger IR-active

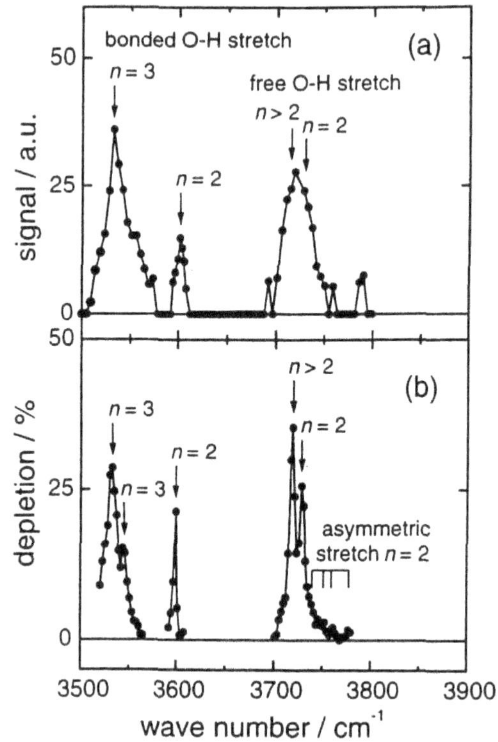

Figure 8: Comparison of spectra recorded for free water complexes (a) and water polymers embedded in large helium clusters (b). Reprinted with permission from Ref. 37. ©1996 American Institute of Physics.

lines is 100 : 93, with the lower frequency being the stronger one,[7] we consider the agreement between experiment and theory as rather good. Similar good agreement is obtained for the position of the free O-H trimer band. Whereas theoretical predictions[7] locate the three components of this band into a small interval redshifted by 9 – 15 cm^{-1} from the corresponding dimer band, a red shift of 11 cm^{-1} is observed in the present study. Unfortunately, the individual components cannot be resolved with the present resolution of our OPO.

Comparing the band positions of free and helium-embedded water polymers, we obtain the following result. For water polymers trapped in helium clusters, the interaction with the environment is very weak, giving rise to rather small red shifts: 1 cm^{-1} for the trimer bonded O-H stretch, 3.6 cm^{-1} for the dimer bonded O-H stretch, 7 cm^{-1} for the trimer free O-H stretch, 5 cm^{-1}

for the dimer free O-H stretch, and 1.5 cm^{-1} for the ν_3 monomer vibration [R(0) line]. As is indicated in the lower spectrum of Fig. 8 (in the vicinity of 3750 cm^{-1}), some weak spectral features are assigned to the asymmetric stretch of the proton acceptor molecule of the water dimer. A more detailed study of this band was presented in Fig. 7.

3.4 Methanol Attached to Water Host Clusters

Very recently we have initiated a series of experiments devoted to the spectral characterization of large water clusters doped with foreign molecules. As our first candidate we chose the system $CH_3OH{\cdot}(H_2O)_N$. Although this system is not so interesting as far as atmospheric aspects are concerned several reasons speak for the choice of this prototype. (i) Free gas phase methanol complexes have been extensively studied in our laboratory in the spectral regions of the C-O [54,55] and O-H [56] stretches, using size-selective techniques. (ii) Similar investigations as the ones presented here for water have already been carried out for methanol embedded in the rare gas clusters Ar_N and He_N. [39,57,58] (iii) Finally, mixed methanol-water complexes $[(CH_3OH)_m{\cdot}(H_2O)_n]$ produced by coexpansion of a methanol/water mixture have been studied before in the C-O stretching region. [59] In this context, the spectroscopic characterization of the methanol-water heterodimer $[CH_3OH{\cdot}H_2O]$ should be mentioned which revealed that the methanol molecule exclusively acts as proton acceptor in this binary complex. [60]

In order to ensure that only single guest molecules were deposited on the water host clusters, the $CH_3OH{\cdot}(H_2O)_N$ complexes were produced with the pickup technique as described in the experimental section. Using the depletion technique with the mass spectrometer tuned to $m = 33$ amu $[(CH_3OH)H^+]$, we have measured the absorption spectrum shown in Fig. 9. Superimposed on a rather broad background which is also present if the water clusters do not contain a methanol molecule, a 13 cm^{-1} broad absorption band is observed at 1040 cm^{-1}. This band is attributed to the excitation of the C-O stretch in single methanol molecules deposited on the surface of large water clusters. Due to our previous studies on gas phase methanol clusters and the methanol-water binary complex, we are in the position to give an interpretation of the observed absorption band.

Our study on size-selected methanol clusters [54,55] has revealed that the methanol dimer C-O stretch spectrum is characterized by two distinct peaks, one centered at 1026.5 cm^{-1}, the other at 1051.6 cm^{-1}. The lower frequency peak is due to the excitation of the proton acceptor molecule while the higher frequency peak has to be assigned to the proton donor molecule. In contrast,

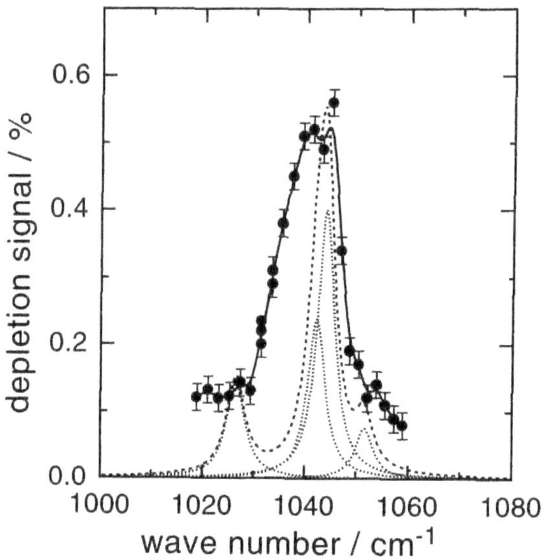

Figure 9: Dissociation spectrum of $CH_3OH{\cdot}(H_2O)_N$ clusters after exciting the C-O stretch of the methanol molecule. The signal was measured in depletion with the mass spectrometer tuned to m = 33 amu [$(CH_3OH)H^+$]. The dotted curves represent results[55] obtained for free methanol complexes [$(CH_3OH)_n$, n = 2, 3, 4].

the methanol trimer and tetramer are characterized by single absorption bands at 1042.2 cm^{-1} and 1044 cm^{-1}, respectively, reflecting the cyclic structures of these polymers. These measured absorption bands are shown in Fig. 9 by the dotted Lorentzian line shape functions, assuming an arbitrary cluster distribution consisting of dimers, trimers and tetramers. The sum of the individual contributions is represented by the dashed curve. On the other hand, our investigation on the methanol-water heterodimer has revealed a single band located at 1027.8 cm^{-1} demonstrating that (i) the methanol molecule adopts the proton acceptor position in this complex and that (ii) the absorption band shifts only slightly to the blue if the proton donor molecule in the methanol dimer is replaced by a water molecule. The latter observation reflects the fact that the interaction between methanol and water is very similar compared to the methanol-methanol interaction.

With this knowledge we may conclude that the methanol molecule adsorbed on the water cluster adopts neither a simple proton donor nor an exclusive proton acceptor position. In contrast, it appears that the methanol molecule simultaneously plays the role of a proton donor *and* acceptor, as is the case if it is engaged in a cyclic structure. The good agreement of the higher frequency feature of the absorption band with the methanol tetramer data suggests that the methanol molecule is hydrogen-bonded with three water molecules in a cyclic structure. The remaining absorption at lower frequencies is attributed to methanol molecules engaged in a third hydrogen bond between the second lone-pair orbital of the oxygen and another proton-donating water molecule. According to the empirical calculations of Tarakanova,[61] this additional interaction causes a red shift of approximately 4 cm^{-1}. Thus, we obtain the result that the methanol molecule is hydrogen-bonded (as proton acceptor *and* proton donor) to two or three water molecules residing on the surface of the water cluster. This implies that the methyl group is pointing away from the surface of the host cluster. It is interesting to note that the same result has been derived from macroscopic investigations of methanol monolayers on liquid water. Using sum frequency generation (SFG) to study liquid/vapor interfaces, Shen and coworkers[62] found that the methanol molecules had their CH_3 groups pointing out of the liquid.

4 Discussion

4.1 Assignments

In the present study, we have measured the O-H stretching frequencies of small gas phase water complexes and of water polymers trapped in helium and argon host clusters. All band positions located in this study are collected in Table 1. Particularly in the experiment devoted to free water complexes, the strong fragmentation of the ionized complexes turned out to be a severe problem when the observed spectral features were to be assigned to specific cluster sizes. However, with the help of the kinematic size selection, by employing a secondary helium beam to disperse the water complexes, it was possible to carry out an unambiguous assignment. For example, in earlier molecular beam studies,[23,24] a vibrational band near 3532 cm^{-1} was attributed to the bonded O-H stretch of the water dimer. Using our size identification method, we could show that this assignment was wrong and that the trimer was responsible for this absorption band.[35] With this reassignment, the band at 3601 cm^{-1} became the lowest frequency dimer band, and therefore, this band had to be attributed to the bonded O-H stretch. On the other hand, it turned out that the symmetric stretch (ν_1) of the proton acceptor molecule, falsely located[23,24] at 3601 cm^{-1}, has not yet been observed. However, this is not surprising since the ν_1 acceptor band should be at least six times weaker than the bonded O-H band, as line strength calculations have revealed.[63–66] This example demon-

Table 1: Intramolecular stretching frequencies (in cm^{-1}) of small water complexes in gas phase and in rare gas host clusters.

size	vibration	free $(H_2O)_n$	$(H_2O)_n \cdot He_N$	$(H_2O)_n \cdot Ar_N$
$n = 2$	ν_3	3745.48[a]	3745	-
	ν_1	-	-	-
	free O-H	3735	3730	3714
	bonded O-H	3601	3597.4	3576
$n = 3$	free O-H	3726	3719	3705
	bonded O-H	3533	3532	3516
$n = 4$	free O-H	3714	-	-
	bonded O-H	3416	-	-
$n = 5$	free O-H	3714	-	-
	bonded O-H	3360	-	-

[a]Ref. 30

strates that the size selection is very important to obtain proper assignments of the spectral features to cluster sizes.

4.2 Comparison with Calculations

Now we want to compare our gas phase results with recent *ab initio* calculations. Kim *et al.*[67] have computed water dimer frequencies at various levels of theory [HF, MP2, CCSD(T)]. Best agreement with the experimental results is obtained for a Hartree-Fock (HF) calculation with a triple-zeta plus double polarization (TZ2P) basis set, followed by the coupled-cluster calculation [CCSD(T)]. The HF/TZ2P calculation yields the following frequency shifts in cm^{-1} from the monomer ν_1 and ν_3 bands: -10 (-10.5)[30] for the ν_3 band of the proton acceptor, -22 (-21) for the ν_3 band of the proton donor (free O-H stretch), and -48 (-56) for the ν_1 band of the proton donor (bonded O-H stretch). Comparing these values with the experimental shifts given in parentheses, it is seen that the agreement is extremely good. The intramolecular frequencies of the larger water polymers [$(H_2O)_n$, $n = 3 - 5$] were first derived by Honegger and Leutwyler [5] and Knochenmuss and Leutwyler [6] in the harmonic approximation at the SCF level. In order to account for the anharmonicity, the authors determined scaling factors for the monomer frequencies which were also used for the larger clusters. Since the scaled frequency of the dimer bonded O-H stretch (3552 cm^{-1}) is significantly smaller than the experimental result (3601 cm^{-1}) it is more meaningful to compare the relative shifts from this dimer frequency rather than the absolute values. This is done in Fig. 10. It is immediately seen that the qualitative behavior of the theoretical curve of Leutwyler *et al.* [5,6] and of our experimental curve is the same. The theoretically predicted red shifts are consistently too large but the overall trend is reproduced very well. If the calculated shifts are multiplied by a factor of 0.76 perfect agreement is obtained. Xantheas and Dunning [7] calculated harmonic vibrational frequencies at the MP2 level with an augmented basis set. At this higher level approach, it was not necessary to adjust the computed values. Plotting the shifts from the dimer bonded O-H stretch as a function of cluster size, it is seen that the agreement between experiment and theory is very good, in particular if one considers that no scaling has been used. The pentamer frequencies have been calculated by Xantheas very recently. [66] Discussing vibrational frequency calculations, the new approach of Jung and Gerber [68] should be mentioned. These authors used the correlation-corrected vibrational self-consistent field (CC-VSCF) approximation and obtained results in fairly good agreement with the experiment. The advantage of the new approach is the possibility to extend the calculations to much larger systems (large water clusters and large hydrated molecules).

4.3 Solvated Water Complexes

As the experiments with water polymers embedded in rare gas clusters have shown, the fragmentation of the hydrogen-bonded complexes upon ionization still remains a severe problem although it is somewhat reduced. Therefore, the size-selective measurements on free water complexes are extremely important to interpret the spectra of solvated species correctly. When the water complexes were embedded in argon host clusters considerable red shifts were found, coming very close to the values observed in macroscopic argon matrices, but without reaching them. From this observation it was concluded that the water polymers were fully solvated in the argon host cluster.

The experiments with water polymers embedded in liquid helium clusters may be summarized as follows. Due to the extremely low temperature (0.4 K) achieved in the host cluster, the widths of the absorption bands were consider-

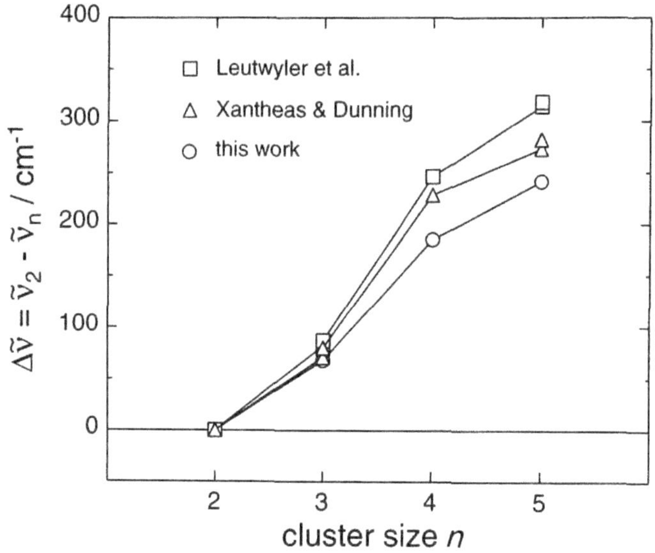

Figure 10: Comparison of the present gas phase results on water clusters with theoretical predictions by Leutwyler *et al.* [5,6] and Xantheas and Dunning. [7,66] The frequency shifts of the bonded O-H stretches of water trimer, tetramer, and pentamer rings, with respect to the bonded O-H stretch of the water dimer, are plotted as a function of cluster size *n*.

ably reduced. Consequently, a splitting in the bonded O-H trimer band, which was predicted by theory as a result of the asymmetric structure of the trimer ring, could be resolved for the first time. Similarly, the free O-H stretching bands of the water dimer and trimer could be separated. It appeared that the interaction between the helium atoms and the water molecule or the water complex was very weak, giving rise to rather small red shifts of the vibrational bands. In addition, it was found that the water monomers and dimers rotate almost freely in the helium nanomatrix. Similarly, it could be shown that the rotation-like tunneling motion of the proton acceptor molecule of the water dimer was not very much affected by the surrounding helium atoms. From these observations it can be concluded that the helium cluster indeed constitutes an almost ideal nanomatrix.

4.4 *Correlation between Frequency Shift and Critical Temperature*

Comparing our results with those of earlier matrix studies, we would like to discuss the possibility of establishing a correlation between the matrix shift and some property of the matrix material. In an earlier paper, Behrens-Griesenbach *et al.*[69] found a linear relationship when the frequencies of the O-H stretching vibrations of methanol monomers and dimers trapped in various rare gas matrices were plotted as a function of the square root of the critical temperature T_c. As has been discussed in more detail in a separate paper,[38] this relationship can be easily rationalized because the strength of the attractive interaction between a chromophore and various matrix materials roughly scales with $\sqrt{T_c}$. The present investigations are the first ones which locate the bonded O-H stretches of water dimers and trimers in the gas phase and, in particular, the first ones devoted to the study of water polymers trapped in the novel matrix material represented by the helium cluster. Therefore, it will be interesting to see how well the present data correlate with previous results obtained in other macroscopic cryogenic matrices.

In Fig. 11 the band positions of the O-H stretching vibrations of water dimers and trimers observed in the present work and in various matrix studies[19–21,27,70,71] are plotted as a function of the square root of the critical temperature. It is immediately seen that for all vibrations and matrix materials considered here the linear relationship between frequency and $\sqrt{T_c}$ is fulfilled very well. The frequencies measured in D_2 and N_2 matrices are somewhat redshifted with respect to those observed in rare gas matrices; but the straight lines connecting the D_2 and N_2 data points are always parallel to the rare gas lines. The extra red shift observed in diatomic matrices is between 8 and 30 cm^{-1} and results from the anisotropy of the matrix material.

The band with the highest frequency is assigned to the asymmetric stretching vibration of the acceptor molecule in the water dimer while the slightly lower frequency band is attributed to the free O-H stretch of the donor molecule. Due to its low infrared activity, the ν_1 dimer band has not yet been observed in the (only weakly interacting) matrix materials neon[21] and helium[37] nor in the free gas phase.[35] In the anisotropic matrices D_2 and N_2, the ν_1 band has gained enough intensity to be located with sufficient accuracy. If now the rare gas data is extrapolated to zero such that the straight line is parallel to the line connecting the D_2 and N_2 data points (and also parallel to the line obtained for the free O-H stretch) we may predict the band position of the symmetric stretch in the proton acceptor molecule of the water dimer with rather high accuracy, namely at 3660 ± 5 cm^{-1}. As already mentioned, earlier molecular beam gas phase studies[23,24] have assigned the band around 3601 cm^{-1} to the symmetric stretch. However, with our size-selective study we could show that the band at 3601 cm^{-1} must be assigned to the bonded O-H stretch. The correlation presented in Fig. 11 nicely supports this interpretation. Furthermore, it is seen that, on the basis of this correlation, a consistent picture for all four O-H stretching vibrations of the water dimer is obtained.

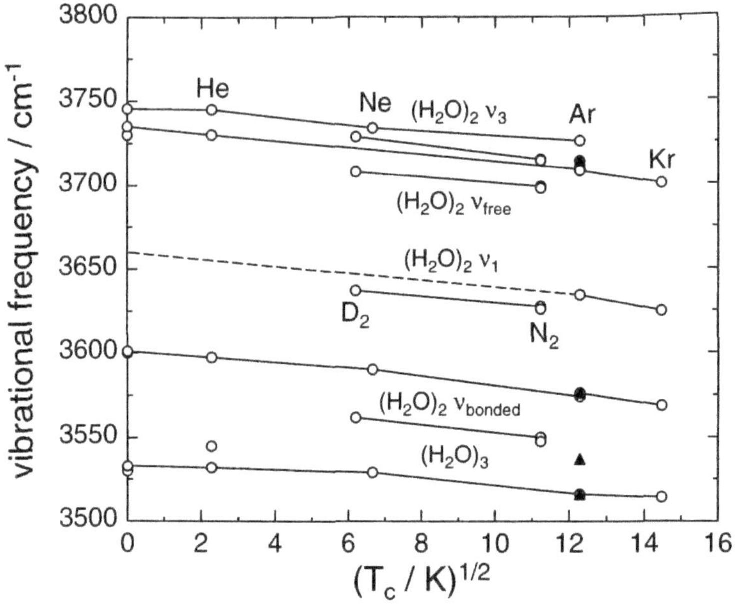

Figure 11: OH stretching frequencies of water dimers and trimers in gas phase, embedded in helium and argon clusters, and trapped in various cryogenic matrices, plotted versus $(T_c)^{1/2}$ of the corresponding matrix material. The solid symbols have been measured for $(H_2O)_n{\cdot}Ar_N$. Reprinted with permission from Ref. 38. ©1997 Elsevier Science.

5 Summary

The intramolecular stretching vibrations of small water polymers [$(H_2O)_n$ with $n = 2 - 5$] have been studied in molecular beams, using the technique of infrared depletion and fragment spectroscopy. Size-specific information was obtained by dispersing the water clusters with a secondary helium beam and measuring the laser-induced depletion as a function of the scattering angle. Three absorption bands could be assigned to the water dimer. The lowest frequency dimer band, located at 3601 cm^{-1}, was attributed to the excitation of the bonded O-H stretch. The symmetric stretching vibration of the proton acceptor molecule could not be observed due to its small transition moment.

In agreement with theoretical predictions, two absorption bands could be identified for each of the larger $(H_2O)_n$ clusters ($n = 3 - 5$), one belonging to the free O-H stretch and the other being due to the excitation of the O-H ring vibration. While the frequencies of the free O-H stretches fall close together the separations of the bonded O-H stretches are considerably larger.

The measurements on free gas phase water clusters were complemented by studies on water polymers attached to large argon clusters by making use of the pickup technique. Such host-guest cluster systems provide the possibility to investigate molecular complexes in a nanomatrix, *i.e.* in an intermediate state between the free gas phase and the bulk matrix. The frequencies measured in the present study for the water dimer and trimer were close to those observed in macroscopic argon matrices, suggesting that the water complexes were fully solvated.

The following study was devoted to water molecules and small polymers trapped in liquid helium nanodroplets. The measured frequencies were only slightly redshifted from the corresponding free gas phase values, indicating that the helium provided an almost perfect matrix material with minimum interaction between the chromophore and the matrix atoms. In addition, it was found that the water molecules rotated almost freely in the helium cluster and that the internal-rotation-like tunneling motion of the proton acceptor molecule in the water dimer was practically not hindered. Due to the extremely low temperature encountered in the helium droplet, the linewidths of the vibrational bands were considerably reduced so that we were able to resolve a splitting in the trimer band and to separate the free O-H stretches of dimers and trimers.

First results were presented for a nanoscopic system, consisting of a water cluster and a single methanol molecule attached to its surface. The spectroscopy of the methanol C-O stretch revealed that the guest molecule was hydrogen-bonded as proton acceptor *and* donor to two or three surface water molecules and that the methyl group was pointing away from the water cluster. While the $CH_3OH{\cdot}(H_2O)_N$ system served as a prototype the forthcoming studies will be concerned with the spectroscopy of water clusters containing guest molecules of atmospheric interest.

Acknowledgments

The author wishes to thank Prof. H. Pauly for his continued interest and support as well as Dr. E. G. Tarakanova and Dr. A. A. Vigasin for very fruitful discussions. The help of Dr. R. Fröchtenicht, Dr. M. Kaloudis, M. Koch, Dr. A. Kulcke, S. Mohammad-Pooran, and O. Werhahn in carrying out the experiments is gratefully acknowledged. This work has been supported by the Deutsche Forschungsgemeinschaft (DFG) in the frame of its Schwerpunktprogramm *Molecular Clusters*.

References

1. O. B. Toon and R. P. Turco, *Scientific American* **264/6**, 40 (1991).
2. S. Solomon, *Nature* **347**, 347 (1990) and references therein.
3. T. R. Dyke and J. S. Muenter, *J. Chem. Phys.* **57**, 5011 (1972).
4. J. E. Del Bene and J. A. Pople, *J. Chem. Phys.* **52**, 4858 (1970).
5. E. Honegger and S. Leutwyler, *J. Chem. Phys.* **88**, 2582 (1988).
6. R. Knochenmuss and S. Leutwyler, *J. Chem. Phys.* **96**, 5233 (1992).
7. S. S. Xantheas and T. H. Dunning, Jr., *J. Chem. Phys.* **99**, 8774 (1993).
8. T. R. Dyke, K. M. Mack, and J. S. Muenter, *J. Chem. Phys.* **66**, 498 (1977).
9. J. A. Odutola and T. R. Dyke, *J. Chem. Phys.* **72**, 5062 (1980).
10. R. C. Cohen and R. J. Saykally, *J. Phys. Chem.* **96**, 1024 (1992).
11. N. Pugliano and R. J. Saykally, *J. Chem. Phys.* **96**, 1832 (1992).
12. K. Liu. M. J. Elrod, J. G. Loeser, J. D. Cruzan, N. Pugliano, M. G. Brown, J. Rzepiela, and R. J. Saykally, *Faraday Discuss.* **97**, 35 (1994).
13. J. D. Cruzan, M. G. Brown, K. Liu, L. B. Braly, and R. J. Saykally, *J. Chem. Phys.* **105**, 6634 (1996).
14. K. Liu, M. G. Brown, J. D. Cruzan, and R. J. Saykally, *Science* **271**, 62 (1996).
15. J. K. Gregory and D. C. Clary, *J. Chem. Phys.* **102**, 7817 (1995).
16. J. K. Gregory and D. C. Clary, *J. Chem. Phys.* **105**, 6626 (1996).
17. P. E. Fraley and K. N. Rao, *J. Mol. Spectrosc.* **29**, 348 (1969).
18. G. P. Ayers and A. D. E. Pullin, *Spectrochim. Acta* Part A **32**, 1629 (1976).
19. R. M. Bentwood, A. J. Barnes, and W. J. Orville-Thomas, *J. Mol. Spectrosc.* **84**, 391 (1980).
20. A. Engdahl and B. Nelander, *J. Mol. Struct.* **193**, 101 (1989).
21. D. Forney, M. E. Jacox, and W. E. Thompson, *J. Mol. Spectrosc.* **157**, 479 (1993).
22. M. F. Vernon, D. J. Krajnovich, H. S. Kwok, J. M. Lisy, Y. R. Shen, and Y. T. Lee, *J. Chem. Phys.* **77**, 47 (1982).
23. R. H. Page, J. G. Frey, Y. R. Shen, and Y. T. Lee, *Chem. Phys. Lett.* **106**, 373 (1984).
24. D. F. Coker, R. E. Miller, and R. O. Watts, *J. Chem. Phys.* **82**, 3554 (1985).
25. J. R. Reimers, R. O. Watts, and M. L. Klein, *Chem. Phys.* **64**, 95 (1981).
26. S. Wuelfert, D. Herren, and S. Leutwyler, *J. Chem. Phys.* **86**, 3751 (1987).

27. L. Fredin, B. Nelander, and G. Ribbegård, *J. Chem. Phys.* **66**, 4065 (1977).
28. A. Engdahl and B. Nelander, *J. Chem. Phys.* **86**, 1819 (1987).
29. B. Nelander, *J. Chem. Phys.* **88**, 5254 (1988).
30. Z. S. Huang and R. E. Miller, *J. Chem. Phys.* **91**, 6613 (1989).
31. U. Buck and H. Meyer, *Phys. Rev. Lett.* **52**, 109 (1984).
32. F. Huisken, *Adv. Chem. Phys.* **81**, 63 (1992).
33. F. Huisken, A. Kulcke, D. Voelkel, C. Laush, and J. M. Lisy, *Appl. Phys. Lett.* **62**, 805 (1993).
34. F. Huisken, M. Kaloudis, A. Kulcke, C. Laush, and J. M. Lisy, *J. Chem. Phys.* **103**, 5366 (1995).
35. F. Huisken, M. Kaloudis, and A. Kulcke, *J. Chem. Phys.* **104**, 17 (1996).
36. T. E. Gough, M. Mengel, P. A. Rowntree, and G. Scoles, *J. Chem. Phys.* **83**, 4958 (1985).
37. R. Fröchtenicht, M. Kaloudis, M. Koch, and F. Huisken, *J. Chem. Phys.* **105**, 6128 (1996).
38. F. Huisken, M. Kaloudis, and A. A. Vigasin, *Chem. Phys. Lett.* **269**, 235 (1997).
39. F. Huisken and M. Stemmler, *J. Chem. Phys.* **98**, 7680 (1993).
40. F. Huisken and T. Pertsch, *Rev. Sci. Instrum.* **58**, 1038 (1987).
41. S. Liu, Z. Bačić, J. W. Moskowitz, and K. E. Schmidt, *J. Chem. Phys.* **101**, 8310 (1994).
42. J. P. Toennies, *Helium Clusters*, Proceedings of the International School of Physics "Enrico Fermi", Course CVII, edited by G. Scoles (North Holland, Amsterdam, 1990), p. 597.
43. D. M. Brink and S. Stringari, *Z. Phys.* D **15**, 257 (1990).
44. M. Hartmann, R. E. Miller, J. P. Toennies, and A. Vilesov, *Phys. Rev. Lett.* **75**, 1566 (1995).
45. M. V. Rama Krishna and K. B. Whaley, *Phys. Rev. Lett.* **64**, 1126 (1990).
46. B. P. Sindzingre, M. L. Klein, and D. M. Ceperley, *Phys. Rev. Lett.* **63**, 1601 (1989).
47. C. Camy-Peyret, J. M. Flaud, G. Guelachvili, and C. Amiot, *Mol. Phys.* **26**, 825 (1973).
48. R. A. Toth, *J. Opt. Soc. Am. B* **10**, 2006 (1993).
49. M. Lewerenz, B. Schilling, and J. P. Toennies, *J. Chem. Phys.* **102**, 8191 (1995).
50. S. S. Xantheas, *J. Chem. Phys.* **102**, 4504 (1995).
51. M. Schütz, T. Bürgi, S. Leutwyler, and H. B. Bürgi, *J. Chem. Phys.* **99**, 5228 (1993).

52. J. G. C. M. van Duijneveldt - van de Rijdt and F. B. van Duijneveldt, *Chem. Phys.* **175**, 271 (1993).
53. L. González, O. Mó, M. Yáñez, and J. Elguero, *J. Mol. Struct. (Theochem)* **371**, 1 (1996).
54. F. Huisken and M. Stemmler, *Chem. Phys. Lett.* **144**, 391 (1988).
55. F. Huisken and M. Stemmler, *Z. Phys.* D **24**, 277 (1992).
56. F. Huisken, A. Kulcke, C. Laush, and J. M. Lisy, *J. Chem. Phys.* **95**, 3924 (1991).
57. A. Kulcke, Ph. D. Thesis, University of Göttingen, 1994, Max-Planck-Institut für Strömungsforschung, Report No. 13 (1994).
58. M. Kaloudis, Ph. D. Thesis, University of Göttingen, 1996, Max-Planck-Institut für Strömungsforschung, Report No. 1 (1996).
59. M. Stemmler, Ph. D. Thesis, University of Göttingen, 1991, Max-Planck-Institut für Strömungsforschung, Report No. 3 (1991).
60. F. Huisken and M. Stemmler, *Chem. Phys. Lett.* **180**, 332 (1991).
61. E. G. Tarakanova, private communication.
62. C. D. Stanners, Q. Du, R. P. Chin, P. Cremer, G. A. Somorjai, and Y. R. Shen *Chem. Phys. Lett.* **232**, 407 (1995).
63. B. A. Zilles and W. B. Person, *J. Chem. Phys.* **79**, 65 (1983).
64. D. J. Swanton, G. B. Bacskay, and N. S. Hush, *Chem. Phys.* **82**, 303 (1983).
65. R. D. Amos, *Chem. Phys.* **104**, 145 (1986).
66. S. S. Xantheas, private communication.
67. J. Kim, J. Y. Lee, S. Lee, B. J. Mhin, and K. S. Kim, *J. Chem. Phys.* **102**, 310 (1995).
68. J. O. Jung and R. B. Gerber, *J. Chem. Phys.* **105**, 10332 (1996).
69. A. Behrens-Griesenbach, W. A. P. Luck, and O. Schrems, *J. Chem. Soc. Faraday Trans.* **80**, 579 (1984).
70. E. J. Murby and A. D. E. Pullin, *Austr. J. Chem.* **32**, 1167 (1979).
71. A. J. Tursi and E. R. Nixon, *J. Chem. Phys.* **52**, 1521 (1970).

www.ingramcontent.com/pod-product-compliance
Ingram Content Group UK Ltd.
Pitfield, Milton Keynes, MK11 3LW, UK
UKHW020358150926
8624IPUK00004B/225

9 789810 232115